# SOUTH AFRICA,

## PAST, PRESENT

## AND FUTURE

## PEARSON EDUCATION

We work with leading authors to develop the
strongest educational materials in Geography,
bringing cutting-edge thinking and best learning
practice to a global market.

Under a range of well-known imprints, including
Prentice Hall, we craft high quality
print and electronic publications which help
readers to understand and apply their content,
whether studying or at work.

To find out more about the complete range of our
publishing please visit us on the World Wide Web at:
www.pearsoneduc.com

# SOUTH AFRICA,

# PAST, PRESENT

# AND FUTURE

## GOLD AT THE END
## OF THE RAINBOW?

ALAN LESTER, ETIENNE NEL
AND TONY BINNS

Longman

*An imprint of* **Pearson Education**

Harlow, England · London · New York · Reading, Massachusetts · San Francisco
Toronto · Don Mills, Ontario · Sydney · Tokyo · Singapore · Hong Kong · Seoul
Taipei · Cape Town · Madrid · Mexico City · Amsterdam · Munich · Paris · Milan

**Pearson Education Limited**
Edinburgh Gate
Harlow
Essex CM20 2JE
England

and Associated Companies throughout the world

*Visit us on on the World Wide Web at:*
http:/www.pearsoneduc.com

First published 2000

ISBN 0 58235626 1

*British Library Cataloguing-in-Publication Data*
A catalogue record for this book is available from the British Library

*Library of Congress Cataloging-in-Publication Data*
A catalog record for this book is available from the Library of Congress

10 9 8 7 6 5 4 3 2 1
04 03 02 01 00

Typeset by 30.
Printed and bound by Grafos S.A., Arte sabre papel, Barcelona, Spain

# CONTENTS

# LIST OF FIGURES

# LIST OF PLATES

# LIST OF TABLES

# PREFACE

Various parts of the world have been selectively thrust onto the centre stage of 'international affairs'. South Africa has merited a considerable proportion of international attention over the last fifty years, owing to its infamous apartheid policies and, more recently, its seemingly miraculous political rebirth. In parallel with the collapse of communism, the fall of apartheid was clearly one of the 'great events' of the late twentieth century. As an emerging champion of democracy and a country positioned between the developed and the developing worlds, South Africa's role is often now seen as one of encouraging democracy, brokering peace deals internationally and, above all, acting as a role-model in a troubled continent.

Dramatic events and extreme social tensions in South Africa have captured the attention of dozens of scholars through the decades. The country's apartheid policies and subsequent post-apartheid initiatives, have been extensively analysed in hundreds of books and articles. Despite the wealth of detailed studies which exist, however, no-one, to my knowledge, has yet attempted a synthesis of what has happened in South Africa in the past as well as in the post-apartheid present, within a single, cohesive volume. The argument for such an approach is obvious: contemporary South Africa differs dramatically from its pre-1990 self, but is also inextricably shaped in countless ways by a social, economic and cultural mould which was cast over centuries. To try to develop such an analysis would have been beyond the capacity of most individual scholars and it was therein that the genesis of the book lay.

I have been privileged to collaborate with two close friends, Tony Binns and Alan Lester, for several years. Our skills and interests are different, but I believe that this enhances our collective insight. Tony has developed a reputation over a long period working on African development issues, and is therefore well positioned to provide a comparative perspective on post-apartheid development endeavours. Alan is an historical geographer of South Africa and comes to the book with a broad temporal and spatial perspective. My own interests lie in contemporary South African development and economic policy. Following a presentation by Tony on South African at the 1997 Geographical Association conference in London, and during subsequent discussions with Alan and Tony about our different but complementary interests in South Africa, the idea for the book crystallized. I was delighted with the enthusiasm which both displayed when I first mooted the idea of a publication.

In writing this book we feel that we have made a contribution to understanding South Africa, past, present and future. In trying to cover such an enormous field in a single work we know that we have not devoted as much attention to all the issues that we perhaps could have done in a larger volume, and we have been selective in the topics that we have addressed. Nonetheless, I would argue that we have provided a framework for understanding contemporary South African society and its economy, together with its development policies, in the light of its past. Hopefully, this book will therefore appeal not only to South Africanists and students of development issues, but to all those who seek to contextualize current development 'problems' in the light of broader histories and geographies.

Etienne Nel, Grahamstown, January 2000
Tony Binns and Alan Lester,
United Kingdom

# ACKNOWLEDGEMENTS

We are grateful to the following for permission to reproduce copyright material:

Academic Press Ltd for Figure 3.14 from 'Labour, capital, class struggle and the origins of residential segregation in Kimberley, 1880–1920' by A. Mabin, *Journal of Historical Geography*, Vol. 12, No. 1, 1986; Africana Museum for Plate 3.2 'Watercolour painting of Greenmarket Square', Plate 3.3 'A trekboer camp', Plate 3.4 'The freed slave' by F. T. I'Ons, Plate 3.5 'The Kat River Settlement in 1838', Plate 3.6 'A Xhosa Chief depicted in the 1820s', Plate 3.7 'Chaka, King of Zooloe', Plate 3.8 'An early portrait of Moshoeshoe, dated 1833', Plate 3.11 'Portrait of Theophilus Shepstone', Plate 3.12 'The Revd. Mr Moffat preaching to the Bechuana' by C. D. Bell and Plate 3.19 'The inmates of a Boer concentration camp'; Blackwell Publishers Ltd for Figure 3.2 from *The Geographical Tradition* by D. N. Livingstone; Butterworth Publishing (Pty) Ltd for Figures 3.4 and 3.8 from *The Eastern Cape Frontier Zone, 1660–1980: A Cartographic Guide for Historical Research* by J. S. Nerghe and J. C. Visagie (1985); Cambridge University Press for Figure 3.5 from 'Settlers, the State and Colonial Power: The Colonization of Queen Adelaide Province, 1834–1837' by A. Lester, *Journal of African History*, 39, 1998; Cape Town Archives Repository for Plate 3.13 'Workers in search of diamonds: E 8615', Plate 3.14 'De Beers Compound, *c*.1900: AG 5997', Plate 3.17 'Alfred Milner: AG 14281' and Plate 3.18 'General Jan Smuts, posing with this commando: AG 11926'; Development Bank of South Africa for Figure 6.1, Development Bank of South Africa 1994; Drum for Plate 5.1 from *Drum*, October 1952; The William Fehr Collection for Plate 3.10 'The Cape Regiment leading an attack against the Xhosa in the Amathole in 1851'; Government Printer for Figure 5A.4 from RSA *Annual Report of the Board for the Decentralization of Industry* (1990); History Workshop Photo Archives for Plate 4.5 'A back street in Pimville in the 1940s'; Independent Newspapers for Plate 3.15 'Woman crossing Joubert Street in 1898', the Barnett Collection, and Plate 5.4 'Hector Peterson picture' by Sam Nzima, 16.6.76; Institute for Contemporary History for Plate 4.3 'Barry Hertzog, Jan Smuts and Nicholaas Havenga: FT69'; Local History Museums' Collection, Durban, for Plate 4.2 'Umvoti Magistrate J. W. Cross explains the implications of a hut tax to African chiefs', copyright Local History Museum; Macmillan Heineman ELT for Figures 2.6 and 2.7 from *New Secondary School Atlas for South Africa* by Shuters-Macmillan (1995); Macmillan Press Ltd for Plate 3.10 from *South Africa: A Modern History* by T. R. H. Davenport (1991); Nasou Ltd for

Figures 2.4a, 2.4b and 2.5 from *Senior Geography Standard: New Syllabus 1987* by C. J. Swanevelder, M. K. R. van Huyssteen and J. C. Kotze (1987); The National Library of South Africa for Plate 3.1 'Khoikhoi family in the early 1700's: INIL 6256V'; Pearson Education Ltd for Figure 4.1 from Southern Africa Since 1800 by D. Denoon and B. Nyeko (Longman, 1984); Reader's Digest Association Ltd for Figure 3.6 from *Reader's Digest Illustrated History of South Africa: The Real Story*, Expanded 3rd edition (1994); E. Nel, C. M. Rogerson for Figure 7.1 from 'Restructuring the post-apartheid space economy', paper presented at the United Nations Centre for Human Settlements Conference on Regional Development Planning and Management of Urbanization, Nairobi, 26–30 May 1997; Routledge for Figures 3.3, 3.13, 4.2 and 5.1 from *The Atlas of Apartheid* by A. J. Christopher (1994); Statistics South Africa for Figures 2.10a, 2.10b and 2.10c from 'The people of South Africa, Population census, 1996', *Census in Brief* Report No. 03-01-11 (1996); Yale University Press for Figure 3.1 from *A History of South Africa* by L. Thompson (1990).

While every effort has been made to trace the owners of copyright material, in a few cases this has proved impossible and we take this opportunity to offer our apologies to any copyright holders whose rights we have unwittingly infringed.

# Chapter 1

# INTRODUCTION

## Economic and social boundaries

This book has been written by one historical geographer (Alan Lester) and two development geographers (Etienne Nel and Tony Binns). There are inevitably differences in our approaches and styles which will become apparent in the separate chapters which we drafted. Partly, our differences stem from different scales of analysis. The treatment of a broad sweep of history in this introduction and in chapters 3, 4, 5 and 9 necessitates the identification of general themes, patterns, connections and arguments. The descriptive 'setting the scene' in chapter 2 and the more detailed treatment of the complexities of government policy over the last six (post-apartheid) years in chapters 6, 7 and 8, however, give rise to a more empirically laden and 'factual' rendition. Nevertheless, through our close co-operation, we have striven to ensure that the book as a whole hangs together, providing a coherent and, we believe, a reasonably comprehensive account of South African society and its geographies in the past, the present and, as far as is possible to predict, the future. By combining our specialisms, we have tried to transcend academic enterprises which are all too frequently pursued in isolation, thus hopefully generating new insights. These are pulled together in particular in the concluding chapter 9. Here, rather than simply summarizing what has gone before, we interpret the current, post-apartheid, transition in South Africa in the light of the country's previous, apparently 'structural' transformations. This long-term analysis suggests that whatever pot of gold the emergence of the 'Rainbow Nation' (as post-apartheid South Africa has popularly been called) heralds, it is unlikely to be distributed very much more evenly than the country's wealth has been in the past.

The most obvious result of our endeavour to link the past and the present in this book is an awareness of the ways in which the legacies of colonialism and apartheid constrain both current and prospective development programmes. One of our central objectives is to explain how the relative deprivation of people currently living in places such as former 'homelands' and townships in South Africa was shaped by the politics of the past, and why it remains intractable. In order to assist those readers who are relatively unfamiliar with South Africa's social and spatial structures, we have included not only a 'setting the scene' chapter (2), but also summary boxes between chapters 3 and 4 and 5 and 6, which distil some of the key inherited features of South Africa's human geographies – features whose origins are discussed at greater length in the historical chapters 3 to 5.

A second, related objective of the book, pursued in chapter 8, is to demonstrate how the new black-led South African government's attempts to deal with the social structures and geographical spaces of exclusion can be seen in comparative context, particularly that of the African continent as a whole. This objective in turn means that we make a particular point of considering South Africa's historic and contemporary connections with other places in the world. Indeed, in the concluding chapter, we argue that such connections have been vital in many ways to the maintenance of South Africa's extreme forms of inequality.

An awareness of South Africa's position within a wider, global political economy, brings us to another, deeper strand of analysis connecting those sections dealing with past, present and future in this book; one which ranges beyond a simple rendition of apartheid's legacies. This concerns our ideas about the underlying processes by which social groups are formed and the means by which they compete and co-operate. All individuals in all societies act within and across certain created 'social boundaries'. These boundaries consist of the differences which mark off one set of individuals from another to create social groups, each with a particular history, sociology and geography of relative privilege and deprivation. Neither historical analysis nor development studies can progress without a realistic appraisal of dynamic boundary-forming mechanisms; that is, the means by which competing or co-operating social groups are demarcated. Shifts in the definition of social groups inform the key changes of the past, and the parameters within which future development will occur. South Africa represents a particularly rich terrain in which to study the kinds of social group boundaries which were originally formed under colonialism, since it was the first part of the African continent to be extensively colonized in the modern era and the last to make the transition from white minority to black majority governance.

Of course, the ways that individuals coalesce and interact within and across social group boundaries are not predictable. Historians are in a process of continuous debate as to the most important forms of co-operation and competition in the past, while development specialists find the identification of contemporary social boundaries and prescriptions for a more benign and inclusive society similarly contentious. Perhaps the greatest criticism of modern, universal prescriptions for 'development', is that they overlook social boundaries which have been constructed differently in different regions. With all their complexity and contingency, however, certain features of social boundaries can be identified – features which inform both the historical and the contemporary sections of this book. The remainder of this introduction sets out what we see as the most significant social group boundaries which have shaped, and which continue to shape, South African society, with a few examples (elaborated in the following chapters) of the ways in which they have done so.

## Race and class

First, we can note the general coincidence between the allocation of scarce economic resources and the boundaries marking perceived social, cultural and physical difference. All societies are materially unequal, and, as we will see in chapter 2, and more particularly in chapter 6, South Africa's is more so than most. Economic resources are concentrated among what is, relatively, a very small group. Since its colonization, the most obvious boundary defining privileged and excluded groups in South Africa has apparently been a 'racial' one. From the nineteenth century (if not before) until recently, the material privilege of South Africa's 'white' colonists and most of their descendants has been more or less sustained, legitimated and consciously understood by them as an inevitable by-product of 'racial' difference.

This 'racial' difference, however, is far from objective. Even prior to extensive colonization, the scientific attempts of European scholars to categorize the world's population rigidly by race were always destabilized by the arbitrary boundaries which had to be drawn between supposedly discrete racial 'types' (Banton 1987). Hall (1987) notes that only about 10 per cent of humankind's genetic variability lies behind those characteristics normally ascribed to 'racial difference', and extensive interchange between regional gene pools, especially over the last five hundred years of colonial and post-colonial 'contact' has broken down many of the genetic distinctions that there once were between isolated concentrations of people. As the case of South African 'Coloureds' (people commonly identified as being of 'mixed race') shows particularly starkly, in a colonial context of imprecise genetic exchange, 'racial' boundaries have had to be constructed and reconstructed deliberately, and 'artificially' through time. This means that both blacks and whites have been socially and culturally, rather than 'naturally' or biologically defined. Correspondingly, in South Africa and elsewhere, it has taken the exercise of social and political power to define and defend privilege according to 'race'. Perhaps the most frequently cited manifestation of the artificiality of South Africa's racial boundaries concerns the reclassifications which occurred on an annual basis during the apartheid period (1948–c.1990). Even under a state system explicitly geared to racial classification and separation, each year hundreds would successfully apply for reclassification to a different and supposedly 'naturally' distinct racial group (West 1988; Unterhalter 1995).

Although 'racial appearance', however arbitrary it may sometimes be, has constituted the most obvious of modern South Africa's social group boundaries, it has never been the only one, and it has always operated in association with other kinds of socially constructed difference. During the early colonial period, in which the Cape was administered by the Dutch East India Company, one can make a case that descent and religion, rather than skin colour, were seen as the most important determinants of status. By the late eighteenth century, however, as we will see in chapter 3, apparently 'racial' distinctions between colonists and darker 'others' were being deployed to defend minority privilege. More significantly for our purposes, since the late nineteenth century, the 'modern' boundaries of class have both reflected and guided wealth production and distribution, and these distinctions of class have become entangled with those of 'race' in various ways (see Bundy 1992).

In our analysis, we avoid the deterministically Marxist view that capitalist class boundaries exist objectively, just as we avoid similar assumptions about 'racial' differences. Rather, we agree with E. P. Thompson, who sees class not as something fixed or given, but 'as something which ... happens (and can be shown to have happened) in human relationships' (Thompson 1980: 8). His point is that relations of production and consumption, relations between the classes controlling resources and those who have been dispossessed of them, emerge not according to some abstract 'logic' of capitalism, but through specific historical struggles within particular societies. As Thompson continues, these relationships 'can only be studied as they work themselves out over a considerable historical period' (Thompson 1980: 11).

While we accept that notions of both class and race difference are socially and culturally constructed in these ways, the interaction between them remains problematic. A superficial impression of modern South African society would indicate a basic divide between a white capitalist class and 'labour aristocracy' on the one hand and a dispossessed black proletariat on the other. But such a pattern is, and always has been, immensely complicated. Three features of South African society in particular have long disrupted any neat coincidence between race and class. First, the continuing existence of a vast African peasantry which has by no means been completely dispossessed. Although, during the late nineteenth and early twentieth centuries, the prosperity of that peasantry was being whittled away by increasingly powerful colonial states and by influential capitalists, often acting in conjunction, it was still resilient enough to remain 'uncaptured'. By this, we mean that Africans were able to force certain concessions on the part of the colonial state and capital, including continued (if diminishing) access to land (Cooper 1981; Lacey 1981; Marks and Rathbone 1982). Many Africans who have been migrant or even apparently fully urbanized workers for some time, still see themselves as members of remnant, rural-based chiefdoms, the identities of which are traced back to the pre-colonial period. They endeavour still to recreate a subsistence in which they will be at least partially autonomous of capitalist employers (Trapido 1971; Beinart 1992; Delius 1996). As Cell argues, such peasants 'are not proletarians or even lumpen-proletarians. Their field of manoeuvre may be narrow and hemmed in. Nevertheless they have something to defend' (Cell 1982: 111). Casting them in the role of a completely proletarianized working class like that of Europe therefore raises considerable difficulties.

Secondly, the coincidence between class and race is complicated by the efforts of blacks to attain a respectable class status within colonial and apartheid society (see Ross 1999). Those efforts have persisted despite all attempts to 'level down' black communities. Indeed, in some cases they were assisted by the white state's need for collaborators, or at least for compliance, among certain black groups. The complexities of social status for these 'interstitial' or 'in-between' groups are explored further in chapters 3 and 4.

Finally, the identification of class with race has been complicated by the failure, despite state assistance, of many whites to attain unambiguous class dominance, and by the tendency of large numbers of whites to slip through the net of material privilege during periods of economic difficulty. While there has been a broad coincidence between class and race in modern South Africa then, it has never been a straightforward one. And when even that broad coincidence was threatened, for instance by the material difficulties of poor whites, it had to be maintained through conscious political action by white governments (chapters 4 and 5).

In recent decades, a new black middle class has emerged in South Africa. It first developed out of those 'interstitial' groups which helped the state to administer black areas, but it became increasingly broad-based as the education system incorporated more blacks and as employers demanded more skilled labour of them, creating a black 'labour aristocracy' during the 1970s and 1980s (chapter 5). Since 1994 and the demise of apartheid, it has swollen and achieved new levels of aggregate wealth as bureaucratic and commercial opportunities have been opened under a black-led government (chapters 6 and 7). Because of these developments, the boundaries of race and class are commonly seen to have been reconfigured quite dramatically in recent years. To what extent is a question that we will be asking in the latter chapters of this book.

## Ethnicity

Conscious political action to protect class and 'race' privilege has been intimately associated with another kind of socially constructed boundary – that of ethnicity. Afrikaner nationalism, based on a vision of 'ethnic power mobilised' (Adam and Giliomee 1979), as we shall see in chapters 4 and 5, became a powerful influence in the creation of modern systems of segregation and apartheid. It can be explained, at least partially, as an attempt to protect the class position of poor, Afrikaans-speaking whites, threatened within a context of imperial conquest and capitalist 'development'. In this case, the ideological construct of ethnicity was deployed 'in the battle to cut the cake of gross domestic product' in a particular way' (Lonsdale 1998: 295).

Despite their longer genealogy, the boundaries of ethnicity between Africans, which are alluded to throughout the text, are also historical phenomena which need to be explained rather than taken for granted (Vail 1989). Immediately before the early nineteenth-century colonization of the South African highveld, African 'tribal' identities were in the process of being dramatically reconstructed through the process conventionally known as the 'Mfecane', discussed in chapter 3. In general, however, 'Politicized ethnicity ... is widely accepted in African studies [as being] a modern phenomenon' (Lonsdale 1998: 295). Very different, politicized ethnicities such as the 'Zulu-ness' propagated by the Inkatha Freedom Party (IFP) in recent years, and the 'Coloured-ness' of the Western Cape, which is now resurgent, have been re-imagined and re-inscribed in response to the particular political circumstances of the twentieth century.

African ethnic boundaries have complicated patterns of resistance to racial structures of dominance in South Africa. Before relatively widespread and overt resistance to segregation and apartheid could be co-ordinated across the country, a black South African identity had to be constructed, even if incompletely, by transcending longer-established and more divisive ethnic and regional distinctions. The persistence of these prior tribal and ethnic affinities meant that Africanist movements were long frustrated in their efforts to mobilize mass opposition to white colonial rule (Cooper 1994). The ANC, which sought to mobilize a united black front against the apartheid government during the late twentieth century, for instance, found itself in conflict with the more particularist notion of Zulu identity propagated by the Inkatha movement in KwaZulu and Natal. South African historians have only recently realized the continuing potential for African ethnic boundaries to be politicized in modern South Africa, as they have been in many postcolonial states to its north (Beinart and Bundy 1987a; Maré 1993; chapter 8).

## Gender

A more recent historiographical and developmental concern than that with either race, class or ethnicity is with the most remarkably persistent and universal boundary of all – that of gender. As Mager has recently demonstrated in a case study of the mid-twentieth-century Ciskei, and as this book all too cursorily indicates, 'historical processes look different when relations between the sexes are seen as integral to them' (Mager 1999: 1). In both African and colonial societies, women have been excluded in varying ways from material and social privilege relative to men. The compromises reached between African and colonial powers in the nineteenth century acted, if anything, to reinforce those gendered boundaries which were common to both cultures. They created 'a patchwork quilt of patriarchies', in which 'everywhere women were subordinate to men' (Bozzoli 1983/1995: 126; Walker 1990a: 1).

These gender systems in turn were modified with the progressive industrialization and urbanization of twentieth-century South African society. As both the colonial authorities and African men attempted to narrow the spaces for African women's manoeuvre in the cities, gendered boundaries were continually redrawn. In the case of black women, the particular form of exclusion which these gendered boundaries defined was usually reinforced by further exclusions on the grounds of class and race (Bonner 1990). As Mager puts it, 'black women stand in multiple jeopardy in classist, racist, sexist societies' (Mager 1999: 10, citing D. K. King).

The significance of race, class, ethnic and gender boundaries lies in the fact that they are both imagined and very real. They are imagined in the sense that notions of difference based upon them provide the main ingredients for people's social identities. Regardless of any biological difference upon which they may be premised, individuals are socialized into groups defined by these socially constructed racial, class, ethnic and gender boundaries. Even though they have never met the vast majority of the members of their 'imagined community' (Anderson 1991), people conceive of themselves as sharing a certain race, class, ethnicity or gender identity, and often excluding those who do not share 'their' characteristics. It is the material consequences of such exclusion, the inequalities to which it gives rise, which make imagined social boundaries 'real'.

'Imagined communities' are necessarily flexibly imagined since their boundaries are capable, as we have hinted, of interacting in various ways. For instance, those white women who struggled to gain the women's vote in South Africa during the 1920s and early 1930s did not do so in the name of women as a whole. Rather, they asserted their claims to citizenship on the basis that white women were allies of white men in the struggle to maintain dominance over Africans of both sexes. In this case, racial boundaries producing solidarity among whites were clearly seen by feminist reformers as being more significant than gendered boundaries, which would have suggested an alternative alliance of white and black women against white and black men (Walker 1990c). White South African women's commitment to 'racial' supremacy has also enabled the burden of their child care to be placed on black women employed as domestic servants. This in turn has often put black women's role as mothers to their own children under severe strain (Jochelson 1995). Such features of South African society indicate that racial boundaries have generally been considered to be of a 'higher political order' than gendered boundaries, despite (or maybe because of) the entrenched nature of the latter. Thus, it has not been the case that, within the common framework of gender, black and white women have sustained tensions over their 'racial' difference. Rather, within the framework of 'race', white men and women and black men and women have waged largely separate struggles over gender.

African nationalists' early politics provides a further example of the complex intersection of different social boundaries, in this case those of race and class. The founders of the African National Congress (ANC) did not initially seek to represent all Africans who were excluded from political rights. Instead, they were most concerned to foster mutual recognition between whites and middle-class Africans, thereby ensuring 'respectability' for the latter. In doing so, they emphasized those characteristics which set them, as members of an elite middle class, apart from poorer and less 'respectable' Africans (Lodge 1983). It was only when avenues for the advancement of a black middle class had been effectively closed off by the white South African state that a black nationalist struggle was built upon both 'race' and class alliances.

These examples demonstrate how gender boundaries can be cross-cut by race and how racial boundaries can be cross-cut by class, but there are many such shifting permutations and combinations which structure politics and influence concrete experiences of wealth or deprivation (see Stasiulis and Yuval-Davis 1995). As Anne Stoler and Fred Cooper note, analysts 'have become increasingly confident that race, class and gender are moving categories', but in order to produce meaningful interpretations, we still have to account 'for how their political saliences shift and which of those affiliations shape political choices in a particular place and time' (Stoler and Cooper 1997: 25). We hope to go some way towards such an account in the historical and contemporary chapters that follow

## Space and power

It is of just as much importance for the planners of the 'new' South Africa as it was for those of colonial and apartheid South Africa that social boundaries, with their implications for economic dominance and exclusion, interact with spatial location.

Socially constructed differences, particularly when reinforced by state intervention, can become manifest in enduring spatial distinctions. A prime example is the removals which modern South African governments inflicted on established black residents in urban and rural areas. Through fear and the desire for control, but also through idealism, as we will show in chapters 4 and 5, these residents were relocated to peripheral townships, just as those defined by their class, their poverty or their race were similarly relocated and (often unintentionally) marginalized in modern European and North American cities.

Such imposed spatial distinctions in turn can help construct new identities – new imagined boundaries. John Western (1996), in his study of Cape Town, indicated that Coloured identity was shaped under apartheid as much by spatial relocation as by anything else, and numerous analysts have identified the ways in which the compression of black South Africans of varying status in fairly homogeneous townships gave rise, unexpectedly, to new forms of racial solidarity and oppositional politics (Smith 1982; Lodge 1983; Robinson 1992; Parnell and Mabin 1995; Evans 1997). The new government in South Africa has inherited not just a sociology of difference and inequality then, but also a geography which has both reflected, and helped to create and maintain that sociology. While it runs throughout the book, this mutually affecting relationship between space and social structure is explored most explicitly in chapters 5 and 6. The summary boxes between these chapters identify those spatial structures which we consider to be the most significant legacies of South Africa's modern past, and which now pose the greatest developmental dilemmas.

The mechanisms which maintain social and spatial distinctions are as complex and ambiguous as the boundaries which define them. They require force, but they also require a degree of active collaboration and passive consent on the part of many of those who are relatively excluded from privilege. In the interpretation of modern South Africa, until recently, brute force has been emphasized, but with the retrospect that the demise of apartheid has allowed, the multiple ways in which power has operated during the modern period have been more readily identifiable. Power takes various forms which may be analytically separable, but which in practice are often indistinguishable. First there is an obviously economic form of power which, following Marx, is known as exploitation. This consists of 'social relations where there is a direct interaction between the economic advantages of some and the economic disadvantages of others'. Secondly, there is a more intangible form of power often referred to as 'domination'. This can be defined as 'control, command or authority in relation to other human beings' (Robinson 1996: 17). Domination, like exploitation, can be effected not only through coercion, but also through consensus, routine or hegemony (the acceptance on the part of the dominated of the cultural norms of the dominant). While power of both kinds in South Africa has relied 'to a large extent on military force to provide space in which it may then operate', it has subsequently become dependent for its everyday functioning on the passive consent of many, or at least on people's self-regulation during the periods when direct force is absent (Foucault 1977; de Kock 1996).

Much of the consent required for the perpetuation of power relations of these varying kinds between social groups in South Africa, however, stems from 'accommodations' which had to be reached between colonists and the colonized. These accommodations dated from the initial, piecemeal European conquest of the region, and they were often the result of European concessions in the face of indigenous resistance (Marks 1986b; Cooper 1994; Lester 1998a). Subsequently, even under well-established colonial and white minority rule, further compromises had to be made in order to secure consent rather than prompt rebellion. Mission-educated Africans who sought the approval and acceptance of the 'civilized' ruling classes, for example, were encouraged to participate, even if as juniors, in the government of the Cape through the creation of a non-racial franchise during the mid-nineteenth century. This concession can be seen as a colonial safety valve – 'a basis for collaboration, for acquiescence, accommodation and consent' (Cell 1982: 19). When the black elite's opportunities for legitimate political action were being closed under the more coherent system of segregation devised during the early twentieth century, its representatives were still 'offered a basis for collaboration that the harder, more inflexible forms of white supremacy could never have permitted or achieved' (Cell 1982: 19; Marks 1986). Even if they were increasingly denied rights of political participation, a small minority of blacks 'might overcome the barriers of class. In their success lay hope, not only for themselves but ultimately for their people' (Cell 1982: 19).

It was not just the representatives of a black elite, but also more 'traditionalist' Africans, who were positioned ambivalently with regard to colonial power. Although their autonomy was undermined during the twentieth century, many African chiefs and headmen were able to enhance or even acquire some aspects of their power through the forms of indirect rule upon which modern segregation and the 'bantustans' or 'homelands' of apartheid were based (Dubow 1989; Evans 1997). In respect of gender, African men, as well as Indian men working on the plantations of Natal, could find their power over women bolstered by colonial authority (Walker 1990a; Beall 1990). The examples of African policemen and court interpreters serve further to demonstrate that everyday life in colonial and apartheid South Africa was 'an arena of complex exchanges rarely reducible to the zero-sum of dominance or resistance' (Comaroff and Comaroff 1997: 34). It is true that throughout South Africa's history since colonization, actual force or the threat of force lurked behind white dominance. We will present evidence for this in abundance throughout chapters 3 to 5. And yet, for extended periods, the ambiguous positioning of certain black groups allowed various models of that dominance, from slavery to apartheid, to be 'largely self-enforcing' (Cell 1982: 19).

## The state

The attempts to identify social boundaries, to examine their implications for the exercise of power, their changes through time and their interactions with spatial arrangements, run throughout this book. As we have indicated already, and as will become increasingly evident in chapters 4 and 5, the modern South African state has reflected, reinforced and helped to create these boundaries and the patterns of privilege which they describe. The modern state is fragmented and differentiated internally, as we will suggest below, but as an institution, it nevertheless possesses a particular power exercised to the same extent by no other group in society – that is, the power to police and punish 'transgression' (Jochelson 1995). At this point then, it is as well to set out our conception of the historical evolution and contemporary functioning of that state.

In chapter 3, the shifting relationships between the Dutch East India Company, which first governed the colonial Cape, and its subjects are touched upon only briefly. The ways in which British colonial states came to represent predominantly the interests of colonial settlers, and the tensions affecting the nineteenth-century Boer republican states, will be set out in the same chapter. From an examination of the advances and reversals experienced by government officials through the nineteenth century, it becomes clear that 'The colonial state, in South Africa as elsewhere, was always an aspiration, a work-in-progress, an intention' rather than the coherent locus of power which its administrators intended it to be (Comaroff 1998: 341). However, power became far more consolidated within the apparatuses of the modern, twentieth-century state, founded in South Africa's case in 1910 with the Act of Union.

Rather than viewing this modern, post-1910 state as an institution prone to being 'captured' and used in the interests of one class or another, we see it as having its own 'particular interests associated with territorial sovereignty'. However, as Robinson insists, it is important to bear in mind that 'it is an institution which is deeply divided within itself' (Robinson 1996: 3; see Bonner, Delius and Posel 1993). The state's relative autonomy from groups in the wider society is based upon its co-ordination of a range of activities extending across the entire territory for which it is responsible. That very responsibility also undermines its coherence since it is split not only into local and central governments, but also into departments which may vie with one  another. As we shall see in chapter 4, a salient feature in the rise of apartheid was the effective 'colonization' of other branches of the South African state by the Department of Native Affairs, while one persistent reason for apartheid's limited effectiveness in securing cities for the white population was the discrepancy between local and central government imperatives (Bonner, Delius and Posel 1993).

Aside from its (fragmented) territoriality, the state's relative autonomy is also guaranteed by its concern with what Foucault calls 'biopower' – the ordering of society 'under the guise of improving the welfare of the individual and the population' (Dreyfus and Rabinow quoted in Robinson 1996: 12). This concern, for all its selective application, means that even the apartheid state was driven by professedly 'neutral' notions of governance and statehood, as well as by the straightforward 'economic logic' of capitalists and the 'racist pressures' of white voters (Robinson 1996: 21).

Nevertheless, the modern state, like individuals, has limits placed upon its autonomy (Evans 1997). We will argue in chapter 4 that a modern system of segregation was formulated by the early-twentieth-century state partly as a compromise with African authorities – authorities which the state did not have the power to mould entirely to its will. By the same token, Western notes that even one of the most callous exercises in apartheid state planning – group areas' removals of blacks from inner cities – was constrained to some extent by the pressure which local groups were able to bring to bear (Western 1996). On the largest scale, the state must seek to avoid both capitalist accumulation crises which would destabilize the national economy, and legitimation crises which would lead to a loss of electoral support among its constituents and, perhaps particularly in modern South Africa's case, the loss of quiescence among its subjects (Habermas 1976). The tightrope walk which these contradictory demands often entail can, at times, lead to a structural transformation, such as that which saw the recent demise of the apartheid state in South Africa and the construction of a new, black-led and democratically elected one. This particular instance of structural state change will, of course, be investigated in far greater detail in chapters 5, 6 and 7.

Finally, there is a further constraint on state autonomy which features particularly prominently throughout this book. This arises from the state's situation within a broader network of power relations. In a modern, global political economy, the South African state has been rendered partially dependent on the support of other states, especially those which are considered 'key players' on a world stage, such as Britain in the first half, and the USA in the second half, of the twentieth century. We will see in chapters 4, 5, 7 and 8 how state policies ranging from segregation through apartheid to post-apartheid development have been conditioned in many ways by the influence of such overseas 'players'.

## Transformations

This book's historical grounding allows the recent abandonment of apartheid and the emergence of the 'new South Africa', to be set in both temporal and spatial context. The post-1990 transition was not the first dramatic shift in social and political boundaries in South Africa, and nor will it be the last. Indeed, the potential and, perhaps more importantly, the limitations, of the current transition, can be appreciated only if it is conceived alongside its historical antecedents. Structures of privilege and exclusion are continually being defended by their beneficiaries and acquiesced in or resisted by the excluded, resulting in both conflict and compromise, but, as we have suggested above, every now and then, and often through some outside intervention, a major transformation in those structures occurs. The abandonment of apartheid is one such transformation, the extent of which will be questioned in the later chapters of the book. But, aside from initial colonization, a number of prior transformations in South African society feature prominently in this account, each of which raises issues concerned with the post-apartheid transition.

First, during the early nineteenth century, British liberal ideas of 'proper' economic and social relations undermined the social order established under the Dutch colonizers. A society founded on slavery and regulated by a patriarchal and paternalistic ethos rooted in the family was, after certain compromises, replaced with one founded on supposedly 'free' labour, and regulated by the British colonial state. As we shall see in chapter 3, this transition prompted many Dutch-speaking farmers to leave the Cape Colony and begin the colonization of the interior. It also laid the foundations for South Africa's 'modern' capitalist development, with its segregationist model of racial exclusion.

This early-nineteenth-century transition was the result of a powerful British metropolitan influence on South Africa's colonial society. During the remainder of the nineteenth century, the colonies in South Africa, like those elsewhere, engaged in an exchange of ideas about racial difference, economic progress and governance with dominant groups in Britain itself. This kind of exchange between colonies and their metropolitan powers is now being widely identified as one of the most important influences shaping global modernity as a whole (Hall 1996b, 1999; Stoler and Cooper 1997; Comaroff and Comaroff 1997; Lester 2000; Nash 2000). One of its most significant outcomes was the idea of Europe 'bringing civilization to the natives' through colonial rule. The fostering of 'civilization' was seen as a real political programme by British humanitarians in the Cape and in the metropolis, and as a convenient justification for a more mercenary exploitation and domination by many colonial settlers, but it was constructed by the British population as a whole as a worthy, 'moral' project.

In two respects, as we shall see in chapters 7, 8 and 9, this colonial trait can be seen as continuing today. First, the new South Africa remains constrained by the dictates of a world economy with its core in the old metropolitan centres as well as in a newer one, the USA. The country's insertion into a network of powerful globalized relations today mirrors that which was brought about by British imperialism in the early nineteenth century. Secondly, and more contentiously, endeavours by the 'West' to encourage the 'development' of 'Third World' countries, including South Africa, along certain lines today, perhaps mirror nineteenth-century efforts to bring 'civilization' (and economic domination) to the 'benighted heathens'. Some would argue that John and Jean Comaroffs' characterization of nineteenth-century colonialism bears a remarkable resemblance to contemporary Western-led efforts at global 'development':

> colonizers set out, on the one hand, to seed modern European techniques and relations of production among subject populations; yet on the other, they often sought cheap, well-regulated labour for purposes of farming, mining and manufacturing – which led them to exploit and perpetuate, not close, the gap between the 'primitive' and the 'civilised' (Comaroff and Comaroff 1997: 258).

A second broad and dramatic transition occurred in South Africa in the late nineteenth century, following the discovery of minerals on an enormous scale. The rapid exploitation of diamonds and gold in the South African interior coincided with a new, more aggressive British imperialism and prompted a tremendous influx of European capital. The region's economic geography was quickly transformed by the growth of industrialized urban centres, its political geography was forcibly reshaped by British armies and a social geography of racial segregation was designed more coherently, if still gradually and haphazardly, by the state and powerful capitalists acting broadly in unison to support more intensive capitalist development.

The particular form that this capitalist development took reinforced 'racial' distinctions between whites and blacks, and between Africans, Indians and Coloureds. As has been indicated in the brief discussion of Afrikaner ethnicity above, however, it also resulted in the redrawing of boundaries within the white population as many less commercially orientated Afrikaners were marginalized relative to English-speakers and capitalists. Full control of the state after 1948, however, gave Afrikaner nationalists the chance to broaden the boundaries of white material privilege so as ultimately to include most Afrikaners in an inter-ethnic alliance with English-speakers. The strategy of apartheid can be seen as a third major reformulation of social and political structures. It was intended to ensure that white privilege as a whole, which was seen to have been undermined by further industrial and urban growth during the Second World War, could be preserved in the challenging conditions of the mid-twentieth century. Aside from its immediate human costs, apartheid state intervention has made the future transformation of social and spatial structures of exclusion vastly more difficult.

Against this deeper context, it becomes clear that the dramatic abandonment of apartheid may represent but a further shift in those dynamic boundaries which have long defined incorporation and exclusion from privilege in South Africa. A glance at today's BMW-owning, black bureaucratic, professional and business elite in South Africa tells us that the social boundaries defining privilege have been redefined so as to embrace, at least, a powerful black middle class. How far they will be further redefined remains to be seen, but this most recent in a series of transformations does not mean that currently privileged groups (of all races) have ceased to define themselves as different from the 'others' who threaten their status. The recent political transition has, however, encouraged a variation in the composition of privileged groups, resulted in a new set of expressed state strategies, including one for overcoming gender domination, and allowed the exercise of privilege in more subtle ways.

The legacies which characterize the 'new South Africa' then, do not just consist of inherited economic constraints and an unequal social and geographical distribution of resources; they also include the underlying mechanisms of inequality by which a modern capitalist society operates – mechanisms based with varying degrees of subtlety on imagined and real distinctions between privileged and excluded groups. The value of an analysis which bridges the pre- and post-apartheid periods lies in its deeper contextualization of recent shifts in these boundaries and in the realization of their implications for 'development'. However, while this analysis attempts to cover such ground, it is important to point out that it is by no means intended to be a fully comprehensive survey of South Africa, nor a definitive prognosis of its future. Issues such as South Africa's shifting foreign policy and its recent educational provision, for instance, are touched upon only briefly in an analysis which concentrates selectively in order to extend itself chronologically. Furthermore, it should be noted that the book is written during a time of rapid transition. While it seeks to define the more structural inherited parameters within which future developments will occur, it does not pretend to predict very far beyond the 1999 general election.

## Organization of the book

Chapter 2 sets the scene for readers who are relatively unfamiliar with South Africa's natural, social and economic environment. It provides some of the basic information which enables the reader who has not studied South Africa before to contextualize what follows. After a brief consideration of the region's pre-colonial history at the beginning of chapter 3, the historical chapters (3 to 5) deal with the period from 1652 when the colonization of the western Cape began, to 1994 when government based on overt racial distinctions finally ended. Chapter 6 consists of a brief 'snapshot' of the patterns of inequality established in South Africa by 1994. This provides the launching pad for the analysis of changes in development practice and policy in the recent post-apartheid period (chapter 7). In chapter 8, South Africa's post-apartheid development initiatives are viewed in the light of other African states' attempts to transcend the inequalities inherited from colonialism, and the country is set more firmly within a regional, continental and global context. Finally, in chapter 9, an overview is presented of the social, economic and cultural forces, operating at scales from the local to the global, that have shaped modern South Africa since its colonization.

Throughout each of these chapters, key interlocking themes, connecting past, present and future, will be intermittently revisited. To separate and define them as clearly as possible, these are:

1. The organization of material privilege and exclusion, especially according to patterns shaped by race, class, ethnic and gender boundaries
2. The mechanisms through which privilege and exclusion have been sustained, legitimated and resisted
3. The spatial distribution of natural, social and political resources
4. The ways in which inherited constraints have affected post-apartheid state initiatives
5. The influence on South African society of its global connections, both historic and current, especially those connections with the metropolises of the world economy.

## Chapter 2

# SETTING THE SCENE

### South Africa's location

Located at the southern tip of the African continent, South Africa covers an area of 1,219,090 km$^2$ between 22° and 35° south (a north–south distance of 1,800 km) and 16° and 33° east (a west–east distance of over 1,500 km). South Africa is Africa's tenth largest country by land area and is more than five times the size of the UK and twice the size of France (Fig. 2.1). With an estimated population of 40.5 million in 1996, South Africa was the fifth most populous African country – after Nigeria (118.4 million), Egypt (64.5 million), Ethiopia (60.1 million) and the Democratic Republic of Congo (48.0 million).

South Africa occupies a dominant economic position within Africa, which far exceeds its position in terms of both physical and population size. South Africa's Gross National Product in 1997 was US$130.2 billion, almost twice the size of the African country with the next largest GNP, Egypt (US$71.2 billion), and more than four times larger than Nigeria (US$30.7 billion), which has the largest population. In terms of Gross National Product per capita, South Africa again had the highest in the continent (US$3,400), and was ranked 45th out of 133 countries by the World Bank

Figure 2.1  South Africa in the world

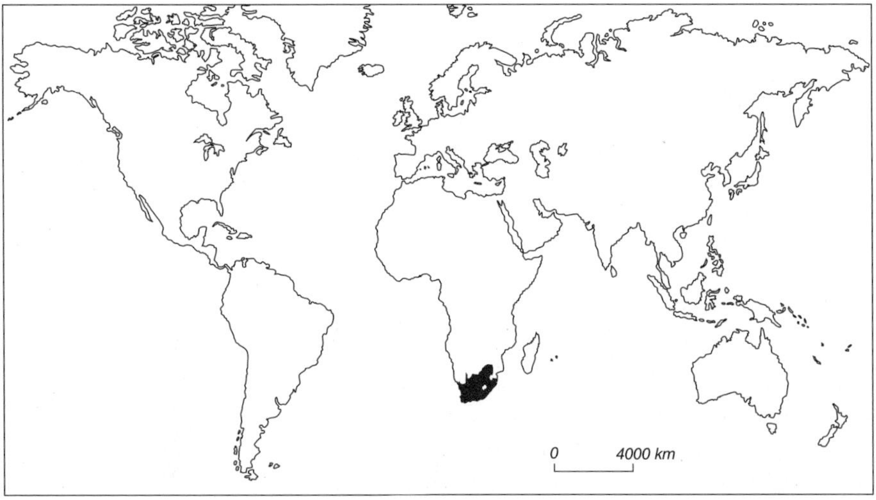

Figure 2.2   South Africa in Africa

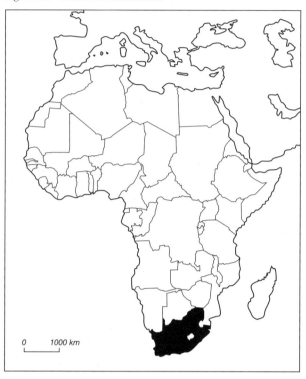

in 1997, not far short of Venezuela (US$3,450 – ranked 44) and Poland (US$3,590 – ranked 43) (World Bank 1998).

South Africa has borders in the northern half of the country with Namibia, Botswana, Zimbabwe, Mozambique and Swaziland, whilst further south the independent kingdom of Lesotho remains as an enclave which is completely surrounded by South African territory (Fig. 2.2). South Africa dominates the Southern African region and has played a leading, if not dominant, role in the Southern African Development Community since it was admitted as a result of its internal reform process, an issue which is considered in more detail in chapter 8. In global political terms, South Africa's strategic importance at the southern tip of Africa has historically had major significance at various times in history, as we shall see in the following chapter. In modern times its strategic position between the Atlantic and Indian Oceans has also been significant, for example when the Suez Canal was closed in 1956.

### Environment and natural resources

South Africa's geological history stretches back over 3.8 billion years, though much of the present landscape has developed relatively recently since the break-up of Gondwanaland some 200 million years ago (Moon and Dardis 1988). The most significant relief feature in South Africa is the Great Escarpment which runs in a wide arc roughly parallel to the coast and between 100 and 200 km inland (Fig. 2.3). The escarpment runs southwards down the eastern side of the country from the Limpopo River on the Zimbabwe border, and then to the south of Lesotho the escarpment continues in a westerly direction, ultimately bending northwards to the Orange River.

The Great Escarpment divides the country into two main physiographic regions – the interior or central plateaux and the plateau slopes and coastal regions. The most impressive part of the escarpment is the Drakensberg, which stretches along the Lesotho–KwaZulu-Natal border and southwards into Eastern Cape province. The highest peaks in this range are Thaba Ntlenyana in Lesotho (3,482 m), Champagne Castle (3,375 m), Giant's Castle (3,312 m) and Mont-aux-Sources (3,298 m). Climatically, the escarpment restricts the inland flow of moist maritime air, whilst historically it was often a major barrier to the movement of people and goods, since road construction across the escarpment was both difficult and costly. From the highlands in the east a number of rivers flow towards the Indian Ocean, for example the Great Fish, Kei, Umzimvubu and the Tugela.

Figure 2.3   Relief, rivers and international boundaries

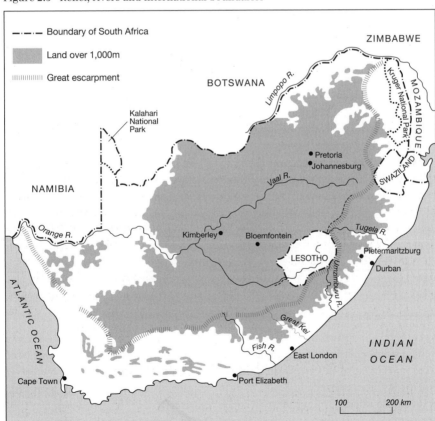

*Source*: Adapted from Christopher (1982)

Plate 2.1 The South-Western Cape coastline

Plate 2.2 The KwaZulu-Natal Drakensberg

The interior plateaux cover about two-thirds of South Africa and rise in a series of steps from west to east, from an altitude of around 1,000 m in the north-west of Northern Cape province to 2,500 m in the Highveld region of the northern Free State and southern Gauteng. The Orange–Vaal river system, which rises in the eastern highlands and flows into the Atlantic Ocean, drains the bulk of the plateaux region, except in the far north where rivers flow northwards and eastwards into the

Plate 2.3  The Highveld, showing power station

Limpopo system and then into the Indian Ocean. The characteristic relief of much of
the central plateaux is subdued, broken occasionally by distinctive flat-topped mesas
and buttes known locally as *koppies*.

The 'plateau slopes' lie below the Great Escarpment. This physiographic region,
which includes the Lowveld (generally below 900 m) in the eastern parts of Northern
Province and Mpumalanga, together with the eastern plateau slopes of KwaZulu-
Natal and Eastern Cape province, is situated to the east of the escarpment.
Fast-flowing streams flow eastwards and through headward erosion have caused the
Great Escarpment to retreat. Also in this broad physiographic region are the Cape
fold mountains in the south of the country, which have a very distinctive landscape
characterized by parallel ridges separated by long valleys, where less resistant rocks
have been removed (see photograph overleaf). The two axes of mountains
(north–south and east–west) meet in the Hex River mountain complex near to the
town of Worcester in Western Cape province, where altitudes of up to 2,260 m are
located. From Worcester, the mountains extend eastwards, eventually linking up with
the Drakensberg. The north–south axis includes ranges such as the Drakenstein and
Cedarberg, whilst the east–west axis consists of the coastal chain (the Langeberg,
Outeniqua and Tsitsikamma mountains) and further inland, the Karoo chain (the
Little and Great Swartberg and the Baviaanskloof mountains) rising to 2,326 m in
the Swartberg, which is situated between the Little and Great Karoo. The latter lies
between the escarpment in the north and the fold mountains in the west and south,
whilst the Little Karoo is situated between the Langeberg and Outeniqua mountains
in the south and the Swartberg range in the north.

South Africa's coastal plain is generally narrow, for example in the south–west in
Namaqualand, where it is also very dry and desert-like. Only in the north–east of the
country, between Richards Bay and the Mozambique border, does the coastal plain
broaden. South Africa's coastline has few natural harbours. On the eastern coast large
lagoons, such as the one at Lake St Lucia, are blocked by sand bars and special

Plate 2.4 Cape fold mountains

measures have to be taken to keep open other lagoon harbours such as at Richards Bay and Durban. The river harbour at East London also needs regular dredging. It might be argued that Saldanha Bay on the south-west coast is South Africa's only good natural harbour.

## Climate

Since most of South Africa lies between the Tropic of Capricorn and the 35° line of latitude, the climate might be broadly described as subtropical with generally warm temperatures peaking in December, January and February and falling to their lowest levels in June, July and August. The country is located within the zone of tropical anticyclones – the Subtropical South Atlantic Anticyclone to the west and the Subtropical South Indian Anticyclone to the east (Preston-Whyte and Tyson 1988). In view of the relative narrowness of the continent compared with the extent of Africa at comparable latitudes in the northern hemisphere, these anticyclones are more stable and have a continuing influence on South Africa's climate throughout the year (Christopher 1982). South Africa's generally dry climate is due to the fact that this is an area of descending air, where the process of rising, cooling, condensation and the formation of precipitation does not readily take place.

Being surrounded by sea on three sides, South Africa's climate is subject to moderating maritime influences. On the west coast the north-flowing Benguela current carries cool water northwards, whilst on the east coast the Mozambique current carries warm water southwards. These currents have a considerable influence on the temperatures of places located at the same latitude, but on different coasts. For example, Durban, situated at 29° 50'S on the Indian Ocean coast, has a mean annual temperature of 20.5°C, whereas Port Nolloth, on the Atlantic coast at 29° 14'S, has a mean temperature of only 14.1°C. Temperatures are also affected by altitude, such that Johannesburg, for example, situated at 26° 12' at an altitude of over 1,600 m on the Highveld, has a lower mean temperature (16.2°C) than Cape Agulhas, the continent's most southerly point (34° 50'), with a mean temperature of 16.8°C. The largest ranges in temperature are experienced away from the coast in the central parts of the country and particularly in the arid northern Karoo. The Drakensberg can experience up to 150 days of frost and most of the interior plateaux experience between 60 and 90 days. Only the coastal belt and the northern part of Northern Province have frost-free periods for more than nine months of the year. The highest temperatures in the country are usually experienced in northern and north-western parts of the country and in the Great Karoo. In January, at the height of summer, the town of Upington in Northern Cape Province is frequently mentioned in weather reports as the hottest place where regular and reliable statistics are collected. Temperatures of well over 30°C are quite common in such places. Unseasonably high temperatures can be experienced on the coastal strip, particularly in late winter and early spring, as a result of Berg winds which may blow for several days or for just a few hours. Differences in atmospheric pressure between the high pressure cell situated over the interior and an approaching low can cause a strong flow of air from the interior towards the sea. As the winds descend from the interior plateaux warming takes place and the hot and dry weather which they produce can be most uncomfortable for humans and animals as well as having a desiccating effect on growing crops (see Figs. 2.4–2.7).

As in so much of the rest of Africa, the availability of water in South Africa is a key controlling factor in both shaping and determining the characteristics of landscapes and human livelihoods. Much of South Africa can be classified as 'semi-arid', with some 70 per cent of the country receiving less than 600 mm of rain per annum. In fact, 20 per cent of the land area receives less than 200 mm – the USA defines an arid climate as one which receives less than 250 mm of precipitation per annum. With the exception of the south-western Cape rainfall totals generally increase from west to east, with the Drakensberg and KwaZulu-Natal coast receiving the largest amounts. Other high rainfall areas are along the windward side of the Cape mountains and the escarpment of Mpumalanga. The smallest amounts of rainfall occur in places such as Namaqualand in the south-west, where the influence of the cold offshore Benguela current means that very little precipitation actually reaches the coast.

Apart from a strip along the south coast, roughly between Cape Agulhas and Plettenberg Bay, South Africa's rainfall is seasonal in character. Whilst the south-western Cape has a Mediterranean regime of predominantly winter rainfall, north of the Cape fold mountains most rainfall occurs in the summer months. In areas of summer rainfall evapo-transpiration rates can be very high, reducing levels of run-off and the availability of surface water for agriculture and human consumption. In the north-west, for example, evaporation losses can exceed 2,750 mm per annum,

Figure 2.4a   Average temperatures for January in degrees Celsius

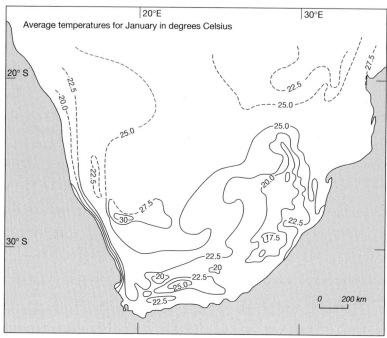

*Source*: Adapted from Swanevelder *et al*. (1987)

Figure 2.4b   Average temperatures for July in degrees Celsius

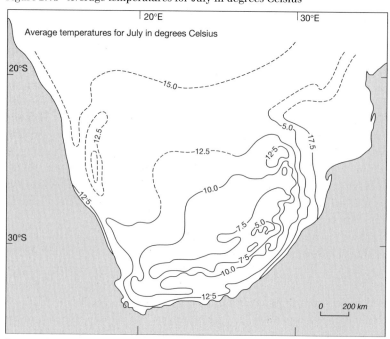

*Source*: Adapted from Swanevelder *et al*. (1987)

Figure 2.5 Average annual rainfall in South Africa

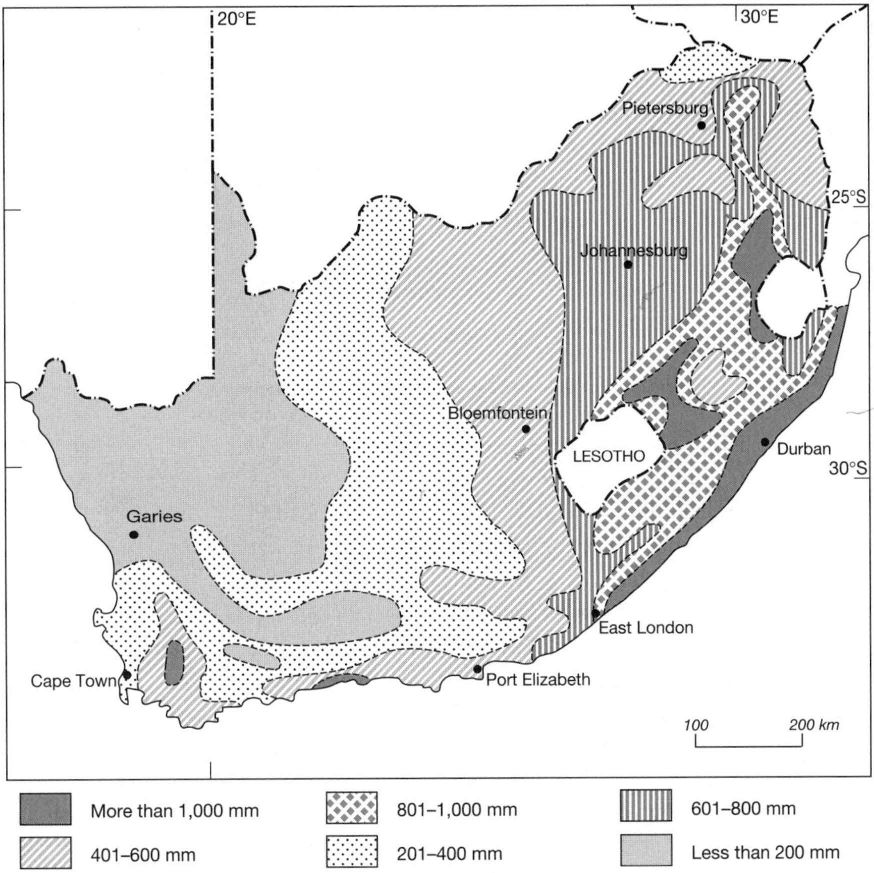

*Source*: Adapted from Swanevelder *et al.* (1987)

whereas along the south and south-east coast such losses are often less than 1,250 mm. As a result, it is estimated that only about one-third of South Africa is suitable for agricultural production without irrigation. Rainfall variability, which concerns the extent to which precipitation at a particular place varies from the annual average, is generally low in the east, but in parts of the north-western Cape it is high, in some years as high as 40 per cent, and in the lower Orange River valley it is frequently 50 per cent above or below the norm.

Throughout its history the provision of an adequate water supply has been one of the key limiting factors in the economic development of South Africa. There have been many attempts to provide more reliable water supplies for rural and urban areas, such as transferring water from the Tugela River system of KwaZulu-Natal into the Vaal River, which is the chief supply source for the economic heartland centred on Johannesburg. More recently, during the late 1990s, the first stage of the ambitious Lesotho Highlands Water Project was completed. This is a joint project between

Figure 2.6  Temperature and rainfall at selected locations

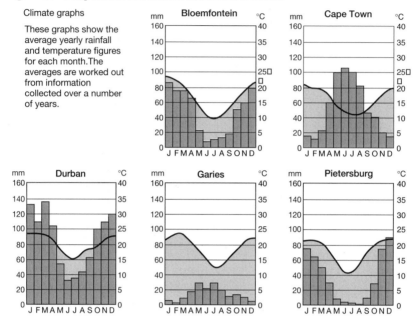

*Source*: Adapted from Shuters-Macmillan (1995)

South Africa and Lesotho designed to move water from the Katse Dam, in the sparsely populated and water-abundant north-eastern region of Lesotho, through a system of tunnels and dams and ultimately to the taps of Johannesburg's rapidly growing population. A further four dams in Lesotho are planned to meet Gauteng's insatiable demand for water. But South Africa is a drought-prone country with below-average rainfall years being much more common than years with above average rainfall. The cyclical nature of the rainfall regime of southern Africa has been studied in detail, revealing that in the period from 1900 to 1988 there have been eight approximately nine-year spells of either predominantly wet years, which average above-normal rainfall during the spell, or predominantly dry years, with below-normal rainfall. It seems that the summer rainfall region of north-eastern South Africa has been most affected by these oscillations, with the most consistently dry spell occurring between 1925/26 and 1932/33 and the most persistently wet spell occuring between 1971/72 and 1980/81. But in general the dry spells have been more persistently dry than the wet spells were wet (Preston-Whyte and Tyson 1988).

It has been suggested that the whole of Namibia and Botswana and more than half of South Africa are under the threat of 'desertification', with large areas of Northern Cape, North West and Northern provinces at high risk (Preston-Whyte and Tyson 1988: 262). However, the issue of desertification is controversial, and since these areas have always been marginal in terms of humans gaining a living from agriculture or pastoralism, livelihood systems have adapted to the rigours of environment over many

Figure 2.7  Climatic regions

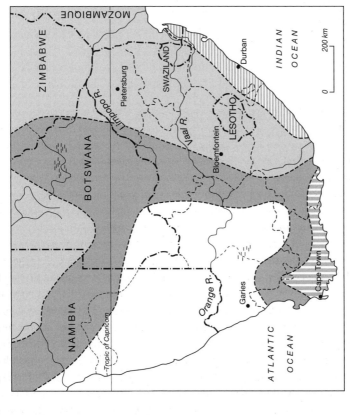

## Climatic regions

### Temperate subtropical

Summers are hot. Rainfall mainly occurs as summer thunderstorms. Winters are dry and clear. Night time temperatures often drop to freezing point especially on the higher ground. Frosts are common and snow is experienced in the mountains.

### Humid subtropical

Summers are warm and humid. Winters are mild. Rain falls throughout the year, although summer is the wettest season. Breezes from the sea keep the coastal areas cool.

### Mediterranean

Summers are hot and dry although there are occasional rainstorms. Winters are cool and wet. In South Africa the Mediterranean climate only extends over a narrow coastal area near Cape Town. Here, hot winds blow from the north in the summer and winds from the sea bring rain in the winter.

### Desert

Very little rain falls here although occasionally there are flash storms. Inland the temperatures can be high, especially when the hot winds blow from the interior. On the coast temperatures are lower because of the cool Benguela current, and fog often occurs.

### Sahelian or semi-desert

This is a region of semi-desert between the tropical region and the true desert. Temperatures are high and the rainfall is low. Rain usually occurs in summer. High evaporation rates reduce the effectiveness of rainfall.

Climatic regions

Temperate subtropical     Mediterranean     Humid subtropical

Desert     Sahelian or semi-desert

*Source*: Adapted from Shuters-Macmillan (1995)

thousands of years (Binns 1990; Thomas and Middleton 1994; Vogel 1994). Vogel (1994) has examined some official responses to drought and concludes that, 'Earlier references to official thinking on drought, clearly illustrate misperceptions of drought impacts and an absence of a holistic understanding of drought that has prevailed in South Africa since the 1920s. Much of this misperception has been rooted in broader socio-political thinking. With the recent political developments in the country, and a growing awareness and motivation to include the rural dimension in drought planning, a dramatic change in drought thinking has occurred' (Vogel 1994: 250).

### 'Human resources'

The October 1996 census was the first time that South Africans were counted as citizens of a democracy and revealed a total population of 40,583,573. This was the first census since 1970 to cover the entire country, since the four so-called 'independent homelands' of Transkei, Bophuthatswana, Venda and Ciskei were excluded after their 'independence' was proclaimed, but since 1994 have been re-incorporated into South Africa. The World Bank estimates that South Africa had an average annual population growth rate of 1.7 per cent between 1990 and 1997, compared with 2.2 per cent between 1980 and 1990 (World Bank 1998). Whilst a growth rate of 1.7 per cent is low by African standards, where most countries still have growth rates above 2 per cent, it is high by North American and European standards. The estimated population in mid-1999 was 43.1 million (EIU 1999).

The 1996 census divides South Africa's population into five groups. In the preface to the census it is stated that '[t]his classification, in common with other countries such as the United States of America which uses a population group-based classification system, is no longer based on a legal definition, but rather on self-classification' (StatsSA 1998). The 1991 England census also used a self-classification of ethnic group with eight categories (White, Black–Caribbean, Black–African, Black–Other, Indian, Pakistani, Bangladeshi and Chinese), and a ninth 'any other ethnic group' category for respondents to enter their own classification. The population groups used in the 1996 South African census were; African/Black, Coloured, Indian/Asian, White, and a fifth category Unspecified/Other. This final category includes the Griquas, 'who preferred to classify themselves as a distinct group (and) are included in the category "other/unspecified" because their numbers were too small for a separate analysis of their situation in this summary publication' (StatsSA 1998). Despite this clear statement in the census preamble, a variety of terms are still used in South Africa to describe the different population groups. For example, most black people are referred to as 'African', whereas others born in Africa but who are not normally referred to as African, are described according to their colour (e.g. white) or ethnic descent (e.g. Indian/Asian). The term 'Coloured', which is used in the census classification refers to mixed race people who, in the past have often been referred to as 'Cape Coloureds'. The Asian group is often in everyday parlance described as 'Indian', yet in fact the group is far more diverse, with origins in such regions and countries as

Malaysia, Indonesia and China. Especially after the rise of the Black Consciousness Movement during the late 1960s and 1970s, the term 'black' has come to be used collectively for Africans, Coloureds and Asians – that is, all 'non-whites'. Finally, the term 'Afrikaner' is still widely used, and whilst it is often used exclusively in relation to the descendants of Dutch settlers, it is more appropriately used to refer to white people whose first language is Afrikaans, some of whom, for example, are descended from French or German settlers. In fact, many Coloured people also regard Afrikaans as their first language, but they are not usually referred to as Afrikaners.

Together, Table 2.1 and Fig. 2.8 show how South Africa's population is distributed across the nine provinces and shows a wide variation in size, density and composition. At the two extremes, KwaZulu-Natal had the largest population (8,417,021), representing 20.7 per cent of the national total, whilst Northern Cape, with less than a million people (840,321), had only 2.1 per cent of the country's people. The overwhelming importance of Gauteng as the country's 'economic powerhouse' is reflected in the fact that it has the second largest population (7,348,423), representing 18.1 per cent of the national total. Since Gauteng is actually the smallest province in terms of land area, its density of population of 432.0 persons per km$^2$ is more than four times greater than that of the next most densely settled province, KwaZulu-Natal (91.4 persons per km$^2$). At the other end of the spectrum, the largest province, Northern Cape, which is predominantly rural and sparsely settled, has a density of only 2.3 persons per km$^2$. Interestingly, the 1996 census revealed that almost 52 per cent of South Africa's population were female. In fact, females outnumber males in all provinces with the single exception of Gauteng, a feature which is presumably due to the in-migration of males for employment in the business and mining sectors.

The relative size, distribution and rates of growth of the different 'racial' groups were crucial variables in many aspects of apartheid policy. Table 2.2 shows the size and distribution of the population according to racial group as revealed in the 1996 census. The African population accounts for well over three-quarters of the total population and in the period 1991–95 had a more rapid growth rate than any other group, and considerably greater than that of the White population (Table 2.3). The distribution of the different racial groups in so many cases reflects aspects of the country's economic history and the legacy of apartheid policies. For example, the largest White populations are in Gauteng and Western Cape, which respectively include Johannesburg and Cape Town, the country's two most important economic poles (Fig.2.9a). The concentration of Indian/Asians in KwaZulu-Natal is a legacy of the large-scale importation of Indian indentured labour in the second half of the nineteenth century to work in the expanding sugar plantations of Natal (see the following chapter). Although the bulk of the workers on these estates are now Africans, many Asian families have remained in the region, engaged in manufacturing and commerce.

The Asian population was subjected to close regulation from the nineteenth century, and the very small numbers in Free State reflects a strong anti-Indian policy during the apartheid era. Provinces with a higher than average African population, such as Northern Province, North West Province, Mpumalanga and Eastern Cape reflect the presence of African 'homelands' during the apartheid era (Fig. 2.9b).

Table 2.1  Population of South Africa by province and gender, 1996

| | Eastern Cape (169,580 km²) | Free State (129,480 km²) | Gauteng (17,010 km²) | KwaZulu-Natal (92,100 km²) | Mpumal-anga (79,490 km²) | Northern Cape (361,830 km²) | Northern Province (123,910 km²) | North West (116,320 km²) | Western Cape (129,370 km²) | South Africa (1,219,090 km²) |
|---|---|---|---|---|---|---|---|---|---|---|
| Male | 2,908,056 | 1,298,348 | 3,750,845 | 3,950,527 | 1,362,028 | 412,681 | 2,253,072 | 1,649,835 | 1,935,494 | 19,520,887 |
| Female | 3,394,469 | 1,335,156 | 3,597,578 | 4,466,493 | 1,438,683 | 427,639 | 2,676,296 | 1,704,990 | 2,021,381 | 21,062,685 |
| Total and % of national population | 6,302,525 (15.5%) | 2,633,504 (6.5%) | 7,348,423 (18.1%) | 8,417,021 (20.7%) | 2,800,711 (6.9%) | 840,321 (2.1%) | 4,929,368 (12.1%) | 3,354,825 (8.3%) | 3,956,875 (9.7%) | 40,583,573 |
| Population density (persons per km²) | 37.2 | 20.3 | 432.0 | 91.4 | 35.2 | 2.3 | 39.8 | 28.8 | 30.6 | 33.3 |

*Source:* StatsSA (1998)

Figure 2.8  Population density by province in 1993

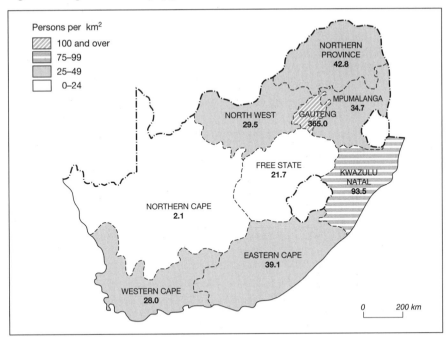

Northern Province, for example, had the homelands of Venda, Gazankulu and Lebowa, whilst North West Province had Bophuthatswana, and Mpumalanga had Kangwane and portions of Kwandebele, Lebowa and Gazankulu. Meanwhile, Eastern Cape had Ciskei and Transkei. The Coloured population is overwhelmingly concentrated in Northern Cape and Western Cape, where the slaves, Khoisan and mixed-race groups from which many of them descended were found, whereas Northern Province has the smallest Coloured population. The 'unspecified/other' group includes minorities such as Griqua and Cape Malay who expressed a strong desire not to be included in the Coloured population (Christopher 1999).

The age and gender distribution of South Africa's population, as depicted in a population pyramid (Fig. 2.10a), displays the broad base and tapering top which is typical of so many developing countries. However, there is some indication of recent changes, with fewer males (12 per cent) and females (11 per cent) in the age category of 0–4 years (bottom line of the pyramid) compared with both the 5–9 and 10–14 age categories. Furthermore, at the top of the pyramid, some 2.2 per cent of women are living beyond the age of 75 (top three lines of the pyramid), compared with 1.1 per cent of men. When population pyramids are constructed for the different population groups, the pyramid for the white population most closely approximates to that of a developed country, whereas the pyramid for the black population has the broad base and sharply tapering top which is typical of many developing countries (Figs. 2.10b–c). This is a feature that was until recently represented to South African schoolchildren as being the result of a 'natural' deficiency of 'development' among

Table 2.2 South Africa, population group by province – numbers and percentages, 1996

| | Eastern Cape | Free State | Gauteng | KwaZulu-Natal | Mpumalanga | Northern Cape | Northern Province | North West | Western Cape | South Africa |
|---|---|---|---|---|---|---|---|---|---|---|
| African/Black | 5,448,495 86.4 % | 2,223,940 84.4 % | 5,147,444 70.0 % | 6,880,652 81.7 % | 2,497,834 89.2 % | 278,633 33.2 % | 4,765,255 96.7 % | 3,058,686 91.2 % | 826,691 20.9 % | 31,127,631 76.7 % |
| Coloured | 468,532 7.4 % | 79,038 3.0 % | 278,692 3.8 % | 117,951 1.4 % | 20,283 0.7 % | 435,368 51.8 % | 7,821 0.2 % | 46,652 1.4 % | 2,146,109 54.2 % | 3,600,446 8.9 % |
| Indian/Asian | 19,356 0.3 % | 2,805 0.1 % | 161,289 2.2 % | 790,813 9.4 % | 13,083 0.5 % | 2,268 0.3 % | 5,510 0.1 % | 10,097 0.3 % | 40,376 1.0 % | 1,045,596 2.6 % |
| White | 330,294 5.2 % | 316,459 12.0 % | 1,702,343 23.2 % | 558,182 6.6 % | 253,392 9.0 % | 111,844 13.3 % | 117,878 2.4 % | 222,755 6.6 % | 821,551 20.8 % | 4,434,697 10.9 % |
| Unspecified/Other | 35,849 0.6 % | 11,262 0.4 % | 58,654 0.8 % | 69,423 0.8 % | 16,120 0.6 % | 12,208 1.5 % | 32,904 0.7 % | 16,635 0.5 % | 122,148 3.1 % | 375,204 0.9 % |
| Total | 6,302,525 100 % | 2,633,504 100% | 7,348,423 100 % | 8,417,021 100 % | 2,800,711 100 % | 840,321 100 % | 4,929,368 100 % | 3,354,825 100 % | 3,956,875 100 % | 40,583,573 100 % |

Source: StatsSA (1998)

Table 2.3   South Africa, population composition and change, 1996

| Racial group | % total population (1996)* | Population change 1991–95 (%)** |
|---|---|---|
| African/Black | 76.7 | 8.5 |
| White | 10.9 | 2.7 |
| Coloured | 8.9 | 5.9 |
| Indian/Asian | 2.6 | 5.7 |
| Unspecified/other | 0.9 | n/a |
| South Africa | 100.0 | 5.7 |

*Sources*: *StatsSA 1998, **EIU 1996

Figure 2.9a   Pre-1994 South Africa showing 'homelands'

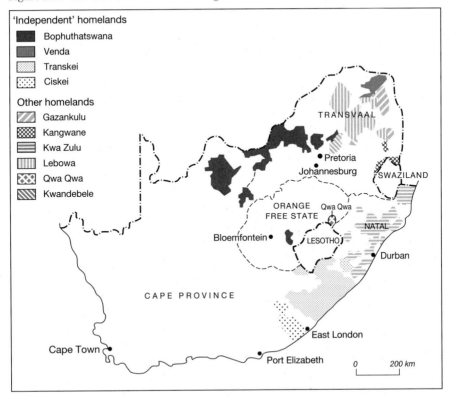

Figure 2.9b  Post-1994 South Africa showing new provinces with former homeland areas shaded in grey

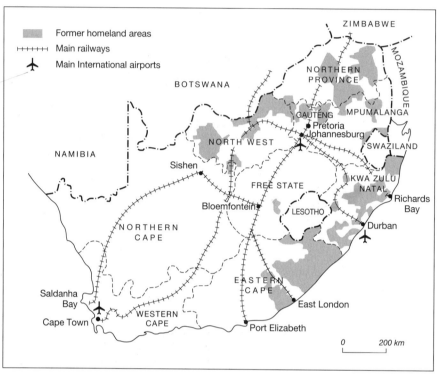

blacks, rather than the outcome of structural inequalities and social and cultural divisions within South African society.

The size and distribution of population is also affected by population movements. Although in the 1990s cutbacks in the mining sector have led to a reduction of workers moving to the mines, migration into the country has increased considerably since 1990. In 1998 the number of official immigrants to South Africa was only 4,371, yet there is a considerable flow of illegal migrants from other African (and to a lesser extent, Asian) countries in search of employment. Some streets in most of South Africa's towns and cities are now occupied by pavement hawkers, many of whom have come from other parts of the continent (Holness, Nel and Binns forthcoming). Interestingly, out-migration is also high, with many whites (and indeed also Coloureds and Indians) leaving the country due to such factors as escalating crime rates.

The 1996 census revealed that 53.7 per cent of South Africa's population was living in urban areas at the time of the census. This figure is high by African standards; for example, neighbouring southern African countries such as Lesotho, Namibia and Zimbabwe have 26 per cent, 38 per cent and 33 per cent respectively of their populations living in towns. Even Nigeria, with the continent's largest population and with more cities with over 1 million inhabitants than any other country in

Figure 2.10a   Age distribution of the population by gender – October 1996

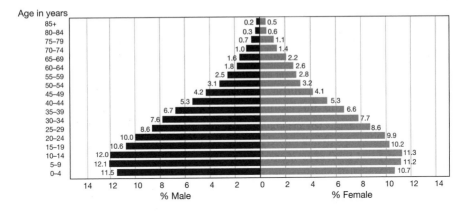

Figure 2.10b   Age distribution of the African population by gender – October 1996

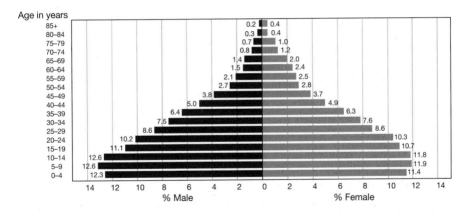

Figure 2.10c   Age distribution of the white population by gender – October 1996

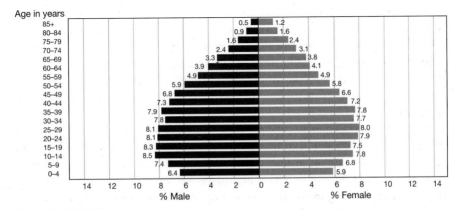

*Source*: StatSA (1996)

Africa, has only 41 per cent of its population living in towns (World Bank 1998). Obtaining reliable figures for the sizes of the main cities in South Africa is difficult, not least because suburbs and African townships are often treated separately from the main urban area. But recent estimates from the Municipal Demarcation Board (MDB 1999) suggest that the size of the five main urban agglomerations are:

| | |
|---|---|
| Cape Town | 2,557,456 |
| Johannesburg | 2,521,352 |
| Durban | 2,519,992 |
| Pretoria | 1,238,127 |
| Port Elizabeth | 942,077 |

## Mineral resources and mining

South Africa has an extraordinarily rich mineral endowment, with the world's largest reserves of gold, manganese, platinum, chromium, vanadium and alumino-silicates, as well as sizeable reserves of iron ore, coal, diamonds, uranium, titanium and nickel (Table 2.4).

Although the mining industry really took off in the second half of the nineteenth century (chapter 3), it was as early as 1685 that the governor, Simon van der Stel, undertook a journey to Namaqualand in the south-west of the country to find out more about the region's copper deposits. However, it took more than 150 years before there was a copper rush to Namaqualand between 1852 and 1854. But the most spectacular developments in the mining sector took place a few years later, when alluvial diamonds were discovered in 1867 in the banks of the Orange and Vaal rivers and then in 1871 a major 'kimberlite pipe' was found in Kimberley (now the 'Big Hole') and other pipes were later found and mined in the same area.

However, it was the development of the gold mining industry which had the greatest effect on the South African economy. Gold was discovered in eastern Transvaal (now Mpumalanga) in the 1860s and 1870s and mined on a small scale in the Barberton area during the 1880s. But it was the discovery of gold in conglomerate reefs on the Witwatersrand in 1885 which led in 1886 to the laying out of the town of Johannesburg (chapter 3). From the early part of the twentieth century exploration of potentially productive gold-bearing reefs spread both west and east from Johannesburg and gold fields in the northern Orange Free State were mined from the 1950s (see Fig. 3.15 on page 109).

In the late 1990s mining and the processing of minerals accounted for some 8 per cent of GDP and 60 per cent of export earnings. Gold mining remains dominant, but the industry is in slow decline and there is much concern within South Africa about the likely impact of this trend on the economy and the problem of finding alternative employment for redundant mineworkers. This problem will be discussed at greater length in chapter 7. In 1998 some 64,000 jobs were shed in the gold-mining industry, whilst gold output fell to a 42-year low of 474 tonnes (Table 2.4). The industry and

Table 2.4a South Africa, trends in mineral production in the 1990s – volume (000 tonnes, unless otherwise indicated)

| | 1990 | 1991 | 1992 | 1993 | 1994 | 1995 | 1996 | 1997 | 1998 |
|---|---|---|---|---|---|---|---|---|---|
| Gold (000 kg) | 605.1 | 601 | 611.1 | 619.2 | 579.3 | 519.8 | 494.6 | 492.5 | 473.7 |
| Iron ore | 30,291 | 29,075 | 28,226 | 29,385 | 32,321 | 32,144 | 30,951 | 33,333 | 29,037 |
| Chrome | 4,618 | 5,100 | 3,002 | 2,827 | 3,599 | 5,130 | 4,982 | 5,794 | 5,419 |
| Copper | 188 | 194 | 167 | 166 | 165 | 161 | 151 | 151 | 171 |
| Manganese ore | 4,402 | 3,146 | 2,464 | 2,507 | 2,851 | 3,165 | 3,254 | 3,095 | 4,292 |
| Diamonds (000 carats) | 8,708 | 8,431 | 10,166 | 10,324 | 10,857 | 9,569 | 10,166 | 10,009 | 8,530 |
| Coal | 174,000 | 178,000 | 174,072 | 182,031 | 195,805 | 203,427 | 208,362 | 218,617 | 218,208 |
| Lime and limestone | n/a | n/a | 18,320 | 18,215 | 19,719 | 18,776 | 18,495 | 18,600 | 19,353 |

Table 2.4b South Africa, trends in mineral production in the 1990s – Gross value (Rm)

| | 1990 | 1991 | 1992 | 1993 | 1994 | 1995 | 1996 | 1997 | 1998 |
|---|---|---|---|---|---|---|---|---|---|
| Gold | 19,063 | 19,358 | 19,458 | 23,169 | 24,953 | 23,335 | 26,482 | 25,077 | 24,280 |
| Iron ore | 1,077 | 1,162 | 1,128 | 1,279 | 1,400 | 1,658 | 1,692 | 2,088 | 2,531 |
| Copper | 1,127 | 1,173 | 1,028 | 1,035 | 1,254 | 1,679 | 1,497 | 1,636 | 1,555 |
| Manganese ore | 849 | 767 | 600 | 549 | 645 | 687 | 784 | 892 | 962 |
| Lime and limestone | 425 | 478 | 494 | 536 | 605 | 688 | 693 | 691 | 719 |
| Total incl. others* | 35,483 | 36,808 | 39,938 | 46,632 | 50,711 | 54,180 | 63,104 | 66,314 | 71,048 |

*Note:* *Others include silver, chrome, coal, diamonds, asbestos and uranium
NB. Exchange rates: December 1995, R3.65: US$1; July 1999, R6.09: US$1
*Sources:* EIU (1996), EIU (1999)

Plate 2.5  Gold mine, Welkom, Free State

economy were also hit by a fall in the price of gold, which in 1980 was $850/oz, but only $250/oz in mid-1999 – a decline which was affected by the announcement from central banks and the IMF that they would sell part of their gold reserves. It is estimated that at mid-1999 exchange rates every $US10 decline in the gold price cost South Africa almost R1billion ($US164 million) in lost export earnings (EIU 1999). In an attempt to reduce costs a number of the largest gold-mining companies, including Anglo American Corporation, the world's biggest gold producer, have been combining their operations.

Diamond production remains important, though output did fall in 1998 to 8,530,000 carats. The South African company De Beers effectively controls the world marketing of diamonds through the London-based Central Selling Organization (CSO). In terms of other minerals, production of coal and chrome rose during the 1990s. South Africa is the world's largest producer of platinum, accounting for four-fifths of total non-Russian supply. Platinum production is likely to increase in the next few years with an upgraded concentrator plant at the Anglo-American Platinum Corporation and the opening of the new Kroondal platinum mine in 1999. With an increasing global demand for platinum and the heavy depletion of Russian platinum stockpiles, there is some optimism in this industry. Whilst the total value of mineral products in rand more than doubled between 1990 and 1998, in fact the value in US dollars scarcely increased, and indeed between 1995 and 1998 actually fell, due to the deteriorating rand–dollar exchange rate.

## Economic performance

The performance of the South African economy during the 1990s was generally disappointing, as we will see in chapter 7. Between 1990 and 1992 the economy contracted, and during the period 1994–98 annual average growth in GDP was

2.7 per cent, peaking in 1996 at 4.2 per cent, but then falling to only 0.5 per cent in 1998. Growth was projected to reach 0.6 per cent in 1999, which is insufficient to generate significant employment and represents a fall in GDP per capita due to higher population growth. It has been suggested that such fluctuations in the health of the economy are due partly to the effects of climatic factors on the agricultural sector and associated agro-processing industries (EIU 1999). Export earnings were much more buoyant and increased by an average of 10 per cent per annum between 1994 and 1998, but the fall in gold prices during 1998 and 1999 considerably reduced export revenues. Encouraging economic trends include a likely fall in the average annual inflation rate from 6.5 per cent in 1999 to 5 per cent in 2000, and a reduction in interest rates in 1999, which started the year at 23 per cent, but fell to 16.5 per cent by July. Nevertheless, as chapter 7 elaborates, the situation with regard to employment creation is not good. As the twentieth century drew to a close, further job losses were expected due to the declining gold price, the restructuring of parastatals and a recession in the construction industry.

Minerals and energy have traditionally dominated the South African export sector. However, unlike most other African countries South Africa has a well-developed manufacturing sector which contributes 25 per cent to the country's GDP – in fact more than any other sector. An important element of manufacturing industry is closely linked to mining and mineral processing, notably iron and steel and heavy chemicals. Many towns have industries linked with food and drink, and milling, together with agricultural processing, remains significant. The annual growth rate of manufacturing output has slowed down since the mid-1970s and in the early 1990s many thousands of jobs were lost. But the cheaper rand, together with increased privatization and liberalization within the manufacturing sector, have improved the competitiveness of South African manufactured goods, particularly in Africa. The effect of this was that manufactured exports increased as a proportion of total exports from 20 per cent in 1989 to 30 per cent in 1998 (Table 2.5).

The contributions of agriculture, forestry and fishing to GDP have been generally declining and in 1998 represented only 5 per cent. However, these occupations do provide considerable employment, particularly in the former African 'homelands'. When account is taken of agriculturally based industries, the sector is very much more significant, with 30 per cent of non-gold export revenue coming from agricultural and processed agricultural products.

The rural landscape of South Africa is varied and has been produced by a complex interaction of environmental, economic, social and political factors (Fig. 2.11).

Apart from agricultural activities being controlled by the usual environmental factors such as climate, topography and soil type, there is in South Africa a marked distinction between the small-scale generally subsistence-oriented agriculture of the former African homelands and the larger and usually better resourced commercial farming activities more typical of the historically white farming areas.

As Christopher observed in 1982 'The major feature of the Black rural areas is their poverty, where traditional agricultural systems have been undermined through competition with a highly organised, and often highly subsidised white agricultural

Table: 2.5   Exports from South Africa (Rmillion; fob)

|  | 1992 | 1993 | 1994 | 1995 |
|---|---|---|---|---|
| Food, beverages and tobacco | 4,746 | 4,937 | 7,455 | 7,769 |
| Inedible raw materials | 11,290 | 12,655 | 12,132 | 16,577 |
| Animal and vegetable oils and fats | 122 | 133 | 132 | 191 |
| Chemicals | 3,720 | 3,843 | 5,602 | 7,144 |
| Manufactured goods | 17,593 | 21,322 | 23,901 | 26,646 |
| Machinery and transport equipment | 4,480 | 5,489 | 6,033 | 9,153 |
| Misc. manufactured articles | 1,353 | 1,825 | 2,377 | 3,627 |
| Total merchandise incl. other exports of which: | 49,135 | 56,988 | 67,352 | 81,387 |
| gold (bullion and coins) | 18,173 | 22,226 | 22,670 | 20,118 |

*Source*: Central Statistical Service (1999)

sector. Costs of production are such that it is more profitable for a Black rural land-holder to work in the White economic sector and retain his land as security, than to attempt to make his living from the land' (Christopher 1982: 72–3). That this has not always been the case will be made clear in chapters 3 to 5. Twenty years after Christopher's comment, and after the demise of apartheid, the rural situation is generally little changed, and only in relatively few areas have black farmers been able to develop a commercially oriented production system which shows some prospect of sustainability (Nel *et al.* 1997; Binns and Nel 1999).

Table 2.6 shows the main types of agricultural production. The decline in maize production, the staple food of many Africans, and the increased output of sugar cane are particularly noticeable. Maize harvests have fluctuated considerably in recent years due to erratic rainfall ranging from heavy flooding to near-drought conditions. At the begining of 1999 there was some concern that after a period of exceptionally hot and dry weather the maize crop might fall to around 6 million tonnes – less than the domestic consumption level of around 7.5 million tonnes. The dry weather and reduced plantings have also resulted in a significant reduction in wheat yields, such that it seemed likely that wheat imports in 1999 would amount to 1 million tonnes. In recent years domestic wheat production has fallen as farmers have shifted to more profitable crops such as sunflower seeds, and cheaper foreign wheat has been imported to make up the shortfall.

Figure 2.11   Landscape regions of South Africa

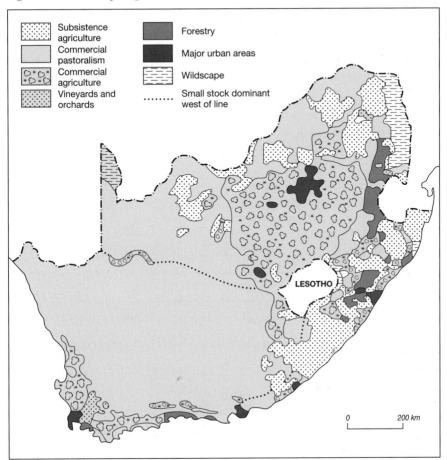

*Source*: Adapted from Christopher (1982)

   The production of wine in the Western Cape in the vicinity of towns such as
Paarl and Stellenbosch, can be traced back to the seventeenth century and particularly to the endeavours of the French Huguenot refugees who developed an export
trade. By the late 1970s the area under vineyards had reached 110,000 ha. While this
figure has remained fairly stable, wine exports between 1993 and 1995 almost tripled
to reach 71 million litres, coinciding with the lifting of sanctions following the
demise of apartheid. The export of food, beverages (including wine) and tobacco
also increased considerably from R4,746 million in 1992 to R7,769 million in 1995
(Table 2.5 on page 38).

Table 2:6   South Africa, agricultural production (crop years; '000 tonnes unless otherwise indicated)

|  | 1994/95 | 1995/96 | 1996/97 | 1997/98 | 1998/99 |
|---|---|---|---|---|---|
| Maize | 13,275 | 4,866 | 10,171 | 10,136 | 7,574 |
| Wheat | 1,840 | 1,977 | 2,711 | 2,294 | 1,469 |
| Sugarcane | 15,683 | 16,714 | 20,951 | 22,155 | 24,460 |
| Citrus fruit | 1,113 | 1,004 | 1,173 | 1,259 | 1,183 |
| Other fruit | 3,845 | 3,885 | 4,237 | 4,461 | 4,150 |
| Fresh vegetables | 265 | 260 | 265 | 270 | 260 |
| Wool | 77 | 68 | 62 | 54 | 54 |
| Cattle and calves slaughtered (no.) | 2,600 | 2,500 | 2,330 | 2,481 | 2,985 |
| Poultry slaughtered (no.) | 311 | 331 | 361 | 356 | 356 |
| Sheep and goats slaughtered (no.) | 11,765 | 11,620 | 11,670 | 11,760 | 11,880 |

*Source*: EIU (Economist Intelligence Unit), 1999, *Country Profile, 1999–2000*, EIU, London

## Tourism

With a combination of spectacular 'wildscapes' and towns which are often steeped in history and generally provide good facilities for the visitor, there is considerable potential for the further development of the tourist industry in South Africa. Indeed, the democratic government regards the development of tourism as a key priority, working closely with private businesses which manage high-quality hotel groups such as Protea and Sun. Better marketing and the falling value of the rand have been instrumental in fuelling the growth of tourism, which has increased considerably since the democratic elections of 1994. The number of visitors from Europe and the Americas doubled between 1994 and 1998 and holiday visits showed a 55 per cent increase in this period (Table 2.7).

The hosting of the Rugby World Cup in 1995 further boosted visitor numbers. But many visitors to South Africa are short-stay, even day-trippers, coming across the borders of neighbouring countries principally to shop or trade their goods. As Table 2.7 clearly shows, Africa is the area of origin of the largest numbers of visitors, with the figure reaching almost 4.3 million in 1998, representing almost 73 per cent of all foreign visitors. One important factor which is holding back the rapid development of the tourist industry is the high rate of violent crime, which has received much publicity both in the South African and the international press, and which is

Plate 2.6  Hotel development on KwaZulu-Natal south coast

Plate 2.7  Umhlanga Rocks near Durban

Plate 2.8  Cape Town's Table Mountain and Waterfront

Plate 2.9  Table Mountain cable car

Table 2:7  Foreign visitors to South Africa ('000)

|  | 1994 | 1995 | 1996 | 1997 | 1998 |
|---|---|---|---|---|---|
| Area of origin |  |  |  |  |  |
| Africa | 3,126 | 3,452 | 3,781 | 3,665 | 4,291 |
| Europe | 463 | 722 | 798 | 878 | 982 |
| Asia | 114 | 158 | 167 | 139 | 138 |
| Americas | 116 | 160 | 177 | 209 | 250 |
| Australasia | 29 | 48 | 63 | 64 | 70 |
| Total incl. others | 3,897 | 4,684 | 5,186 | 5,170 | 5,898 |
| Purpose of visit |  |  |  |  |  |
| Holiday | 2,625 | 3,564 | 3,938 | 4,002 | 4,731 |
| Business | 761 | 619 | 621 | 601 | 676 |
| Contract worker | 158 | 125 | 123 | 108 | 85 |
| Work | 70 | 71 | 118 | 86 | 81 |
| Other | 283 | 305 | 386 | 373 | 325 |

*Source*: EIU (Economist Intelligence Unit), 1999, *Country Profile, 1999–2000*, EIU, London

discussed at greater length in chapters 5 to 7. A World Health Organization (WHO) study in 1995 revealed that South Africa has one of the highest murder rates in the world, as well as having some of the worst figures for rapes, robberies, hijacks and burglaries (EIU 1999).

## Transport and telecommunications

South Africa's transport infrastructure is without doubt the best in Africa. There are some excellent stretches of roads, with motorways in and around the major cities. There is also a good internal air system linking most of the main urban centres as well as important tourist destinations such as the Kruger Park in the north-east. Cape Town, Durban and Johannesburg are the main hubs of the domestic air network, as well as having international flights (see Fig. 2.9b). During 1998–99, Johannesburg International Airport underwent a major redevelopment to cope with the increasing numbers of flights and throughput of passengers. The government has strongly encouraged privatization and investment from outside South Africa. For example, in March 1998 a company controlled by Italy's Aeroporti di Roma won the bid for the privatization of South Africa's airports. In June 1999, it was announced that Swissair had won a bid to acquire 20 per cent of the national air carrier, South African Airways (SAA), which competes with smaller airlines such as Airlink, Comair (a subsidiary of British Airways) and Sun Air.

South Africa has over 21,000 km of railways, representing the densest network in Africa. This network, which developed from the 1880s onwards, was initially a colonial creation to facilitate the exploitation and export of minerals, to strengthen white (and particularly British) political domination and to enable the penetration of European capital (see chapter 3, Griffiths 1995). Between 1892 and 1895 railways were constructed from Johannesburg to key ports such as Cape Town, Durban, Port Elizabeth and Lourenço Marques (Maputo) in Mozambique. At key positions on the growing railway network 'railway' towns developed to service the system, for example Touws River, Noupoort, Waterval Boven and De Aar, an important railway junction now in Northern Cape province. At the beginning of the twenty-first century the railway system is looking in need of urgent and considerable investment. In 1997 the Rail Commuter Corporation announced plans to sell railway coaches to the private sector for refurbishment, which would be leased back to the rail utility. The government is considering granting rail concessions on metropolitan rail services, which are presently controlled by the parastatal Transnet, which has a rail monopoly and some road interests. The privatization of Transnet is high on the political agenda, but is being hindered by the considerable debts in the pension funds of such parastatals. In spite of considerable 'pruning' of the railway system during the 1980s and 1990s, it seems likely that in the next few years many more lines will be closed and the network will represent a small fraction of its former extent and complexity.

Unlike the railways, there has been a considerable expansion in telephone coverage since 1994, with about 1.3 million additional homes being connected, helping, if only in a small way, to redress the highly unequal situation of 60 telephones per 100 whites and only 1 per 100 among the African population. In April 1997 a deal was signed involving the purchase of 30 per cent of the state-owned telecommunications group, Telkom, by Telekom Malaysia and a US-based company SBC Communications. The government has earmarked about US $1 billion of its receipts from this US $1.3 billion deal, towards supplying 3 million new lines before 2004.

Having established some of the most basic features of the South African physical and human environment in this chapter, we turn now to three chapters which seek to explain how South Africa's social boundaries have been created, and how its enduring material and spatial inequalities have been forged, over the last three and a half centuries.

# Chapter 3

# PRE-INDUSTRIAL SOUTH AFRICA, 1652–*c*.1900

## Pre-colonial society

The interlinked processes of colonialism and capitalism, and the cultural notions and practices which have guided these processes are integral to this book. They help explain the ways in which the social groups comprising modern South African society have been created. But neither Europe's culture nor its capitalist practices were simply imported into South Africa, through colonialism, subsequently to be reproduced there in some pristine form. What colonists and capitalists were able to achieve in southern Africa, and indeed what they themselves became in the southern African environment, was the result of accommodation, adaptation and the partial absorption of indigenous social and cultural forms. Resilient African societies, well established in the subcontinent, powerfully shaped the nature of colonialism and local forms of capitalism, even as these societies were themselves affected by the encounter. One has to envisage a mutual, complex interaction between the pre-colonial and the colonial, the precapitalist and the capitalist if one is to attempt to render contemporary South Africa intelligible. This interaction resulted in new social and cultural forms displaying in some respects the universal attributes of 'modern' colonial and capitalist processes, but manifesting them in very particular ways according to the varied strategies adopted by local people.

### Hunter-gatherers, pastoralists and farmers

Until recently, it was generally thought that the region which is now South Africa had long been occupied by hunter-gatherers known as San (or 'Bushmen' to Europeans) and by pastoralists known as Khoikhoi (more properly 'Khoena', but 'Hottentots' to Europeans), sharing a Stone Age technology, but that these original inhabitants were displaced across the eastern, if not the western half of the subcontinent during the first few centuries of the first millennium AD. In orthodox accounts, their displacers and successors were Iron Age, Bantu-speaking farmers migrating in coherent waves from the north. However, the picture of early occupation suggested by recent archaeology is more complex. It is one in which it is far less easy to identify rigid social and cultural boundaries between San, Khoi and Bantu-speaker.

Hunter-gatherers, who could be called 'San', had indeed been established over much of the region now known as South Africa for at least 20,000 years before pastoralism was introduced. Indeed, their ancestry could possibly be linked to some of the earliest humanoid remains in the world, dating from 1.5 to 2 million years ago, which were recently discovered in the area around Sterkfontein near Pretoria (*Mail and Guardian*, 19 Dec 1997). The San organized themselves into small bands, each consisting of 20 to 50 people. These bands were highly mobile, hunting game and foraging for wild fruit and vegetables over a wide area, and their social structure was egalitarian, given the need to share the food products obtained. Evidence from their rock paintings suggests a considerable degree of interaction between San groups across southern Africa.

In the early years of the first millennium at the latest, bands of hunter-gatherers in what is now northern Botswana acquired sheep and goats, and subsequently cattle, probably after acting as low-status herders for farming communities of the type discussed below. Partly through the migration of members of these bands to the south and east, and partly through the adoption of livestock by hunter-gatherer groups *in situ*, some of the hunter-gatherer bands in what is now South Africa had become 'Khoikhoi' nomadic pastoralists by about AD 400. There was no rigid genetic boundary between Khoikhoi pastoralists, numbering some 50,000 people by the mid-seventeenth century, and San hunter-gatherers, whose numbers are impossible to determine (Elphick and Malherbe 1988). Both were mobile and the former could

Plate 3.1  Khoikhoi family in the early 1700s
*Source*: South African Library, Cape Town

resort to San lifestyles when they lacked sheep and cattle, while the latter could effectively become Khoikhoi if they acquired livestock. It was therefore possible for the same individuals to be both San and Khoikhoi during different stages of their lives. It is for this reason that the collective term Khoisan is often used to describe all of the indigenous inhabitants of the Western Cape at the time that Europeans arrived there.

However, there is evidence that some kinds of social boundary were invoked to separate hunter-gatherers and pastoralists, ensuring the dominance of the latter when both were in proximity. Hunter-gatherers were often absorbed within pastoral communities in subordinate roles and it may not have been quite as easy as was once thought to move between hunter-gatherer and pastoralist identities and lifestyles. There are certainly strong indications that pastoralists and later farmers upheld an ideology in which hunter-gatherers were 'lesser beings' than themselves by virtue of their lack of livestock (Boonzaier et al. 1996). Cattle especially acquired symbolic value in Khoikhoi communities. They became indicators of both material status and political power, and the word 'San' itself was derogatory, applied by Khoikhoi pastoralists to those who lacked cattle. (The San have no indigenous term of their own to describe the collectivity of hunter-gatherers and so a derogatory appellation, either the Khokhoi word 'San' or the colonists' word 'Bushmen', is unavoidable.)

Despite the recognition of differing degrees of status and wealth, however, the 'idea that individuals could gain exclusive control of land or water was alien' to all groups in pre-colonial society. While outsider groups had to seek permission from local Khoikhoi chiefs for the exploitation of water or grazing resources, the Khoikhoi's 'experience with other Khoikhoi or San would have done little to prepare them for the [later] remorseless privatisation of these resources by [European] settlers' (Guelke and Shell 1992: 805).

As we have suggested, Khoikhoi herders in the Cape traced their ancestors to those who had first acquired livestock from farmers to the north. These farmers had not simply swept south through the continent, carrying with them a 'package' of Iron Age technology, Bantu languages and cultivation techniques as was once imagined. Rather, their agricultural practices, involving both crop cultivation and livestock herding, their Iron Age technology and their Bantu dialects had been diffused through the continent from south of the Sahara through a combination of fragmented and intermittent migratory movements and their adoption in situ by formerly hunter-gatherer and pastoral groups situated in favourable areas. The dependence of farmers' sorghum and maize crops on summer rainfall had led to their occupation of the eastern side, but not the more arid and winter rain-fed west of what is now South Africa, by the early centuries AD. By the eighth century, Bantu-speaking farmers were using iron implements to clear the forests of the eastern coastal strip for cattle grazing and were holding livestock around larger settlements on the more sparsely vegetated highveld (Elphick 1985; Hall 1987).

Unlike hunter-gathering, both pastoralism and farming allowed for the accumulation of storable, surplus products, or wealth. This generated significant social effects among the Khoikhoi and those who adopted cultivation. It meant that households could engage in reciprocal relations, one lending stock to another in times of the latter's need, with the relationship between the two potentially being reversed in the future. Larger-scale reciprocal networks and alliances thus developed. As a result of the progressive and more secure accumulation of surplus products, elites could construct patron–client relationships, leading to the partial displacement of the sharing ethos of hunter-gatherer societies by ideologies founded on social stratification.

The heads of certain lineages were able to create chiefdoms in which the allocation of land and other resources was the hereditary prerogative of the chief (a member of the dominant lineage), and in which subservient groups were more clearly marked. While such stratification generally took a less pronounced and fairly fluid form among constantly mobile pre-colonial Khoikhoi chiefdoms such as the Cochoqua and Gorinhaiqua, settled farming communities tended to have more rigid social boundaries. Nevertheless, chiefly authority could still be moderated or challenged by the headmen of other lineages within the farming chiefdom who gathered to form councils, and chiefdoms across the region were subject to fission as groups broke away to establish a more autonomous political existence (Elphick 1985, Hall 1987; Boonzaier *et al.* 1996).

Along what would later become the eastern frontier of the Cape Colony, around the Kei River, farmers were established by about 1500 at the limits of the summer rainfall zone. These particular farmers, speaking Xhosa dialects, were mingled with Khoisan pastoralists and hunter-gatherers. Transactions between the groups were generally based on accommodation, with trading contacts allowing eastern Khoikhoi to distribute the narcotic dagga (marijuana), iron and copper goods and clay pottery, which they acquired from the Xhosa, among other Khoikhoi tribes to their west. The Xhosa acquired cattle and herding service in return. Some Khoisan subsequently became clients within Xhosa chiefdoms and could ultimately be accepted as Xhosa subjects in their own right. Sexual interaction over an extended period generated frontier clans and chiefdoms that were 'hybrid' in the sense that they were defined clearly neither as Xhosa nor Khoikhoi (Hall 1987; Boonzaier *et al.* 1996). On the highveld too, interactions between pastoral and farming communities were commonplace. Because of these extensive interactions, the Khoikhoi and Bantu-speakers who lived along the farming frontier can be described as components of 'a single interlocked system of food producing – a "division of labour" more than two distinct societies' (Hall 1987: 60).

### *Social stratification and the construction of states*

Stratification within African chiefdoms was most obviously marked along the lines of chief–commoner, elder–youth and male–female. Degrees of power were reflected in differing access to resources such as crops, sheep and cattle. A set of rituals of deference marked the superior status of chiefs over commoners, elders over juniors and men over women. Accompanying gendered status differentials, there was generally a division of labour between men and women. While men tended the livestock, hunted, raided other chiefdoms for cattle and defended the chiefdom's own stock, women cultivated the crops, distributed the milk and meat produce obtained from the livestock and maintained the home.

Although the male exploitation of women's labour power was a fundamental feature of farming society, and women were generally subordinate to men, women's control over the household economy gave them a counteracting, if limited form of power (Guy 1990). Their reproductive role could also make them politically significant since marriages between members of different chiefdoms cemented alliances, while the necessity for the groom to pay bridewealth to the bride's father in cattle meant that young men were forced to seek the patronage of older, wealthier men. In eastern and highveld farming communities, the practice of bringing married women into the group from outside, and allowing them to retain their connections with their own lineage, constituted a form of protection from the worst excesses of male dominance within the new household, while women could influence succession disputes and develop powerful roles as diviners mediating between the spiritual and material worlds (Hanretta 1998). In the western Khoikhoi chiefdoms, women could own stock in their own right, trade alongside the men and occasionally became regents in the absence of a legitimate male chief (Boonzaier *et al.* 1996; Wells 1998).

Nevertheless, certain people in pre-colonial pastoral and farming chiefdoms always constituted the 'structural poor', and women tended to be especially vulnerable (Iliffe 1987). Such people included those unable to perform their own tasks or to render labour service to another, and those without family support. Adaptable forms of reciprocity and kinship precluded the more extreme kinds of inequality brought about later by colonialism and capitalism. However, as Bundy puts it, while pre-colonial 'forms of clientship staved off want', they also 'perpetuated differentiation' (Bundy 1992: 27). Age and gender stratification, aside from contributing to inequality, also had its own geography. It was made manifest, for instance, in the layout of farming settlements, with the wives and offspring of the headman being allocated huts either side of his own hut according to their seniority. Thus 'the plan of a settlement is not simply a pragmatic or random arrangement of architectural features, but is rather a map of the relative status and interrelationships of members of the community' (Hall 1987: 72).

By the ninth century, the stratification generated by surplus wealth creation in farming communities was being intensified in some areas of the subcontinent. Along the Limpopo River, ruling elites developed extensive trading networks with the east coast, allowing them not only to accumulate wealth in certain local products, but to trade those products with Arabic merchants. Rare commodities could thus be brought into the chiefdom from outside, and control over their distribution enabled chiefly lineages to exercise enhanced military and cultural power over their subjects. Through these wider transactions, certain favourably located chiefdoms were transformed into states, in which a ruling class exercised extensive territorial control and in which escape from the king's authority through fission was more difficult. As Unterhalter explains, 'The emergence of these kingdoms was associated with the assertion by dominant groups of a national identity, a common founding father, a common language, new religious rites and new forms of social organisation. A national identity, containing within it notions of superiority to neighbours, was advanced by the dominant clan to legitimate seizure of land from neighbours and demands of labour from subjects' (Unterhalter 1995: 210).

The first of these enlarged states to be relatively well documented was Mapungubwe, dating from the eleventh century and situated just to the south of the Limpopo River. Within this kingdom, wealth was generated and retained by an elite group through exchanging locally produced ivory and animal skins for the iron and copper ware, glass beads and textiles brought to ports such as Kilwa (in modern Tanzania) by Islamic traders. The products of the kingdom were then distributed across the Islamic Middle East. The 'relations of production in the Mapungubwe state would have created a pattern in which cattle, military service and other forms of tribute would have flowed inwards to the major centres of power, while beads, cloth and other valued signifiers of high status would have moved outwards to regional centres and to local chiefs who acknowledged the suzerainty of the Mapungubwe kings' (Hall 1987: 89).

Mapungubwe declined as the primary product demanded by Islamic states shifted to gold during the fourteenth century. The exchange of gold thereafter favoured the nascent Zimbabwe state, located further to the north and nearer to goldfields. It was the shift of the hub in trade networks to this state which enabled the construction of the impressive stone-built capital at Great Zimbabwe. Although the Zimbabwe state had itself long disintegrated by the time that Portuguese traders began to dominate inland trade routes during the sixteenth and seventeenth centuries, the exercise of power by dominant African elites, and attempts at state creation continued within southern Africa. To the north and west of Zimbabwe, the successor states of Monomotapa and Torwa were constructed partially around trade relationships with the Portuguese, but more fundamentally on relations of dominance with surrounding, tributary chiefdoms. On the South African highveld by 1600, substantial Sotho-Tswana-speaking chiefdoms were based on large settlements located around scarce water supplies. These centres, some of which contained populations numbering 15,000 to 20,000, acted as nodes in trans-subcontinental trade networks as well as centres of political authority. On the eastern coastal strip, smaller chiefdoms speaking Nguni dialects of Zulu and Xhosa were scattered across the grazing land (Fig. 3.1). A violent process of state formation conventionally known as the 'Mfecane', was to further transform each of these chiefdoms during the late eighteenth and early nineteenth centuries, as we will see below (Hall 1987; Hamilton 1995; see summary box 2).

### The creation of colonial categories

The southern tip of Africa was first 'discovered' by Europeans when Portuguese navigators, in search of new trading routes to the east, rounded the Cape from the Atlantic into the Indian Ocean during the late fifteenth century. The oceanic trade route made Far Eastern sources of silks and spices (essential to disguise the taste of unpreserved food) directly accessible to European merchants for the first time. This sea route would not be disrupted by the warfare that plagued the ancient overland trade routes between Europe and Asia, and the fact that middlemen could be avoided and goods shipped directly allowed for significant savings. Subsequently, European crews called in at the Cape, strategically positioned halfway on the voyage, between the Atlantic and Indian Oceans, in order to replenish supplies (Figs. 3.2 and 3.3).

Figure 3.1  Major linguistic and political groups in the sixteenth century

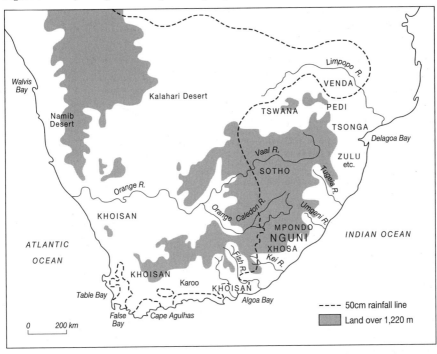

*Source*: Thompson (1990)

By the seventeenth century, the local Khoikhoi had gained a reputation for savagery among these European visitors, a reputation associated with the resistance that
Portuguese sailors had encountered when they had tried to impose more favourable
terms of trade. Nevertheless, during the mid-seventeenth century, the directors of
the Dutch East India Company – one of the most powerful of the European mercan-
tile capitalist enterprises which circulated trading fleets between Europe and the
Indian Ocean – ordered that a supply station be established by a group of Company
employees at the Cape. In the sheltered harbour at what is now Cape Town, sailors
could take on board essential fresh produce, grown at the company station, and
receive medical attention while their ships underwent repair. Jan van Riebeeck, the
commander of the outpost established in 1652, was instructed to develop amicable
relations with the local Khoikhoi so that they would supply cattle in exchange for
European products. Passing ships could then be well provisioned with meat from the
Company stores (Elphick 1985; Elphick and Shell 1988).

By the end of the eighteenth century however, Cape colonial society was by no
means characterized by the harmonious exchange of resources between a European
enclave and a politically independent, Khoisan-populated interior, as envisaged in
van Riebeeck's instructions. Rather, it was marked by relatively clearly defined status
groups distributed across the entire western Cape and a broadly capitalist economic

Figure 3.2 Trade routes developed by the Portuguese

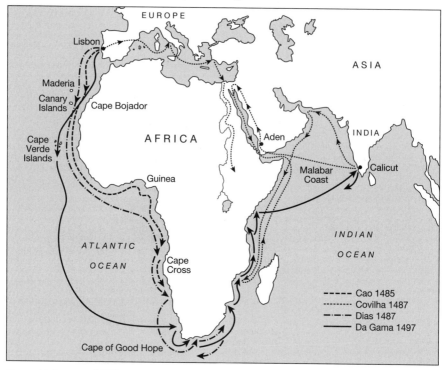

*Source*: Livingstone (1992)

structure which embraced both Khoisan and colonists, but on uneven terms. Most of the colony's population, consisting of 20,000 Europeans, 25,000 slaves and 15,000 Khoisan and people of mixed descent by the late eighteenth century, would have understood where they were positioned relative to others in the social hierarchy. That position, by and large, reflected their command of material resources. Critical features in the construction of this colonial society and its hierarchy were the importation of slaves by the Dutch East India Company, the expansion of European colonial settlement and private land ownership into the interior, and the corresponding dispossession and subjugation of the Khoikhoi and San. The material inequalities generated by these processes were associated with, and consolidated by, the construction of racial and religious social boundaries. In the next section, we explore these processes in more detail.

### Slavery, colonial expansion and the subjugation of the Khoisan

Soon after the establishment of the Cape outpost, the Company authorities decided to encourage first former Company employees and then other Europeans, known as burghers, to settle nearby. They were instructed to develop intensive agriculture, selling their surplus exclusively to the Company so that passing trade fleets could be more effectively supplied. Between 1657 and 1701 plots comprising the peninsula Khoikhoi's best grazing land along the Liesbeeck River were distributed to individual colonists (Fig. 3.3). Local Khoikhoi chiefdoms, who had previously both traded with and occasionally raided the Company post, combined to mount armed resistance to this incursion, but they were ultimately forced to acquiesce in expanded European settlement (Elphick 1977).

Slavery had long been established in European discourse as an 'acceptable' solution for Europeans located in colonies where land was abundant, but the labour to work it scarce. In the East Indies the Company relied on slaves traded with local elites to produce the goods which it carried to Europe. During the early years of colonial settlement, however, the enslavement of the surrounding Khoikhoi would have

Figure 3.3   Early colonial Cape Town and its immediate surrounds

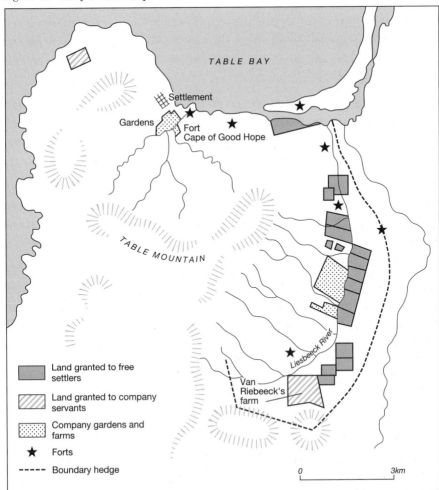

*Source*: Christopher (1994)

jeopardized the colonists' vital trade in livestock with them. A slave workforce was acquired instead from India, Indonesia, East Africa and Madagascar. By 1808 some 63,000 slaves had been imported from these diverse regions (Shell 1994).

It was not only their different origins which made the Cape's slaves a heterogeneous population, but also their different material standing and location. Despite their mutual lack of freedom, there were stark differences in living standards, especially between those living in towns and those in the countryside. Skilled, town-based artisans were often hired out by their owners to other employers in return for cash wages. They generally escaped daily coercion and could even accumulate some private capital. Field and domestic labourers on more isolated inland farms, however, were frequently brutalized by their masters and remained propertyless (Shell 1994). In the larger settlements, slaves could have greater recourse both to the Company's protective laws and to what Dooling (1994) describes as a 'moral community' of owners. He was referring to a collective effort on the part of masters to regulate paternalistic relationships with their slaves. In return for the slave complying with the duties which were generally expected of him or her, the master would be careful not to impose burdens which were too onerous, and would ensure that the slave had a certain degree of self-determination. Such paternalism could be described as the 'ideal' system of slavery as far as masters and the Company authorities were concerned. It was a form of domination resembling that of the male head of household over his wife and children – one requiring neither the constant use of physical coercion (although this did exist within the family too), nor the Company's continuous oversight (Crais 1994; Shell 1994; Scully 1997).

As far as most slaves themselves were concerned, whether in town or countryside, there was nevertheless an acute awareness of the masters' and the Company's ability to employ violence if it was felt to be necessary. Punishments inflicted by the Company for slave escapes, and particularly for attacks on the bodies or the property of masters, were harsh – far harsher than the punishments which masters were liable to incur, for instance, for the murder of a slave (Ross 1983). They involved gruesome torture and often death. The public enactment of such punishments symbolized, for the population as a whole, that the power normally held latent by the dominant group could be manifested at will (Foucault 1977). The Company's selective use of exemplary violence was mirrored in the approach of individual masters. As the slaveholder Samuel Hudson wrote, 'I have seldom occasion for punishment and they are well aware of its being regularly inflicted when they deserve it' (Shell 1994: 221). Although there were occasional slave revolts, slaves' awareness of the Company's and masters' potential for violence rendered paternalism, where it was available, generally preferable to overt resistance.

The colonial burghers were initially former company employees, recruited mostly in Holland and what is now Germany. In 1688 they were joined by 156 Huguenot refugees fleeing religious persecution in France. The burghers' claims to individual ownership and occupation of land at the Cape were driven by their imperative to produce at above-subsistence level, as far as conditions allowed. They came from a Western Europe which was completing a long, varied and complex, but fundamental transition from feudal to capitalist social structures. This transition entailed two particularly significant and interlinked changes: from communal access to land to privately owned and enclosed farms, and from limited material aspirations among the vast majority of the population to expectations of the accumulation of wealth among a significant minority (Hilton 1976). Despite the constraints on free enterprise imposed by Company monopolies at the Cape, 'the ethos of the colony's elite was certainly that of competitive capitalism', and all of those Europeans who emigrated there, 'whether they produced grain, wine or stock, must be seen as tied to the market [and generally in search of accumulation] to a greater or lesser extent' (van Duin and Ross 1987: 3).

A system of private landholding was a critical feature of transplanted European capitalist society in the Cape, distinguishing it from pre-existing Khoikhoi and San social structures. On the one hand, it facilitated the accumulation of tremendous levels of wealth among an elite now known as the 'Cape gentry', who were able to seize the most valued land nearer the Company port at Cape Town (Guelke and Shell 1983). By 1731, colonists of this stature, comprising 7 per cent of the European population, owned over half of the colony's landed wealth (Guelke 1989). On the other hand, private landholding guaranteed a corresponding relative poverty among those who were relatively or absolutely deprived of access to the land. Such class stratification was to be an enduring feature of colonialism across broad swathes of southern Africa, as settlers pushed capitalist practices inland (Bundy 1992).

From 1702, burghers were able to acquire 'loan farms' from the Company, averaging 6,000 acres, in exchange for a small annual fee. Soon they were moving beyond the boundaries recognized by the Company, dispossessing the western Cape Khoisan of their land, as we will see below, but were still being granted effective freehold of the land they occupied because the Company continued to rely on their produce. During an economic boom in the late seventeenth and early eighteenth centuries, the

Plate 3.2  Cape Town in 1764
*Source*: Africana Museum Johannesburg

*Company allowed illegal settlement for Trade*

western Cape gentry developed intensive wine and wheat production. Other colonists, striving to acquire land of their own, pushed yet further inland, ignoring the protestations of an increasingly nervous Company (Guelke 1988). On the fringes of European settlement, only livestock could be transported profitably to the Cape Town market. Where colonial settlement pressed into the semi-arid region to the north and east of Cape Town, crop cultivation became almost impossible. Extensive pastoralism was therefore adopted as colonists moved inland. The interior, especially the southern coastal strip, offered poorer colonists 'pasture in abundance, unparalleled hunting opportunities in a land teeming with game and opportunities for trade and robbery in cattle and sheep' (Guelke 1991: 15).

As the burgher population grew from 259 in 1679 to 10,500 in 1780, colonial territory expanded with it, each rising male generation expecting to lay claim to its own land (Guelke 1988; Shell 1994; van der Merwe 1995; Fig. 3.4). So too did the reach of the burghers' commercial system. This was not a 'modern' capitalist system in which most items, including labour, are commodified. Frontier farmers continued to hunt and grow their own food for instance. But it was nevertheless one which was reliant on exchange, and on a developing money economy. Even the remotest farmers produced commodities such as meat, soap and candles so that they could be exchanged for luxuries like tobacco and coffee, and essentials such as wagons, guns and ammunition. Such exchanges occurred initially as the result of long, intermittent treks across rough tracks back to Cape Town. But once a frontier district had been sparsely settled by burghers, itinerant traders known as *smouse* would visit, connecting farmers directly to the colony's money economy.

With the foundation of small market towns such as Swellendam and later Graaff Reinet, and the founding of banks, local commercial transactions became more firmly rooted. By the mid-eighteenth century, an expanding commercial nexus was well established. It had its local hub in Cape Town and the surrounding wheat and wine districts dominated by the gentry. But beyond the port of Cape Town, it stretched right back to Europe, from where manufactured items were imported by the

**Plate 3.3** Frontier colonist with Khoisan servants, livestock, wagon and tent, 1700s
*Source*: Africana Museum Johannesburg

Figure 3.4  The Cape Colony in 1790

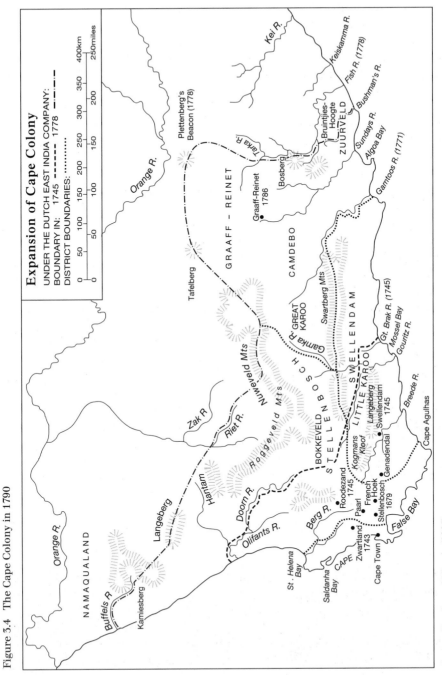

**Expansion of Cape Colony**

UNDER THE DUTCH EAST INDIA COMPANY:

BOUNDARY IN: 1745 — · — · — 1778 — · · — · ·

DISTRICT BOUNDARIES: · · · · · · · ·

*Source:* Berghe and Visagie (1985)

Company (Neumark 1957; Ross R. 1986, 1988; Newton-King 1992). Although the Company attempted to maintain a variety of partial monopolies and monopsonies within the colony, there was still scope for private enterprise, not least on the part of Company officials themselves. While the Company attempted, for instance, to control all sales of wine to European merchants, private entrepreneurs on the estates of the western Cape were able to secure good prices in their own right. Having produced 660,000 litres of wine in 1725, colonists were selling over 3 million litres by 1775. Aside from benefiting from this wine export, the gentry of the western Cape also engaged in the large-scale commercial production of wheat. Increasing domestic consumption and an expanded foreign trade stimulated gains in production from 15,000 hectolitres in the 1720s to over 50,000 by the 1770s (Ross 1988). The Company, then, 'could only seal off the colony to a very limited extent from the world market' (Ross 1988: 248).

However, before extensive colonial settlement and commercial exchange could develop in the interior, European military power had to be demonstrated more extensively. By 1700, having been given permission to barter livestock inland, colonists frustrated by the Khoikhoi's reluctance to part with sufficient cattle were already using their horses and guns to raid Khoikhoi herds. The effect was devastating. In 1705 the Landdrost (Company-appointed magistrate) at Stellenbosch reported that the surrounding Khoikhoi had been deprived of all their stock. He continued, they 'have nearly all become Bushmen, hunters and brigands, dispersed everywhere' (Penn 1989: 5). While such colonial raiding prepared the way, European settlement itself was extended through small groups of colonists co-operating in commandos, groups of armed and mounted men acting under the authority of their own appointed leaders (and later, in some instances, under the authority of Company officials). They concentrated first on monopolizing access to the most valuable land in a semi-arid environment – that surrounding a spring or a watercourse – and getting their claim to it recognized by the Company. Control over water supplies, and the game which they attracted, then enabled dominance over a far more numerous Khoikhoi population, often without the need for direct conquest. By such means, dramatic colonial expansion was effected during the eighteenth century by just a few thousand colonists and their families (Guelke and Shell 1992; van der Merwe 1995).

Khoikhoi groups responded to European intrusion in various ways, ranging from 'open resistance' through 'cautious withdrawal' to 'eager co-operation' (Elphick and Malherbe 1988: 16). But those who wished to retain some of their own cattle had either to displace other Khoikhoi groups further inland (leading to a general exacerbation of conflicts between clans), ally themselves with the Company authorities, or work for the colonists. Many of the interior Khoikhoi, remote from the nearest Company officials, became the clients of individual frontier colonists (Marks 1972; Elphick 1977). Colonial domination over such Khoikhoi and San 'clients', like that over the slaves, worked through more subtle as well as overtly violent means. For Khoikhoi threatened by San and colonial hunters, incorporation on a colonial farm was a relatively secure option and, as Guelke and Shell point out, 'accommodation even to virtual serfdom would have made a good deal of sense to a people with hierarchical traditions in which clientage and even slavery was practised' (Guelke and Shell 1992: 819).

While colonial farmers offered protection, food, tobacco, 'dagga' (marijuana) and alcohol to their Khoikhoi clients and servants, and allocated grazing for their cattle, the employment of Khoikhoi labour was in turn vital to the colonists (Marks 1972; Ross *et al.* 1993). Without it, as the traveller Otto Mentzel noted, 'no European or Afrikaner [a colonist born in the Cape] could live in these regions' (van der Merwe 1995: 115). Grevenbroeck, a burgher from Stellenbosch, pointed out that Khoikhoi men not only acted as herdsmen to the sheep and cattle, they also trained oxen for ploughing, drove wagons, broke horses, and chopped wood. Meanwhile, Khoikhoi women and children 'wash plates and dishes, clean up dirt, gather sticks from the fields ... light the fires, cook well, and provide cheap labour for the Dutch' (Guelke 1991: 13). The subordination of Khoikhoi labour was the only thing which allowed the remotest colonists, who often lacked slaves, to engage in a 'primitive' form of capital accumulation, connecting them with the colony's broader commercial system.

The extension of individual colonists' authority and economic dominance in the western Cape was assisted greatly by a devastating smallpox epidemic brought by a European ship in 1713. This alone killed the vast majority of Khoikhoi in the region, reducing the survivors to a greater dependence on Europeans. But by the late eighteenth century, the northern and eastern frontiers of the colony were beginning to hinge on a different balance of power. In the late 1770s colonial expansion to the east ground to a halt as the vanguard colonists entered lands densely settled by Xhosa-speaking chiefdoms around the Fish River. To the north, when colonists pushed in greater numbers into the boundary between summer and winter rainfall zones around the Roggeveld mountains, they were entering an area of particularly scarce resources. The competition for these resources sparked by the colonists' arrival, precluded the kind of paternalistic relations with Khoisan clients which developed in some other parts of the colony. In the face of overt Khoisan resistance, commandos had to be employed continuously in order to protect stock and access to grazing, and burghers were frequently repulsed from the lands to which they laid claim. The region's numerous San, who lacked any tradition of clientage, put up particularly formidable resistance to conquest. Frontier colonists responded by instituting a 'reign of terror', constructing the 'Bushmen' as being only half-human and exterminating the men wherever possible (Newton-King 1994: 228). They also captured and brutalized local Khoisan women and children, forcing them to provide labour – a practice which was tacitly endorsed when the Company allowed the legal indenture of captives and other Khoisan born on colonial farms, at first until the age of twenty-five and later for a total of twenty-five years. To the east, Xhosa resistance to colonial encroachment proved just as effective in halting colonial advance and necessitating more forceful responses. Colonists on that frontier were compelled to evacuate their farms on more than one occasion during the late eighteenth century.

Together, Khoisan resistance to the north and Xhosa to the east halted colonial expansion and guaranteed an unstable and often violent equilibrium in frontier districts throughout the last thirty years of the eighteenth century. Indeed, when Xhosa reprisals were combined with Khoisan labour rebellion on the eastern frontier from 1795 to 1802, colonists were forced to retreat *en masse* (Marks 1972; Penn 1989).

### *Religious and racial boundaries*

Slaves and Khoisan were obviously subordinate groups within colonial territory in the eighteenth century. But status differentials were always in danger of becoming blurred. There were no hard and fast natural distinctions to elevate the colonists above those whom they dominated and so, as in other colonial situations, social boundaries were constructed to take their place (Stoler 1989). The rituals which were developed to symbolize these boundaries were especially necessary where, as under slavery, there was a great deal of intimate contact between superordinate and subordinate groups. Slaves were housed on the property of their master, often living within the same building, and they were in constant interaction with the master's family. Status distinctions were tested almost daily, and had constantly to be upheld through displays of deference. Thus, if slaves were to avoid punishment, they had to accept facetious names, like those which would be given to family pets, to look meek and speak respectfully to the master, and to go barefoot and without wearing a hat in public. At the larger scale, within colonial society as a whole, religion and later race served as more robust boundary markers, bolstering the social privilege of the colonists (Mason 1994; Shell 1994; Worden and Crais 1994).

According to the doctrines of the Dutch Reformed Church (DRC), in which most burghers placed their faith, baptism allowed one to inherit property, marry, be buried in a Christian graveyard, bear legal witness and fill official posts. It therefore marked a rite of passage to full citizenship. At the global scale, Europeans had been identifying the possession of Christianity as one of the major distinctions between themselves and the other peoples, described as 'heathen', whom they had encountered since the Renaissance conquest of South America (Todorov 1984; Thomas 1994). Although the DRC had ruled that 'heathen' slaves should be converted and baptized for the sake of their souls, the Dutch East India Company was too preoccupied with commerce to embark on any widespread evangelization. It therefore left the conversion of the colony's slaves to their masters, most of whom remained resolutely opposed to it. For individual masters to confer the rights of baptism on their slaves would be to erode one of the most basic distinctions which ensured their social privilege. This, and the fact that Christian slaves could not legally be sold, meant that 'most slaves were systematically excluded from the Christian community' (Shell 1994: 332). The same fear of a potential equalization in status lay behind the general reluctance, indeed explicit resistance, to the full conversion of the Khoisan. To the relief of many colonists, the distinction between Christian burghers and heathen slaves was enhanced during the eighteenth century as many slaves turned to the more accommodating belief system of Islam (Elphick and Shell 1988; Fredrickson 1981).

Although Christianity was a 'jealously guarded group privilege', it was insufficient by itself to insulate burghers from their subordinates' demands for social inclusion and equality (MacCrone 1937: 256). A certain imprecision remained inherent in the early colonial hierarchy, and it was revealed most clearly in the case of the offspring resulting from sexual intercourse between members of different status groups. Clearly, a double standard applied to such relationships. It was extremely rare for a

European woman to engage in intercourse with a man who was not European – even a slave who had been freed. Burgher women, like female colonists elsewhere, were bound by particularly confining, patriarchal conventions (Stoler 1989). Their role was constructed as one of 'marriage and submission to the authority of their husbands, the supervision of the household, the bearing of children and the inculcation of the norms and values of their society into the next generation' (Walker 1990c: 317). Thus, for a European woman to encourage a relationship with a slave was to transgress two significant boundaries: that between free and unfree, and more importantly, the gendered boundary protecting European men's control over what they considered 'their own' women (Fredrickson 1981; Elphick and Shell 1988).

Yet, during the early years of colonial settlement, sexual 'transgression' was common among burgher men. With the relative shortage of European women, sex with, and indeed marriage to, freed slaves was a practical and attractive option for male colonists. For slave women such marriages might represent the opportunity of freedom for themselves and their children (Elphick and Shell 1988; Ross et al. 1993). During the early eighteenth century, the Cape's European community, like that in other Dutch East Indies possessions, was too small and vulnerable to govern without the incorporation of such children as burghers with full rights of citizenship (Fredrickson 1981). Some 1,000 marriages between burghers and ex-slave women were therefore tolerated by the Company, mostly in the late seventeenth and early eighteenth centuries, as a means of generating a 'loyal and useful group of colonial subjects' (Guelke 1989: 41).

However, less formal sexual contact, including frequent rape, was also conducted between European men and relatively powerless slave and Khoikhoi women. The offspring of such informal unions might assume an ambiguous position within seventeenth- and eighteenth-century colonial society. If their mothers remained slaves, they too would be born slaves. However, if the mother was a freed slave or Khoikhoi, the offspring would be legally free, but were distinguished as 'Bastaards'. This designation was, at times, all that set them apart from the minority of poorer European colonists who were not themselves granted full burgher status – mostly company servants or those recently released from their contracts. Although they could be baptized, own slaves, land and livestock, and carry weapons, 'free blacks' like the 'Bastaards' had to abide by the Company's slave curfew and were discriminated against not only in public office, but also in the practice of certain occupations (Elphick and Shell 1988; Giliomee 1988). In the case of those 'free blacks' born to a Khoikhoi mother and a slave father, known as 'Bastaard-Hottentots', it was customary practice, and from the late eighteenth century, a legal requirement, that children be indentured to serve the colonial master on whose property they were born. It was the clarification of the indistinct status of those 'Bastaards' born to European fathers and freed slave and Khoikhoi mothers however, which prompted the creation of the colony's first explicitly 'racial' boundaries.

In many colonial societies, 'the arrival of European women ... was particularly significant in reinforcing class and racial distinctions, in part because of the emergence of new sanctions against intermarriage between indigenous women and European men' (Stasiulis and Yuval-Davis 1995: 14). Guelke (1989) suggests that, as the proportion of European women increased in the late-eighteenth-century western Cape, the position of 'Bastaards' was defined more explicitly in order to ensure their exclusion from the burgher community. The reason was that a more balanced European gender ratio reduced the imperative for intercourse with slaves and indigenes in order to maintain the dominance of colonists. It could now be made 'more difficult for non-Europeans to be admitted to full membership' of the privileged group (Guelke 1989: 43). This local concern for a less 'diluted' form of European dominance meshed well with a contemporaneous, European-centred shift in discourses of 'racial' difference. During the eighteenth-century Enlightenment, the Renaissance taxonomy of the human race, based on the divide between those human groups who possessed, and those who lacked Christianity, was being adjusted. A secular categorization of humanity became increasingly dominant, and it depended not so much on a religious world view, but on a more rational, 'scientific' classification of peoples within a 'natural' hierarchical order known as the 'Great Chain of Being'. The new discourse encouraged more precise distinctions to be made between discrete racial groups on the basis of their appearance and culture, including their political and religious organization and their modes of agriculture and warfare (Pratt 1992; Thomas 1994). Within late-eighteenth-century European racial discourse, differences in pigmentation assumed much of the cultural significance that differences in religious orientation had acquired during the seventeenth century.

In the late-eighteenth-century Cape, unions between 'white' colonial men and 'brown' or 'black' women of mixed or slave descent were increasingly frowned upon, perhaps especially by 'white' European women seeking to defend their exclusive position. In 1765, for instance, the colony's 118 free black women (mostly former slaves and 'Bastaards') were condemned by the Cape Town authorities for imitating the clothing and appearance of 'respectable burghers' wives', and barred from wearing silk skirts, lace, bonnets, curled hair or earrings in public (Shell 1994: 323). The Company also banned marriages between Europeans and "pure-blooded" persons of other races', that is Khoikhoi and most slaves (Guelke 1989: 43). In contrast to the relative fluidity of relations between Europeans and those who were of mixed descent in the early eighteenth century, by 1798, Lady Anne Barnard could write that 'the word slave-born or half-caste' was enough to dispel any notion of 'quality' or 'virtue' (Shell 1994: 324). The desire of 'respectable' Capetonians to preserve certain boundaries was matched by the concerns of poorer colonists, especially non-burghers, who felt that they had increasingly to compete with free blacks for jobs and social status (Elphick and Shell 1988). In the late-eighteenth-century western Cape then, there was agreement across the spectrum of European classes on the need for tighter racial boundaries.

In the northern and eastern Cape, a burgher consensus was also being generated on the need for a clearer distinction between colonists and their subordinates on the grounds of 'race'. With the escalation of warfare on the northern frontier during the late eighteenth century, Penn (1989) notes that the 'Bastaards' were identified by burghers as potential military allies. For the first time, it became essential to define precisely what their duties and rights should be. The resolution reached by the local authorities was that although they were eligible for military service, they should continue to be excluded from the full rights of burghers. Simultaneously, on the eastern frontier, available land was becoming scarce as colonial farms abutted Xhosa settlements. As Giliomee (1988) points out, the consequence was increasing stratification within colonial society, with burghers pressing the priority of their claims to resources more fervently against their own and other colonists' 'Bastaard' offspring.

Decisions to exclude the 'Bastaards' from the material, occupational and official opportunities available to burghers across the frontier districts, prompted many men and women of 'mixed race' to emigrate over the northern colonial boundary where they carved out new polities of their own. Under missionary influence, these polities along and beyond the Orange River were labelled collectively as the Griqua. Although they had escaped the increasing racial constraints of the colony, members of Griqua communities maintained their commercial links with colonial burghers, enabling them to acquire both horses and European weaponry. During the early nineteenth century, as we will see below, some of them played a significant role in the extension of slave raiding into the interior, on burghers' behalf (Marais 1957; Ross 1981; Legassick 1988). By then, a 'trend towards white exclusiveness' was well established in the Cape as a whole (Fredrickson 1981: 121).

Within a largely commercial colonial society, oriented to wine and wheat production in the west and extensive semi-commercial pastoralism in the north and east, religion and colour were being developed as reinforcing, and more impermeable social boundaries during the late eighteenth century. The wealth generated through production and exchange with Cape Town, and through that port to the wider, European-dominated Western world, was being channelled in ways defined by these social boundaries. A western Cape gentry, well connected with the locus of political and legal authority – the Dutch East India Company – had greatest access to material and ideological power, but white burgher citizens as a whole were able to deploy formal and informal mechanisms to preserve relative social and material privilege over non-burgher colonists and dependent slave, Khoisan and 'Bastaard' subjects.

## The transition to liberalism

During the first half of the nineteenth century, the practices by which Cape colonial society had been regulated under the Dutch East India Company were to be reformulated. This reformulation can be ascribed to the interventions of a new and more powerful metropolitan centre within the global capitalist economy – Britain. British imperialism, based on a different conception of citizenship from that of the Cape colonists, saw to the end of slavery and the legal 'liberation' of the Khoisan. Paradoxically, however, it also brought about the construction of new ideologies and practices through which racial inequality was entrenched and extended. In order to comprehend the changes, we have to examine the shifting boundaries of privilege in Britain as well as the Cape, since it was connections between the two sites which shaped official policy and socio-economic practice in the colony.

British officials took control of the Cape from the waning Dutch East India Company for the first time in 1795. They had identified it as a strategically vital port during Britain's war with Revolutionary France. It was briefly handed back to the Dutch government in 1803, but the threat of Napoleon's military expansion in Europe led to renewed British control after 1806. Concerned only with the Cape's strategic role as naval base, the government in London concentrated initially on administering the colony at minimal expense. Accommodations with the existing Dutch elite were essential to the effective economic management of the Cape and during the 1800s and 1810s, conservative policies were adopted (Peires 1988). In fact, these policies reflected the British ruling class's own domestic concerns.

This was a period in which the landowning aristocracy and gentry who dominated British political life felt besieged. They were fighting not only Napoleon's armies on the Continent, but also political radicalism, partially inspired by the French Revolution, and popular unrest driven by an economic downturn, at home. Tory governments, fearing revolt, were reacting with repression, cracking down on public meetings and political agitation with a series of legislative and judicial measures (Evans 1996). The British elite's emphasis was on stabilizing fraught relations between established social groups, and British governors' first, limited attempts to intervene in Cape colonial society generally had the same aim (Macmillan 1927; Atmore and Marks 1974). In the Cape, status divisions, as we have seen, had come to correspond quite closely with 'racial' difference. Accordingly, early British policies in the Cape consolidated the racial boundaries which had been forged during the late eighteenth century. For example, in 1809, Governor Caledon issued the 'Caledon Code' which, in return for outlawing the extreme abuse of Khoikhoi servants, ensured their continued subservience. In particular, it required them to obtain a colonial farmer's pass before leaving their 'place of abode'. Otherwise they could be arrested and jailed. Although such measures were aimed at stripping the power to punish Khoikhoi away from individual colonists and placing it in the hands of the new colonial government, they were certainly not intended to remove the inequalities of status between colonists and their servants (Fredrickson 1981; Crais 1992).

On the colony's eastern frontier, the new British authorities showed as little sympathy for the Xhosa as they had for Khoikhoi servants in the west. Describing the 'Kaffirs', as the Xhosa were universally called by colonists, as a 'nation' imbued with 'a thirst for plunder and other savage passions', Governor Cradock implemented a more forceful and separationist frontier policy from 1809 (Lester 1997: 641). In 1812, Xhosa chiefdoms were forced east from contested land along the frontier by British regular soldiers, British-officered Khoikhoi troops and colonists. Xhosa who resisted were shot in the process (Maclennan 1986).

### Freeing the labour force

British influence in the Cape, however, was soon to become more ambivalent. British society itself underwent a dramatic transition in the aftermath of the victory over Napoleon in 1815, and as the industrial revolution gathered pace. The most significant change was the rise of middle-class, or bourgeois, economic, social and political influence. Intrinsic to bourgeois identity was evangelical, nonconformist religion, orchestrated originally in opposition to the exclusive and conservative Anglicanism of the ruling class. However, during the early nineteenth century, the aristocracy increasingly accepted the need for an alliance with the bourgeoisie, so that both might defend their property, whether it was landed, commercial or industrial, against the potentially radical working classes. Through personal friendships with government ministers, their influence on committees and, ultimately, through parliamentary reform, bourgeois reformers and evangelists were given greater access to the levers of power (Elbourne 1991; Colley 1992; Stoler and Cooper 1997).

In this context, colonial missionaries, most of whom were from lower-middle class origins, were able to forge a political alliance with reformists in Britain (Comaroff and Comaroff 1991). While missionaries concentrated on 'saving the souls' of the heathen within the empire and on getting the slave trade and then slavery itself abolished (achieved in 1807 and 1834 respectively), metropolitan political reformers focused on a range of domestic issues including securing the vote for the middle classes, regulating factory conditions and providing poor relief so as to ameliorate the harshest conditions suffered by the working classes, thereby pacifying and 'improving' them (Keegan 1996). Despite the wide range of their concerns and the contradictions between them, during the 1820s and 1830s, humanitarian reformers thus created a politically powerful network spanning the British empire (Lester 2000).

The first success of this global humanitarian network in the Cape involved the treatment of Khoisan servants. Circuit courts were established in 1812 to punish colonial masters if they inflicted brutality on their labourers (Macmillan 1927; Peires 1988). The position of the Cape's slaves was also being addressed as part of a much broader campaign. First, the Atlantic slave trade, which evangelicals represented as grossly inhumane, and a direct contravention of God's commandments, was targeted. Popular agitation against slavery in Britain secured the end of the trade in 1807 (Walvin 1982; Turley 1991; Colley 1992). Thereafter, only the slaves captured by the Royal Navy from other nations' ships were brought to the Cape. With the transatlantic trade in slaves abolished, reformist zeal was brought to bear next on the conditions of those in Britain's colonies who were already enslaved. Missionaries in the Cape and elsewhere agitated during the 1820s to ameliorate the slaves' punishment, food, clothing and working hours (Holt 1992). Within the evangelical discourse which dominated British public life in the early nineteenth century, such measures were intended to secure atonement in God's eyes for the British nation as a whole (Hilton 1988).

Despite some fundamental points of disagreement, these religiously inspired and humanitarian efforts against slavery were assisted by a new economic orthodoxy being consolidated by British theorists. Adam Smith, in particular, interpreted Britain's economic growth as the outcome primarily of 'free trade'. The lesson was that metropolitan and colonial produce should be both manufactured and exchanged without government interference. In the metropolitan context, that meant the removal of landed aristocratic protection by opening the market to cheaper grain imports. In the colonies, the practice of slavery was an obvious government-imposed constraint on the efficient use of labour. It meant that workers were not able to sell their labour power within a free market, and employers were not free to hire and fire as and when labour was needed. It also meant that labour was not optimally productive: labourers were forced to work for others by law, rather than encouraged by individual incentive to 'improve' themselves. They were thus not motivated to contribute their full potential to the broader society's prosperity. Thus philanthropy and economics became interwoven as reinforcing rationales for the abolition of slavery (Keegan 1996).

By the 1830s, the anti-slavery campaign in Britain was reaching a crescendo, with a mass petition to parliament signed by one and a half million Britons manifesting widespread moral repugnance and economic opposition (Colley 1992). Within this atmosphere, the British government overrode the material interests of the West Indies planters and other colonial slave-owners. All slaves within the British empire were emancipated in 1834, although they generally remained with their former masters as 'apprentices' for a further four years. During this four-year 'preparation for freedom', bourgeois reformers intended the slaves to learn the responsibilities of sober, productive citizenship, just as they felt that those responsibilities had to be learnt by Britain's own radical and potentially rebellious working classes (Holt 1992; Worden 1994).

Plate 3.4 Slave celebrating his forthcoming emancipation, 1833
*Source*: Africana Museum Johannesburg

John Philip, local Director of the London Missionary Society (LMS) and his son-in-law the newspaper editor, John Fairbairn, were key figures in the global humanitarian network, based in the Cape. Riding the back of the campaign against slavery during the late 1820s, they were able to draw attention to the continuing plight of those Khoisan who were unable to seek legal assistance, deploying the economic argument that the Khoisan would be better-motivated and more docile workers if freed from their legal bondage. Philip and his allies in the British parliament succeeded in 1828 in getting Ordinance 50 ratified by the Cape and British governments. Although the ordinance still classified 'hottentots and other free persons of colour' differentially from the white colonists, it abolished the pass laws which had restricted the Khoisan's mobility under the Caledon Code. The ordinance also released the Khoisan from the legal requirements which bound them and other free blacks to serve the colonists and explicitly recognized their right to own land. For humanitarians, Ordinance 50 was hailed as the Khoikhoi's Charter of Freedom (Macmillan 1927; Ross A. 1986).

In the space of six years, a humanitarian and economic reform movement spanning the British empire had thus accomplished the abolition of the legal measures which kept both Khoisan and slaves in the Cape in subservience. As Crais has pointed out, the emancipation of these groups 'threw into question ... the extent to which freedmen [and women] would [subsequently] be included in the body politic' (Crais 1990: 190). What would prove decisive for South Africa's nineteenth-century social structures was not so much the act of emancipation itself, but how this critical question would be answered.

### Disciplining the freed labour force

The extent of the freed slaves' and Khoisan's incorporation within the colonial 'body politic' was in practice to be limited, not only by resistance from colonists, but by the prospects which their own 'saviours', the missionaries and humanitarians had in mind. While they generally supported Ordinance 50 and universally backed the abolition of slavery, colonial humanitarians, like their bourgeois counterparts in Britain, believed fervently in the necessity for the Khoisan and former slaves (and on the frontier, the Xhosa too), to behave in particular ways and use their newly won freedom to prescribed ends.

The prescriptions offered by missionaries and humanitarians derived from the metropolitan context, and from the broader Enlightenment understanding that all societies were 'arranged on a vertical scale of civilization', with European and especially British society, at its apex (Porter 1997: 376). Both indigenous peoples and former slaves were to aspire to the conduct of particular elements within this society. For the few Khoikhoi who were granted land along the Kat River on the eastern frontier, and the Xhosa beyond the colony's borders who still retained their land, the prescription was systematic. They were to become a prosperous and diligent peasantry. Integral to this envisaged transformation was dramatic cultural change. They were to supplant their own forms of gender discrimination with ones which had recently been forged in Europe, depriving their women of responsibility for agriculture and confining them to domestic duties within their own and settlers' homes (Cock 1990; Lester 1998b). They were to replace round huts with square European houses, semi-nakedness with European clothing, the hoe with the plough, common grazing land with the cultivated enclosure, and rainmaking rituals with irrigation (Etherington 1978, 1989a; Comaroff and Comaroff 1991; Comaroff 1997).

For the freed slaves and Khoikhoi who lacked their own land, the humanitarians' prescribed role was not so much that of the bucolic peasant as that of the dutiful worker. The colonial programme for the emancipated colonial labourers was an extension of that which the bourgeoisie envisaged for the working classes in Britain. Fairbairn expressed the approach succinctly: bonded labourers must be freed from legalized oppression, but their emancipation should be achieved with 'least shock to property and vested rights' (Fairbairn quoted in Keegan 1996: 113). This was why the colonial government, with the blessings of the humanitarians, was soon able to enact legislation imposing different kinds of constraints over the colonial workforce (Worden 1994; Lester 1998c).

Despite many freed labourers' attempts to reconstruct their lives in urban areas and on mission stations away from colonial farms, opportunities for ex-slave and Khoikhoi workers, who were now classed together for the first time as 'Coloured', remained limited (Marais 1957; Ross R. 1986; Crais 1990). Most lacked access to productive land and capital with which to pursue an independent livelihood. Even those seeking only residential plots in the towns were discriminated against by new

Plate 3.5  Kat River Settlement, 1838
*Source*: Africana Museum Johannesburg

municipal councils run by colonists, and were effectively forced into 'squalid locations' (Macmillan 1927: 270). In Cape Town, a class of bourgeois slum landlords
emerged, hastily building property with the money acquired from the British government as compensation for the freeing of their slaves, and renting it out to freed slaves
themselves (Meltzer 1994). Aside from the tiny minority of Khoikhoi granted fertile
land along the Kat River (partly in order that they might form a buffer between the
colony and the Xhosa), former slaves and bonded labourers were eventually obliged
to seek poorly paid employment.

The freed labour force had even less opportunity to evade working on terms laid
down by white employers once the colonial government passed a new Masters and
Servants Ordinance in 1841. This centrepiece legislation was intended to promote
the kind of work ethic which officials and humanitarians, as well as colonists,
required of the emancipated labourers, and it was based on equivalent legislation
approved by parliament to control workers in Britain (Thompson 1980). As a result
of metropolitan insistence, the Masters and Servants Ordinance was phrased so as to
make no racial distinctions. However, in the Cape, any distinction between a 'Master'
and a 'Servant' was effectively racial, regardless of whether it was made explicit. The
ordinance prescribed contracts between employers who were overwhelmingly white
and labourers who were overwhelmingly Coloured, imposing criminal penalties for
the labourers who broke their terms and more gentle punishments for masters.
Among its requirements for contracted labourers were the acceptance of corporal
punishment for desertion or 'misconduct', despite generally long working hours and
poor working conditions as well as low wages (Ross 1994). For other offences, imprisonment and formal flogging would now replace the master's individual punishment.

Aside from reinforcing class differentials between employers and servants and
consolidating the power of the state over and above that of the individual colonist,
labour legislation such as the Masters and Servants Act also bolstered gendered
boundaries within the subservient population. It recognized women as the legal
minors of their male partners, thereby enabling their subjection to the conditions
'agreed' by those partners, as well as to the domestic duties expected of them within
the family home (Scully 1997). Primarily as the result of the Masters and Servants
Ordinance, and other informal mechanisms of control, productivity within the western Cape was less affected by emancipation than it was in many of Britain's other
former slave colonies (Ross 1994).

The first major transition within colonial society, marked by the abolition of slavery then, meant simply a shift from one system of dominance to another. As Worden
and Crais argue, '[t]he goals of emancipation were clearly revealed: the replacement
of slaveowner arbitrary tyranny by a more powerful state regulation of labour, but one
which would continue to ensure the maintenance of social hierarchy and inequality of
race and class'(Worden and Crais 1994: 13). Indeed, in a world without legal
bondage, measures such as superior white schooling and segregated religious services
assumed new significance in order to ensure that the Coloured workforce's inferior
social status was perpetuated (Fredrickson 1981; Trapido 1980a; Keegan 1996).

## British settlers, capitalist expansion and segregation

Along with humanitarianism and the emancipation of bonded labour came other important consequences of the Cape's integration within the British empire. The first of these was the relative decline of the old colonial elite – the Dutch-speaking Afrikaner gentry – and its partial absorption within a new dominant economic class. Brought thoroughly within the orbit of a world economy shaped by industrializing Britain, the Dutch-speaking elite was ultimately deprived of the protection, including guaranteed markets for its wine exports, that the Dutch East India Company had provided (Rayner 1986; Ross 1988; Ross *et al.* 1993). At the same time, it was joined, and soon outnumbered, by an English-speaking merchant community seeking new colonial outlets from which to accrue profit.

English-speaking merchants were now advantaged in Cape Town due to their personal connections with, and financial backing from, the City of London, the greatest centre of finance in the world. Their commercial hegemony was reinforced when the colonial currency, the rixdollar, was replaced by British sterling. While it was establishing economic dominance, the bourgeois English-speaking elite was also carving out a new colonial identity at the Cape, and especially in Cape Town. Opposed to what was seen as the lethargic, religiously fundamentalist backwardness of the stereotyped 'interior Boer' (Afrikaner), this identity was explicitly rational, commercially driven and in touch with the currents of scientific thought being generated within the British and particularly Scottish Enlightenment of the late eighteenth and early nineteenth centuries (Dubow 1999; Ross 1999). Many of the old Dutch-speaking elite around Cape Town soon adopted the same values and pursued the same economic strategies as their new British counterparts. During the first half of the nineteenth century the predominantly, but not exclusively, English-speaking merchant elite would act as the conduit for financial resources from the metropole, resources which allowed a more aggressive and exploitative form of colonialism to be consolidated and extended (Keegan 1996).

The Cape's new merchant community remained, for the time being at least, broadly sympathetic to the humanitarian aim of generating a stable and prosperous black peasantry, not least because profits were to be made from trading with such a group (Trapido 1980a). But a further consequence of the colony's incorporation within the British empire was the emergence during the 1830s of another colonial group with connections in metropolitan Britain, one which became vehemently opposed to humanitarian programmes. This was comprised of a diverse array of settlers from all parts of the British Isles, whose first representatives had arrived as a coherent group on the eastern frontier in 1820. Whereas humanitarians stressed the universal qualities of human nature and the potential for every individual, regardless of current cultural differences, to be 'redeemed' by Christianity and a version of bourgeois 'civilization' modelled on that of Britain, settler discourse came to emphasize instead the 'inherent' and 'immutable' differences between 'civilized' Europeans and the 'savages' encountered in the colonies. In the following section, we examine ways in which British settlers and merchants in particular, sometimes conflicting and sometimes colluding, moulded a capitalist and racist colonial society from the 1830s to the 1850s. Together, they initiated, as Keegan puts it, 'the take-off of a new phase in colonial development' (Keegan 1996: 116).

### The 1820 settlers and the colonization of the Xhosa

Some 4,000 Britons were brought to the eastern Cape in 1820, one year after the Xhosa chief Ndlambe had retaliated against colonial interference by attacking the British garrison at Grahamstown. The town, in the heart of fiercely contested territory – the country from which Ndlambe had earlier been expelled – was now to become the new settlers' capital. These settlers were sent largely because the metropolitan government was in the midst of a revolutionary scare and emigration was seen as a way of relieving pressure on resources and easing political tension. But they had been requested by the Cape authorities in particular in order to help stabilize the eastern frontier of the colony. They came from all parts of the British Isles and from a spectrum of classes, ranging from the gentry to the labouring poor. Landing at what is now Port Elizabeth, they were allocated farming land in the district of Albany surrounding Grahamstown. Despite early class, regional and religious tensions, within

Plate 3.6 Xhosa chief, thought to be Ndlambe, c.1820
*Source*: Africana Museum Johannesburg

twenty years, the diverse array of settlers had forged a new identity as 'British settlers' which transcended many of their differences. The first condition which united the settlers was the desire for material gain within their new colonial environment. All came from a Britain in which capitalist aspirations to turn resources into profit, and to live at a level well above subsistence, were entrenched. All saw colonial emigration as a means of enhancing their standard of living (Lester 1998d).

Once established in the Cape, continued colonial control over Khoikhoi labour was seen as being essential if the settlers' material aspirations were to be fulfilled. Within a few years even those settlers who had been labourers in Britain were able to accumulate capital, partly through cross-frontier trade with the Xhosa but also through their domination of a cheap Coloured workforce. Humanitarian interference to free the Cape's Khoikhoi of their legal constraints was immediately resented by most settlers. Thomas Stubbs, for instance, described Ordinance 50 as 'that abominable false philanthropy which made [the Khoikhoi] free and ruined them ... They were a people that required to be under control, both for their own benefit and the public; the same as the slaves in this country' (Lester 1998d: 523). Ultimately, the settlers' attitudes towards the Khoikhoi were extended, in more virulent form, to the Xhosa. While colonial and metropolitan humanitarians wished to see the frontier Xhosa 'brought under the mild and transforming power of the Gospel' to the status of an improving and contented peasantry, by the mid-1830s settlers had visions of their lands being appropriated as profitable settler farms ('Justus', quoted Lester 1998c: 17).

Settler hostility towards the Xhosa was aroused first when those settled on land from which Xhosa chiefdoms had been expelled experienced retaliatory cattle raiding. It was intensified among the propertied classes when a realization of the eastern Cape's great potential for commercial sheep rearing coincided with a massive increase in demand for raw wool in Britain (Le Cordeur 1981; Peires 1988; Crais 1992; Mostert 1992). Instead of a reasonably equitable trade with an independent people across the frontier, settlers now found that sheep farming provided the greatest opportunities for wealth generation. In order to develop a prosperous woollen industry, however, elite settlers realized that there were three preconditions (in addition to reliable and cheap labour): a market, capital and land. The proliferation of textile factories in Britain provided a lucrative and seemingly endless market. Much of the compensation money awarded to the Cape's slave owners for the freedom of their slaves was now circulating in the hands of British merchants and newly established banks, providing the essential loan capital (Davenport 1982; Meltzer 1994; Keegan 1996). It was land hunger that remained to be fulfilled. During the early 1830s, settler spokesmen such as Robert Godlonton, who edited the newspaper *The Graham's Town Journal*, tried to gain more land by exaggerating the Xhosa's 'depredations'. Proclaiming the vulnerability of the British subjects who were forced to confront their 'savagery' directly, he hoped to provoke the authorities into seizing the frontier Xhosa's territory (Keegan 1996).

But most settlers developed an enduring antipathy towards the Xhosa only after they themselves became victims of a Xhosa counterattack on the colony in 1834–5. During the war which ensued, the Cape's governor himself adopted the phrase

Figure 3.5   Albany and the eastern Cape frontier, 1778–1865

*Source*: Lester (1998b) and (2000)

'irreclaimable savages' to describe the enemy. Despite the fact that some Xhosa chiefdoms had assisted the colony in the war, settler spokesmen increasingly adopted the
phrase to describe the Xhosa as a whole (Lester 1998a). Indeed, they went further,
enthusiastically backing pseudo-scientific attempts to prove that the Xhosa were
inherently inferior beings by analysing the shape and characteristics of their skulls
(Bank 1996; see summary box 1, pages 117–20).

In the wake of the war, the colonial government, backed by the settlers, intended
to extend the colonial boundary so that it encompassed much of the frontier Xhosa's

land, renamed Queen Adelaide Province (Lester 1998b). However, John Philip and other sympathizers with the Xhosa utilized the powerful humanitarian network to criticize the governor's plans, and the British government was persuaded, both by its own desire to cut colonial expenditure, and by these representations, to reverse the annexation of the province (Galbraith 1963; Macmillan 1963; Elbourne 1991; Lester 1998c). The British humanitarian vision of a potentially flourishing and Christian African peasantry, still in possession of its lands, had won out in this instance over the alternative vision of settler capitalist advance and indigenous dispossession (Keegan 1996). But the humanitarian victory, as we will see, was to be short-lived. Capitalist British settlers on the frontier continued to constitute a powerful force within the region – one which persisted in agitating for the expansion of colonial wool farming into Xhosa territory (Lester 1998d).

### The 'Mfecane' and the trekker republics

We will return to the struggles which ensued between British humanitarians and settlers, but first we must address the vast extension of a different order of colonial authority carried out by Dutch-speaking colonists from the eastern Cape. This extension was itself partly in reaction to humanitarian reforms. Through a mass emigration subsequently mythologized as the 'Great Trek', some 10–15,000 Afrikaner colonists referred to as 'trekkers' established new colonial communities in the interior between 1834 and the mid-1840s. Their reasons for pushing beyond the Cape frontier were diverse, as was their economic standing, but they commonly arose from the prospects that could be anticipated in the interior compared to those under continued British control of the frontier zone (see summary box 3, pages 124–6).

Most of the trekker leaders aspired to the kinds of commercial enterprise that the British settlers were engaging in on the eastern Cape frontier, but they believed that they had less chance of success within the colony (Ross R. 1986; Keegan 1996). Given that the British governors and the new officials who had displaced Afrikaner representatives clearly favoured British settlers in the allocation of further land grants, their assumption was apparently well founded (Lester 1998b). The final overcoming of San resistance and the development of better technology to extract groundwater in the form of wind-driven pumps, now raised alternative possibilities inland. A long pent-up move to the semi-arid north as well as to the more fertile eastern coastal strip, both of which had already been well traversed by colonial hunters and traders and partially settled by the Griqua and small groups of colonists, became possible (van der Merwe 1995).

Some trekker leaders probably had strong commercial motivations for seizing land in the interior, but it is not clear that most of the emigrants shared their leaders' imperatives (Delius and Trapido 1982; Etherington 1995; Keegan 1996). Many of the emigrants simply wanted new land at a time when it was rapidly running out on the frontier, and most were poorer farmers, less able to compete in a 'free' market for labour and alarmed by their potential loss of control over slave and Khoikhoi workers under British authority (Ross 1986; Giliomee 1988; van der Merwe 1995; Keegan

Figure 3.6   Routes taken out of the eastern Cape colony and onto the highveld by Dutch-speaking emigrants (trekkers), 1834–1840s

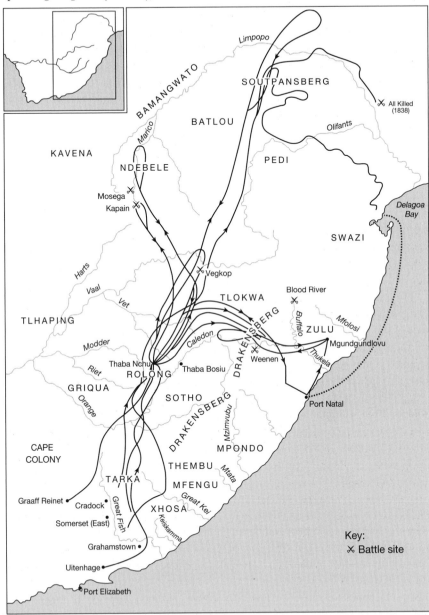

*Source*: Reader's Digest (1994)

1996). On the frontier in particular, there was little faith in the colonial state's ability to discipline a recently increased population of Coloured workers, so the undermining of the individual master's authority associated with emancipation was widely resented (Crais 1990, 1994). Many trekkers probably felt that humanitarian-inspired reforms held the potential to overthrow the 'natural distinction of race and religion', as one trekker put it, which maintained their superiority and privilege on the frontier (Macmillan 1927: 81). In collusion with British settlers, a trekker leader, Piet Retief, publicized a manifesto of the trekkers' grievances, directed largely against humanitarian 'interference' in the relationships between colonists, slaves, Khoikhoi labourers and independent Xhosa chiefdoms. Those Afrikaners who emigrated took their subordinate workforce with them into the interior, relying on their own methods of labour coercion and patronage to maintain established status distinctions.

On the highveld and on the eastern coastal strip, the trekker immigrants were initially able to establish a number of small, fractious polities only through a series of accommodations with local African chiefdoms, backed, when necessary, by demonstrations of military force. They found two such demonstrations in particular unavoidable. These were directed against the powerful Ndebele and Zulu polities, which attempted to resist their incursion into 'buffer territory' which each kingdom had established to safeguard its territorial core. Before we return to the conflicts and accommodations between trekkers and interior Africans, however, we need to understand a set of processes that had been transforming all African polities on the highveld and along the eastern coastal strip since the end of the eighteenth century.

As we saw above, the formation of hierarchically organized kingdoms, each with a strong monarchy, had been responsible for ethnic differentiation among Africans in parts of southern Africa since the tenth century. The assertion by dominant groups of an ethnic identity and a paramount lineage had long helped legitimate control over land and labour where supplies of each were scarce (Unterhalter 1995). But during the early years of the nineteenth century, Nguni-speakers to the east of the Drakensberg mountain range (modern KwaZulu-Natal) experienced increased inequality and violence, both within and between chiefdoms. This was probably associated initially with an ecological crisis including severe drought, but it was exacerbated by competition over the trade routes which linked the region to the Portuguese trading post at Delagoa Bay (modern Maputo, Mozambique), and subsequently, by externally organized slave raids.

In the midst of drought, the power differential between the beneficiaries of trading and those lacking independent control of the activity facilitated the conquest of weaker polities by stronger neighbours (Eldredge 1995). The most powerful polity which ultimately emerged from the conflicts between chiefdoms on the eastern coastal strip was that of the Zulu, marshalled by the paramount chief Shaka from the late 1810s to 1828. Through conquest and absorption, Shaka was able to considerably enlarge the population adhering to a Zulu identity, tracing a common, if often mythical ancestry, and speaking a Zulu dialect (Wright and Hamilton 1989; Wright 1995). While the Zulu kingdom was being constructed to the east, on the other side of the Drakensberg, conflict among highveld Tswana and Sotho-speaking polities had been mounting independently, caused by similar structural changes: environmental degradation, and increased competition for the control of long-distance trade and cattle (Parsons 1995a, 1995b; Manson 1995). A similar process of ethnic differentiation and centralized state power was occurring in response. During the 1810s and 1820s, the migration of Nguni groups escaping the violence to the east of the Drakensberg onto the highveld brought the two regions, and numerous, partially ethnically defined polities, into an interlinked and particularly intense cycle of warfare conventionally known as the 'Mfecane' (Omer-Cooper 1966; see summary box 2 pages 121–3).

This period of generalized and intensified strife may have been similar in origin to preceding episodes in the region, but in this case it was perpetuated and exacerbated by the intervention of colonial groups. Some Griqua, using the military technology which they had acquired from within the Cape Colony, engaged extensively in slave raiding across the Orange River during the 1820s, utilizing the turmoil of the interior to gather captives and sell them, illegally, to colonists in the Cape. At the same time, the Portuguese base at Delagoa Bay was itself a centre for slaving expeditions into the interior. From there, slaves were sold on to the plantations of the Americas (Cobbing 1988; Eldredge 1994a, 1994b). Despite the illegality of slave trading in the Cape, some English- and Dutch-speaking colonists also capitalized directly on the instability of

Plate 3.7 Romanticized portrait of Shaka
*Source*: Africana Museum Johannesburg

African polities by taking captive labour as and when the opportunity arose (Cobbing 1988; Kinsman 1995; Peires 1995).

The collapse of established means of subsistence in many parts of the interior, caused by further drought, insecurity and raiding, led thousands of Africans who were not enslaved, but nevertheless destitute, to migrate during the late 1820s. They sought clientship among the followers of relatively powerful chiefs such as Moshoeshoe's Sotho, or employment with colonists in the northern and eastern Cape, providing the latter with a source of cheap labour to supplement that of their Coloured servants and of those Xhosa admitted across the eastern frontier to work (Peires 1981). Some 16,000 of these refugees, subsequently known as Fingos or Mfengu, were absorbed as clients among the Gcaleka Xhosa, and were co-opted onto the colonial side during the 1834–5 frontier war (Marks and Atmore 1980; Webster 1995; Lester 1998b).

On the highveld and the eastern coastal strip, the need for extra defensive measures during this period of heightened insecurity persuaded both the Ndebele and the Zulu states to maintain partially depopulated zones around their centre, within which were located smaller tributary chiefdoms. Beyond these zones were further stretches of land that were frequently contested. The trekkers, arriving from the eastern Cape

Plate 3.8  Romanticized portrait of Moshoeshoe, 1833
*Source*: Africana Museum Johannesburg

during the late 1830s, accordingly saw much of the interior as 'empty land', over which they were determined to establish control (Etherington 1995; Parsons 1995a). Unlike the chiefs of more disrupted and fragmented highveld polities, however, the Ndebele and Zulu kings felt able to resist the trekkers' incursion, and attacks were launched upon them as they entered each kingdom's area of control. The trekker parties, with their horses and military technology, had to establish their armed superiority in pitched battles. The decisive battle with the Zulu, taking place on 16 December 1838, was subsequently known as the Battle of Blood River and mythologized by Afrikaner nationalists from the late nineteenth century as evidence of divine intervention on behalf of the trekkers (see chapter 4).

Such instances of dramatic military conquest were, however, not the norm. The preferred means by which the Cape colonial emigrants laid claim to land was negotiation backed by only an occasional and small-scale deployment of military power (Thompson 1982a; Delius 1986). After all, the concentrated firepower of the commando (the armed and mounted unit of men that had helped conquer the western Cape from the Khoisan and which also sustained trekker military supremacy on the highveld) could be mobilized only intermittently. Keegan (1996) stresses that the kind of society that many trekkers initially created could co-exist with post-'Mfecane' African polities. Like Africans, most trekkers were cattle herders utilizing resources to maintain subsistence and engage in the relatively small-scale exchange of surplus. Those in the far northern trans-Vaal region relied for some time largely on hunting, the exchange of ivory and slaving for their subsistence, as did many of their African neighbours (Wagner 1980). Like African men, trekker men exercised a patriarchal authority over the women and children of the household and some were initially willing to live under the authority of African chiefs, or at least forge necessary alliances with them against more threatening polities. Finally, like Africans, the trekkers initially formed a society where degrees of material privilege and deprivation were far less extreme than they were in the more capitalist society of the Cape (Trapido 1980b; Bonner 1983; Delius 1983).

The trekker parties' settlement was easiest in the central highveld, where African polities had been most disrupted during the 'Mfecane' (Thompson 1982a). Here, trekker authority, backed by the usually latent military power of the commando, was simply superimposed on the remnants of African chiefdoms which had already been in a relationship of subordination to more powerful African neighbours (Delius 1986). Most of the trekkers lacked capital to buy their own livestock and draft oxen, and so they negotiated sharecropping arrangements with the African homestead heads already settled on the land. For many African patriarchs, the turning over of a proportion of their crop and of their family's labour to a new white 'landlord' was not far removed from the accustomed payment of tribute formerly demanded by more powerful African states. Sotho and Tswana-speaking sharecroppers were joined on highveld farms by 'Oorlams' – Africans originally taken captive from, or traded with, surrounding polities as children. They and their descendants were brought up as acculturated Afrikaans-speaking farm servants, providing an intermediate and supervisory stratum between the owner and sharecropping tenants. A similar role was played by landless whites known as 'bywoners', whose numbers increased as landholding became more concentrated (Delius and Trapido 1982).

On the highveld sharecropping farms, relationships between landlords and tenants were generally mediated through a discourse of paternalism akin to that associated with slavery in the Cape, with the white farmer acting out the role of father to his various 'dependants'. While such an arrangement 'represented a compromise, a modus vivendi which enabled … working relationships' through most of the nineteenth century, it was also reliant on the regular infliction of violent punishment on 'dependants' who stepped out of line (Keegan 1986: 157; Trapido 1978; van Onselen 1996, 1997). Such informal methods of control were especially vital in the early years of trekker settlement, when political authority was dispersed between a number of small, far-flung trekker groups claiming allegiance to particular leaders.

It was not until the 1850s that power on the highveld became centralized in coherent trekker republics. On the eastern coastal strip, the first trekkers' constitution was drawn up within the republic of Natalia in 1839. Like subsequent trekker constitutions, it restricted citizenship, including the franchise, to white adult males. Natalia soon succumbed to British governance, as we will see below, and its main successor state, established in 1844, was located at Potchefstroom in what became the Transvaal. Among the diverse articles of government laid down in Potchefstroom was the prohibition on 'bastards … down to the tenth degree' becoming members of the governing body or volksraad. Further centres of authority were created during the 1840s by individual trekker leaders at Orighstad in the eastern Transvaal, Zoutpansberg in the north and Winburg near the Modder and Vet Rivers (Davenport 1991; see Fig. 3.7). As we will see below, the region between the Orange and Vaal Rivers known as trans-Orangia was occupied by British authorities in 1848, depriving those among its trekker community who sought autonomy of the chance to establish an independent government until 1854.

Plate 3.9 Transvaal farm, *c.*1860
*Source*: National Cultural History and Open-air Museum

By the early 1850s though, when the British government recognized the independence of the trans-Vaal and trans-Orangia trekker republics, the private accumulation of land had enabled a degree of political centralization among a broader-based trekker elite (Trapido 1980b). After 1849, members of this elite forged a federal entity in the Transvaal, bringing local leaders at Mooi River and the Magaliesberg, the Zoutpansberg, Lydenburg (the former Orighstad polity) and Marico under a united volksraad and dividing the official roles of state between them (Davenport 1991). To the south of the Vaal, in trans-Orangia, the Orange Free State (OFS) was forged as a separate republic in 1854 by commercially oriented trekkers who maintained trading contacts with British settlers in the eastern Cape (Keegan 1988; Fig. 3.7).

Within the new highveld state structures, the paternalism of the farm was progressively supplemented by state-imposed safeguards on white privilege. Regardless of their ownership of property, African men and women were denied citizenship in both the Transvaal (known as the South African Republic (SAR) after 1859) and OFS and subjected, not always successfully, to a pass system designed to prevent them leaving farms without the white owner's permission. In the Transvaal, blacks were also unable to own land, although missionary societies held patches of land on their behalf and small areas were protected from the trekkers' scramble to register ownership of land by their designation as government locations. In these areas, Africans could farm independently of direct white control. The OFS's constitution, devised in 1854, was modelled on the USA's, but with an explicit Calvinist religious framework, the franchise again restricted to whites only, and a prohibition on blacks owning land in any form. Land already owned by the Griqua in the west and the Rolong of Thaba 'Nchu in the east was bought up by trekkers, and particularly by the elite farmers who had access to the volksraad (Ross 1981; Delius and Trapido 1982; Bonner 1983; Keegan 1988; Davenport 1991). The alienation of Griqua land ultimately led to a mass emigration to an area that became known as Griqualand East, located beyond the southern border of Natal, but this area too would succumb to white encroachment during the 1870s.

With all the similarities between the trekker and African social systems of the 1830s and 1840s noted above, there was one profound difference which would prove decisive, especially after the establishment of the central republican governments. This was the trekker elite's expectation, and progressive realization, of private land ownership. It was this which allowed stratified landed and landless classes, with differential access to state power and resources to emerge within the trekker republics (Trapido 1980b; Keegan 1991). Once a capitalist landowning elite had asserted itself, it was almost inevitable that it would seek to expand its territorial base, just as British settlers had already done in the eastern Cape. A greater intensity of conflict with surrounding African chiefdoms would follow. This was the case in Natalia and trans-Orangia, established nearer the Cape with its commercial markets, before it was in the Transvaal. From the 1840s, in liaison with eastern Cape British settler entrepreneurs, the trekker leadership of trans-Orangia launched an extended struggle with Moshoeshoe's Sotho over fertile land along the Caledon River, while the Natalia trekkers' threat of displacing local polities was the act that prompted official British intervention in the region (Eldredge 1993; Keegan 1996).

Figure 3.7   The trekker states (later known as Boer Republics)

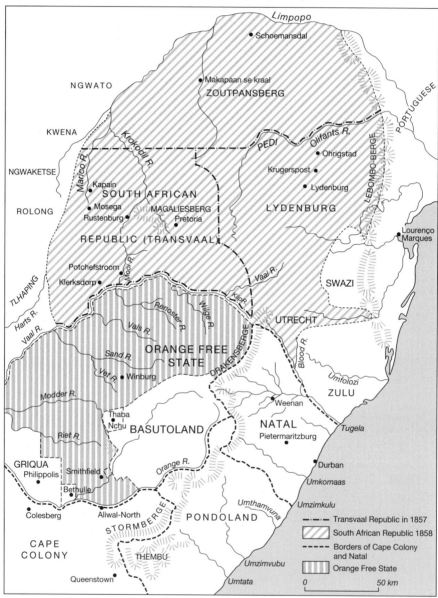

*Source*: Boeseken *et al*. (1922)

## Settler capitalism and segregation

In the eastern part of the Cape Colony, British settlers had continued to agitate for the dispossession of the Xhosa after the abandonment of Queen Adelaide Province. In 1846, frontier Xhosa who had resisted further land seizure by raiding colonial farms were again defeated by colonial forces in an all-out war. The following year their territory was once again annexed by the British authorities and renamed the colony of British Kaffraria (Peires 1981; Fig. 3.8). British Kaffraria was later to be absorbed into the Cape Colony.

Throughout these struggles on the eastern Cape frontier, British imperial expenditure on the local military constituted a significant source of capital for the colony as a whole. It was overshadowed only by wool revenues. Wool production in the east had soared since the mid-1830s from 114,000 lb. to about 4 million lb. and the commodity was now by far the greatest provider of revenue for the colony (Le Cordeur 1981; Keegan 1996; Lester 1997). If wheat and wine had initially secured the Cape's place within a wider, European-dominated world economy, that role was now fulfilled by wool, and the fastest growth in wool production was occurring on lands seized from the Xhosa. As Ross points out, with the combination of imperial military expenditure and woollen production, 'the colony was partly kept in business by its eastward expansion' (Ross R. 1986: 75).

While colonial forces were still insufficiently powerful to expel the Xhosa completely from their lands in British Kaffraria, their chiefdoms were confined to 'locations' and much of their former land parcelled out to settler farmers during the late 1840s and 1850s. A patchwork of colonial and African agriculture developed. Xhosas under their chiefs in the 'locations' (later known as 'reserves') were 'ideally' to be taxed in order to ensure their dependence on cash wages. This in turn would stimulate men and women from the locations to find work on surrounding white farms, partially satisfying settler sheep farmers' demands for labour as well as land (Peires 1981; Crais 1992). Although, in practice, most Xhosas successfully resisted the payment of tax until much later in the century, the scheme envisaged for the administration of British Kaffraria was to prove an influential and enduring one within the region (du Toit 1954). It contained principles of racial land division and African labour flows which lay at the heart of successive state attempts to reconcile African chiefdoms to the practices of settler capitalism.

The colonial government's willingness to support settler capitalist expansion in the eastern Cape resulted, however, in unprecedented levels of conflict. During the late 1840s, aside from dispossessing the Xhosa, settler commercial farmers were attempting to undermine Coloured peasant communities, such as that along the Kat River, largely through local, settler-dominated government restrictions on the inhabitants' economic activities (Kirk 1980; Mostert 1992). The humanitarian ideal of a landed and flourishing commercial peasantry, manifested for a time in the Kat River settlement, was effectively undermined through such endeavours. Between 1850 and 1852, a combined Coloured and Xhosa uprising marked the last-ditch attempt by black communities under assault to reverse the advance of acquisitive settlers and the colonial state in the eastern Cape (Crais 1992; Legassick 1993; Keegan 1996).

Once the Khoikhoi rebels' and Xhosa chiefdoms' resistance had been crushed in an unprecedentedly prolonged and brutal conflict, and with their cattle herds newly afflicted with lungsickness, thousands of Xhosas turned in desperation during 1856–7 to millennial prophecies of a new era of prosperity and purity. The mass

Figure 3.8   British Kaffraria showing a) districts and b) English county names adopted within its districts

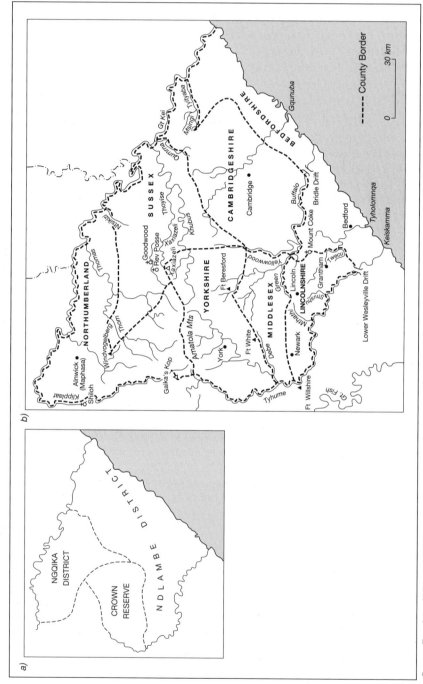

*Source:* Bergh and Visagie (1985)

Plate 3.10 Khoi soldiers serving in the British Cape Regiment launch an attack on the Xhosa during the 1850–52 Frontier War
*Source*: William Fehr Collection

slaughter of cattle and destruction of crops, intended as a sacrifice to the ancestral spirits, left some 30,000 starved and an equal number absorbed into the colony as cheap colonial labourers (Peires 1989; Bradford 1996). The Cattle Killing movement gave the new Cape Governor, Sir George Grey, the opportunity of consolidating colonial control over the frontier Xhosa's partially depopulated territory. He was assisted by the settlement of some 4,500 German military settlers (Davenport 1982; Peires 1989). Grey's programme for the 'civilization' of the Xhosa, informed by his experiences among the Maori of New Zealand, was to be the most coherent yet. It would involve 'employing them upon the public works which will open up their country [and] establishing institutions for the education of their children and the relief of their sick'. By such means, Grey hoped that the Xhosa would be won over from 'unconquered and apparently unconquerable foes, into friends who may have common interest with ourselves' (Schreuder 1976: 290). However, for the foreseeable future at least, these 'friends' would be overwhelmingly docile labourers rather than fellow citizens. With the Cape apparently secure for British settlement and British capital after 1857, over 30,000 new immigrants arrived in the colony from Britain (Beinart and Delius 1986).

While an ambitious form of domination was being engineered for the Xhosa in British Kaffraria, a more tenuous project of British colonization was unfolding in the former trekker republic of Natalia. A small British trading community had been agitating for the British annexation of the area around Port Natal (now Durban) since the mid-1820s. When, in 1842, Cape and London officials became concerned that trekker attempts to expel Africans to the south would further destabilize the Cape frontier, the territory had been taken under British control. Its administration as a separate colony was intended to be strictly self-financing (Thompson 1982b; Brookes and Webb 1987; Fig. 3.9).

Figure 3.9   Natal in the 1860s

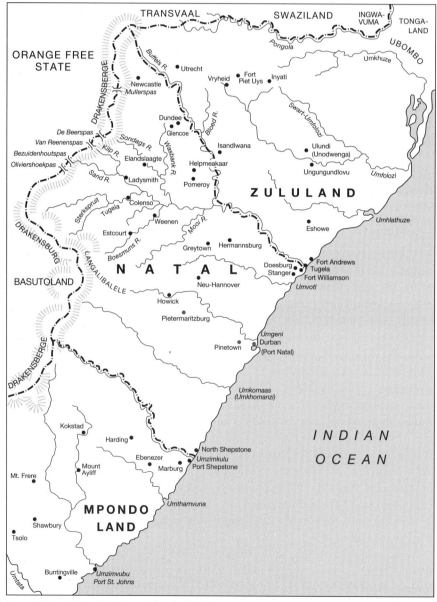

*Source*: Adapted from Berghe and Visagie (1985)

British authority having been extended to Natal, settlers were swift to follow. Most of the Afrikaner trekkers left the colony for the highveld as some 5,000 British immigrants arrived in one scheme alone between 1849 and 1852, purchasing the best land from the new colonial authorities. Many of the arrivals, from the Cape as well as from Britain, were speculators who took possession of huge farms but were content to rake in profits from African tenants while they awaited a price increase. Others developed smaller-scale commercial farming or engaged in trading and commerce from the towns of Durban and Pietermaritzburg. Some British entrepreneurs began sugar production in plantations along the coast. However, with the resort to tenancy still open to those local Africans who lacked access to communal land, commercial farmers complained constantly about the lack of labour supplies, while plantation owners were forced, from the 1860s, to rely on Tsonga migrants from the north and indentured workers 'imported' from British India (Slater 1980; Lambert 1995).

The indenture of Indians and their 'export' to labour-intensive enterprises enabled the partial replacement of emancipated slave labour in many parts of the British empire during the mid-nineteenth century. Between 1860 and 1866 the Natal government arranged for some 6,000 labourers to be shipped in from Madras and 300 from Calcutta for five-year periods of service, with the option of remaining in Natal or returning home thereafter. Indentured labourers were subsequently joined by those who emigrated without binding contracts, known as 'passenger' Indians, many of them Muslims and mainly from Gujarat. These families were generally escaping natural disasters and imperial taxation in India and seeking new craft and trading opportunities abroad. By 1900, much to the dismay of white settlers, the bulk of the African trade in Natal was in their hands (Thompson 1982b; Richardson 1986; Brain 1989).

In Natal as in British Kaffraria, a form of spatial segregation between settler and African landholding was developed. Thousands of Africans continued to practise an established form of homestead production on absentee-owned estates, on land now owned by the government, or on mission land. Others remained more firmly under chiefly authority in communal 'reserve' lands similar to the 'locations' in British Kaffraria and the Transvaal. Since the Natal government was dependent largely on the taxation of productive African farmers in order to raise its finance, it resisted settler farmers' agitation for the further extension of commercial agriculture into their 'reserve' lands, and tried to moderate demands that they be forced to render greater labour service (Welsh 1971; Slater 1980; Lambert 1995).

Demands for more Africans to be prised off the land and into service were additionally resisted by the Natal government, because it judged that it was in all the colonists' interests to maintain accommodations with the surrounding African chiefs. Even by 1871, there were only eleven magistrates from a white population of 17,000 available to 'oversee' an African population of over 300,000 (Welsh 1971; Marks 1985a). As the settler newspaper the *Natal Witness* explained, since they were vastly outnumbered by Africans, '[t]he safety and welfare of the colonists are intimately connected with the proper management and control of the native population' (Welsh 1971: 24). To undermine chiefly authority by attempting the complete dispossession of Africans and their reduction to 'civilized' labour (as was being attempted to a certain extent in British Kaffraria), would be to risk a rebellion which the government could not contain. Since it lacked the finances and the might to impose its own structures of judicial and military administration over the African population of the region, the Natal government compromised by co-opting chiefly authority. Given the turmoil which had recently been experienced with the 'Mfecane' and the arrival of the trekkers, and the additional problems of establishing control over 'Mfecane' refugees returning from the Zulu kingdom to the north and elsewhere, reciprocal accommodation with the colonial authorities was a relatively secure option for many African chiefs and kraal heads themselves. Indeed, colonial recognition could assist in maintaining their authority (Lambert 1995).

The official associated most directly with the system of administration in Natal was Theophilus Shepstone. The son of a missionary and a fluent Zulu-speaker, he was appointed 'Supreme Chief' and expected to co-ordinate the activities of the lesser African chiefs. As would be the case under subsequent systems of indirect colonial rule in South Africa and in other parts of the continent, an individual like Shepstone assumed great responsibility for the local success or failure of colonial power. Through the mid- to late nineteenth century, he appointed chiefs (many of them from commoner backgrounds) where there was no evident incumbent, punished those whom he considered had pushed their autonomy too far, and oversaw the implementation of a flexible body of 'native law' by which Africans in the colony would be governed. Until the late nineteenth century, the malleable system of segregation which he devised, resting as it did upon indigenous 'concepts of sovereignty' and a relatively high degree of local African autonomy, worked to sustain settler-capitalist enterprise whilst avoiding massive resistance of the kind which had been encountered in the eastern Cape (Hamilton 1998: 128; Welsh 1971).

Plate 3.11  Theophilus Shepstone, diplomatic
agent to the 'native tribes' in Natal
*Source*: Africana Museum Johannesburg

While British settler capitalism was imposing or insinuating itself into the eastern Cape and Natal during the late 1840s, it was also thrust into trans-Orangia. Spurred on by speculators in land, in 1848, the Cape's new governor Harry Smith (the former Commander of Queen Adelaide Province and British Kaffraria) annexed the nascent trekker republic for Britain, with reluctant metropolitan approval. It was renamed the Orange River Sovereignty. Smith seized lands which had been leased by trekkers from the Griqua and the Sotho, and those still settled by Tswana- and Sotho-speaking polities. They were then allocated to co-operative trekkers and to new British arrivals (Marais 1957; Macmillan 1963). Major Henry Warden was instructed to arbitrate between the new Sovereignty and Moshoeshoe's Sotho over possession of the fertile Caledon valley lands. The colonial boundary which he drew incorporated over one hundred Basotho villages and thousands of inhabitants, allocating their land to a dozen or so white farmers (Davenport 1991; Fig. 3.10). Smith also decided to support tributary chiefs on the fringes of Basotho territory in an endeavour to undermine Moshoeshoe's authority (Keegan 1996). Even with official appeals to the racial sentiment of the colony's Afrikaner trekkers, however, too few were willing to fight the Sotho in order to enforce the new British claims. Ultimately, both trekker and Sotho resistance convinced the British government that Smith had misrepresented the costs of running the Sovereignty and, after Moshoeshoe had defeated one colonial force and parlayed effectively with another, a withdrawal was ordered in 1854. During the brief extension of British colonial control in trans-Orangia, however, certain colonial and capitalist foundations had been laid. A political elite comprising both British settlers and Afrikaner trekkers had been established and allowed to accumulate land and wealth. Regardless of the official withdrawal, this elite would ensure the region's continuing involvement with British commercial enterprise even when it reverted to being the independent OFS republic (Atmore and Marks 1974; Keegan 1988, 1996).

Despite the great differences between colonial practices in the Cape, Natal and the short-lived Orange River Sovereignty, there was no distinction between these British administrations in principle. Each colonial government, and the majority within each community of settlers, saw the more effective dispossession of Africans, and the encouragement of settler enterprise on their land as an ideal policy. Each, however, was deterred from pursuing that policy to its conclusion by the potentially disruptive and costly consequences of African resistance. Local Africans' partial spatial confinement to 'reserves' or 'locations', accompanied by attempts varying in intensity to render them 'civilized', servile and dependable labourers, comprised a second choice, a compromised policy. The greater concessions made to African authorities in Natal compared with the Cape stemmed from different financial and security considerations, rather than a different philosophy of colonial government. In each of southern Africa's colonial territories, systems of segregation were constructed, under British authority in particular, as ways of inserting pockets of colonial-capitalist endeavour between still relatively powerful African chiefdoms. Segregation thus reflected 'the limits [that] capital encountered in trying to tame Africa's labor power' (Cooper 1994: 1524).

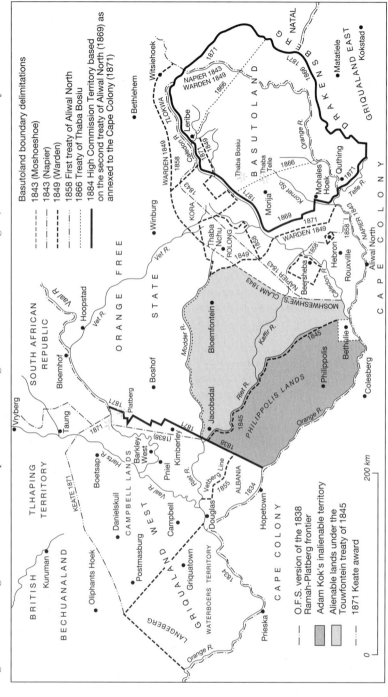

Figure 3.10   Shifting boundaries in the Griqualand West–Orange Free State–Basutoland region, mid-nineteenth century

*Source:* Davenport (1991)

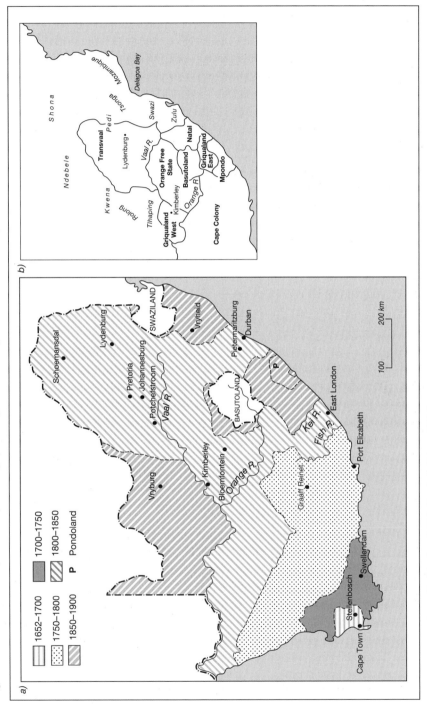

Figure 3.11   South Africa by the 1870s, showing a) the sequence of European annexation and b) states

*Sources:* Adapted from Christopher (1994), Reader's Digest (1994)

Nevertheless, British settler capitalism, founded above all at this stage on wool production, relied on a different formulation of power, and different types of social boundary, from any which had been experienced previously in southern Africa. Colonial subordinates in the eighteenth-century western Cape were slave and Coloured labourers who had little resort to pre-colonial social and cultural resources. Many (though not all) of the Africans incorporated as sharecroppers on farms in the interior had also experienced the extensive disruption of their pre-colonial polities during the 'Mfecane'. While they were backed up by new state structures, it was possible for many whites in the Transvaal to control relatively 'atomized' African subordinates within the household, through an overarching Dutch–Afrikaans culture incorporating religious and informal social boundaries. The nineteenth-century capitalist British settlers, however, were aware that their subordinates in the eastern Cape and Natal generally belonged to more intact and resilient African societies (Marks and Atmore 1980). These societies' residual power led to Africans being psychologically conceived and treated by most British settlers as a threatening, indeed a contaminating presence, to be kept as far as possible outside of the settlers' domestic and communal space, and admitted into it solely in their labouring function. The British settlers' sense of social distance from African subordinates was also probably connected with their aspiration to replicate the British metropolitan bourgeoisie's conscious distancing from Britain's own working classes (Scully 1994; McClintock 1995; Sibley 1995; Dagut 1997, forthcoming).

For these reasons, it was largely British settlers who introduced what Cell describes as a 'horizontal' arrangement of power, partially replacing the 'vertical' arrangement developed by the Dutch colonists and practised by most Transvaal farmers (Cell 1982: 50). By this, Cell means that settler capitalists relied for their privilege on a territorial separation of groups defined by their race, class and culture, rather than the rituals of deference between status groups which had characterized the more intimate system of slavery, and which continued to play a significant role in highveld sharecropping relations. In adopting a segregationist compromise between capitalist acquisition and fear of its destabilizing consequences, British settlers laid the foundations for modern systems of citizenship and state planning in South Africa.

### Self-government and the decline of humanitarianism

Settler-capitalist schemes in southern Africa were given a considerable boost during the 1850s by two related developments. First, colonial governments were freer to act in the interests of local settlers due to the waning of humanitarian influence in the colonies and in Britain. Secondly, and partly as a result of that waning influence, settlers themselves were given a say in the running of the colonial governments.

Both humanitarians and settlers drew upon the concept of 'civilization' to explain their respective programmes for Africans. However, the immediate implications which they attached to the word were very different. Following Ordinance 50 and the abandonment of Queen Adelaide Province, a political contest arose between British interests, a contest connecting settlers, officials and humanitarians in the Cape with powerful groups in Britain (Lester 2000). A similar contest arose in Natal with some missionaries trying to defend Africans from the settlers' demands for labour (Welsh 1971). The eastern Cape settlers' newspaper, *The Graham's Town Journal*, referred to a struggle between the settlers' sense of 'stern justice' and the humanitarians' 'mistaken philanthropy' (Macmillan 1927: 282). By the 1850s, the struggle was turning decisively in the settlers' favour. In order to understand why, we need to broaden our focus, situating the politics of the Cape alongside those of other settler colonies and the British metropolis itself.

In the decades following the abolition of slavery, settlers in the Cape and elsewhere tried to persuade the British Colonial Office based in London that freed slaves and indigenous peoples had to be rigidly controlled for the good of society as a whole. In the West Indies, plantation-owners were particularly successful in generating persuasive propaganda. They blamed the post-emancipation decline of the local sugar industry, not on the attempts by former slaves to escape planter control and construct autonomous family lives for themselves, but on an inherent laziness which led blacks to avoid any kind of labour unless they were coerced. Many members of the British establishment themselves had plantation interests, and the planters of the West Indies were well connected in London. Through such connections, and through the powerful and popular books and treatises of metropolitan allies such as Thomas Carlyle, plantation owners were able during the late 1840s to contribute to increasing humanitarian disillusionment. Even Henry Taylor, one of the key protagonists of the abolition of slavery, stated in 1846 that 'negroes, like children, require a discipline which shall enforce upon them steadiness in their own conduct and consideration for the interests of others' (Holt 1992: 285).

In the West Indies and in the other formerly slave-holding colonies, including the Cape, settler propaganda was assisted by the importance that humanitarians themselves had placed on 'proper' and 'virtuous' conduct of slaves once they were free. Given that many slaves did not become the sober, diligent and docile Christian peasants and labourers that the humanitarians had wished for – that instead, the landless chose as far as possible to escape labouring for whites by moving to the mission stations and urban centres, or by becoming wandering 'vagabonds' – the humanitarians were made to seem naive. Settlers were also able to point out that the humanitarians themselves had always believed in the necessity for European intervention to 'improve' both slaves and Africans, by acculturating them and incorporating them as a dependent class. If they would not improve themselves voluntarily under humanitarian guidance, settlers reasoned, they would have to be improved more forcibly under settler tuition (Bank 1995; Lester 1998c). When the Xhosa went to war again in defence of their lands in 1846, the Colonial Secretary in London was persuaded by the settler argument. He stated that the Xhosa had 'abused' their independence and thus now deserved to have it taken from them (Keegan 1996: 217).

During the 1850s, with an anti-humanitarian stance already becoming widespread, further 'evidence' was supplied to metropolitan audiences of the 'irreclaimability' of black peoples within the colonies as a whole. As the British settlers in the Cape were advancing their brand of capitalism on the colonial frontier, and facing the

Plate 3.12  A mission station at Thaba Nchu in 1834
*Source*: Africana Museum Johannesburg

consequences of resistance, so were settlers in New Zealand and Australia. Each set of settlers was diligent in ensuring that their compatriots 'back home' knew all about the 'savagery' of their respective foes, whether they were Xhosa, Maori or Aborigine. When the infamous 'Mutiny' broke out in northern India in 1857, even the relatively well-respected Indian civilization seemed to many in Britain to be inherently prone to unpredictable acts of barbarism. In 1865, the humanitarians' failure to have Governor Eyre of Jamaica prosecuted for his brutal repression of a black uprising confirmed that racism was becoming an embedded part of popular discourse in Britain (Belich 1986; Holt 1992; Bank 1995; C. Hall 1996a; Lester 1998c).

In the Cape, perhaps the most striking indication that the tide had turned against the humanitarians was John Fairbairn's change of tune. Fairbairn had been the staunchest of humanitarian propagandists through his editorship of the *South African Commercial Advertiser*. He had also been the London Missionary Society Director, John Philip's most valuable ally in the struggle against Khoikhoi oppression and the colonization of the Xhosa in the 1820s and 1830s. By 1847 though, Fairbairn had announced his belief that the Xhosa had to be 'put down or expelled' as a result of their renewed resistance. He continued, 'not victory but conquest is to be the end of this [1846–7] outbreak' of Xhosa resistance (Keegan 1996: 216). Fairbairn was joined in this opinion by an increasing number among the Cape's merchant trading elite. While they had previously generated profits through trade with Xhosa peasant producers, many merchants now saw the potential for far greater accumulation in colonial 'military advance and dispossession to the benefit of [wool-producing] settlers' (Keegan 1996: 129). Metropolitan humanitarians continued to be concerned about events in the colonies through the late nineteenth and early twentieth centuries, maintaining philanthropic organizations such as the Aborigines Protection Society, but they never again had the kind of political clout which domestic bourgeois reformism had earlier given them. During the late nineteenth century, the necessity for colonized peoples' Christian conversion and progressive 'civilization' was still upheld, but in a more 'muscular' manner, more in tune with colonial settlers' expansionist schemes.

While a shifting colonial discourse allowed greater metropolitan sympathy for settlers, it was specific policies of free trade which allowed them to be 'given their head' with representative government. Free trade, as we have seen, was a central demand of bourgeois merchants who stood to profit from it. By the late 1840s, it had been extended to allow cheaper foreign corn imports into Britain, and the removal of protection against foreign shipping. Colonial reformers argued that along with free trade went self-governing communities of colonial settlers, established first in Canada. If they were free to pursue their own material gain within their own political structures, it was reasoned, colonists would both increase their trading transactions with metropolitan firms and find their loyalty to the British empire as a whole repaid. All parties, metropolitan and colonial, would thereby benefit (Cain and Hopkins 1993). In addition, the British government came to hope that the settlers in southern Africa, having been granted self-government, would contribute more towards the costs of their own defence as they pushed the frontiers of agrarian capitalism forward into African lands (Trapido 1980a).

In the Cape, a political crisis caused by metropolitan plans to send convicts to the colony, in order to serve out their sentences, prepared the way for more vehement local demands for self-government from 1849. Both Afrikaans- and English-speaking commercial interests forged a wider alliance of classes in Cape Town to protest at the use of their settlement as a penal colony. Despite securing the allegiance of eastern Cape British settlers whose capitalist expansionism he supported, Governor Harry Smith found that he could not successfully govern the colony as long as the Cape Town-based interests opposed him on the convict issue. While the convict transport ships were eventually sent to Van Diemen's Land instead, resolving the immediate issue, the political crisis had given many Cape colonists the confidence, and the determination to secure greater powers for themselves from the metropolitan government (Keegan 1996).

The form that representative government in the Cape took was a compromise between local interests. While the British settlers generally argued for a franchise qualification that would include only wealthier capitalists such as themselves, effectively excluding many Afrikaners as well as Coloureds and Africans, the constitution actually adopted set the relatively low voting qualification of £25 worth of property, regardless of race. This was the result of an alliance between those liberals who still held to non-racial principles and western Cape merchants on the one hand, and Afrikaner politicians on the other. While the former wanted the vote for 'respectable' Coloured artisans, partly in order to encourage them in their economically profitable activities, the latter were determined that poorer Afrikaans-speaking whites should also be included within the body politic. This alliance ensured the political incorporation of a broad upper social stratum, including the wealthiest Coloureds as well as the poorest Afrikaners. In doing so, it allowed scope for the future enfranchisement of a landowning and commercially orientated African peasantry in the eastern Cape (Trapido 1964, 1990; Keegan 1996). Within two years of the Cape precedent, the colony of Natal also received representative government, but due to the predominance of British settlers among the colonial population there, a much higher franchise qualification was adopted, excluding all but a handful of blacks (Trapido 1963; Marks 1987).

In practice the non-racial constitutions of the two British colonies acted as a kind of 'safety valve' to release tensions among the colonized. Some white politicians believed that the 1850–2 rebellion in the eastern Cape had demonstrated the need for some form of assimilation to be offered, even if only to an elite among the emergent Coloured and African peasantry. The non-racial constitution could act as a counter to those destabilizing effects of settler expansionism which had caused the rebellion. As the Cape's Attorney General put it, 'I would rather meet the Hottentot at the hustings voting for his representative than meet the Hottentot in the wilds with his gun upon his shoulder' (Trapido 1980a: 262). A similar function was served by Natal's constitution, devised under a legislative council in 1856. Here, a very few Africans were allowed to apply individually, by virtue of their property, housing, education, monogamy and Christianity, for exemption from the 'native law' which governed the African population as a whole. If exempted they might then qualify for the franchise, but this was by no means guaranteed (Welsh 1971; Lambert 1995). Natal's Lieutenant Governor was quite explicit about the purely legitimating function of the new arrangements. In 1864, he explained that:

> If this scheme of exemption was likely to throw a large number of Native voters on the colony, so as really to influence any election, I should be amongst the first to oppose it, but this ... cannot be so for generations to come, and certainly not before the Colony will have become densely peopled by white inhabitants ... [therefore] I see no necessity for any illiberal ... barrier ... to a Native becoming a voter (Welsh 1971: 62).

The vast majority of Coloureds and Africans in both the Cape and Natal, of course, like those in the Afrikaner republics, would remain disqualified from the franchise. In the British colonies though, they would be excluded by their lack of property rather than their race *per se*, and they would continue to be controlled through class legislation like the Masters and Servants Law.

## The mineral revolution

By the time that mineral exploitation in the interior initiated South Africa's industrial revolution, the Cape Colony's and Natal's ostensibly open frameworks of citizenship had proved their utility in regulating unequal class and race relations for some fifteen years. They had done so without the need for explicitly discriminatory legislation like that which existed in the Transvaal and OFS. But a mineral-driven economic revolution, consequent upon the discoveries first of diamonds and then of gold, brought unprecedented challenges to established systems of governance.

The expansion of colonial markets brought about by the mineral revolution meant new opportunities, not only for those of white complexion, but for anyone who had the resources to engage in commercial production and exchange. During the late nineteenth century, the economic participation and educational advancement of a Coloured, African and Indian elite threatened to erode the informal class boundaries protecting white colonial privilege, rendering legal non-racialism insufficient for the colonial elite's protection. At the same time, within an economy increasingly dominated by large-scale mining, it became imperative that unprecedented numbers of African labourers be brought within the colonial ambit. First the separate Afrikaner republics and the British colonies, and then a new overarching South African state, grappled with the problem of how to exclude elite blacks from citizenship while exploiting the labour of the black population as a whole. As was the case with the earlier, post-emancipation reformulation of social boundaries, necessary change was managed in such a way that the composition of the social elite altered marginally, but that material privilege remained concentrated within its hands.

### Diamonds, confederation and commercialization

European prospectors had been exploiting African knowledge of mineral resources in the interior for decades before 1868, when diamonds were discovered just across the Cape's northern border in Griqualand West, an area in which Griqua and Tswana polities had retained some independence from OFS control. As a result of recession, capitalists who had made their money primarily out of wool in the Cape were already searching for new outlets for investment when the first claims were staked, so there was a ready supply of credit, channelled to prospectors through the Cape's recently expanded banking system (Mabin 1984). A dispute over control of the diamond diggings between Tswana Tlhaping chiefdoms, a Griqua polity and the Afrikaner republics was resolved swiftly by the Cape government's annexation in the name of the Griqua. Despite the Griqua claim, however, the new administration sold land in the territory to white speculators and corralled the Tlhaping into locations like those in British Kaffraria. The crushing of a combined Tlhaping and Griqua uprising in 1878 demonstrated the British determination to hold on to the first major source of economic power that had been discovered in the region since the expansion of the woollen industry (Shillington 1982, 1985).

In the event, the expectations of future wealth that the diamond discoveries occasioned would prove well founded. By 1900, total exports of wool, still the Cape's most significant agricultural product, amounted to £14 million, but diamonds accounted for £89 million, making the difference in establishing a favourable trade balance (Kubicek 1991). As the result of mineral development, an expanding network of trade, finance, communication and transport quickly tilted the region's economic core away from the coastal areas of the Cape and Natal and into the interior (Mabin 1984; Lester 1996). This shift in economic geography was manifested most obviously in the extension of a railway network centred on the mineral finds. From the late 1840s, Britain's own industrial revolution had been driven more by heavy industrial growth, in coal, railways, iron and steel, than by the initial generator of economic growth, the textile industry (Hobsbawm 1968). Large-scale capital within the imperial metropolis was now ready to invest in similar heavy industrial enterprises overseas (Atmore and Marks 1974). The Cape government was thus able to secure £20 million from private investors in London between 1874 and 1885 to connect the harbours of Cape Town and Port Elizabeth to Kimberley by rail (Fig. 3.12). European immigrants, amounting to 23,000, swelled the white working class of the mineral fields over the same period (Marks 1985a).

Combined with cycles of boom and bust as the diamond market fluctuated and the costs of mining rose, investment from European shareholders resulted in a progressively greater concentration of mine ownership during the 1880s. Individual prospectors were bought out by joint stock companies, the whites among them often being taken on as overseers of African labour. With finance raised in Britain and France, Cecil Rhodes' De Beers company was able to buy out rival concerns on the diamond diggings, creating a monopoly by the 1890s (Kubicek 1991). In the twentieth century, De Beers would go on to monopolize not just local diamond production, but world distribution of the gems. Anglo-American, founded by Ernest Oppenheimer using American capital in 1917, assumed control of the company in 1929 and became South Africa's largest multinational concern (Innes 1984). Along with a few other giant corporations, the much-diversified Anglo-American ensures that capital

Figure 3.12   South Africa's rail network by *c*.1900

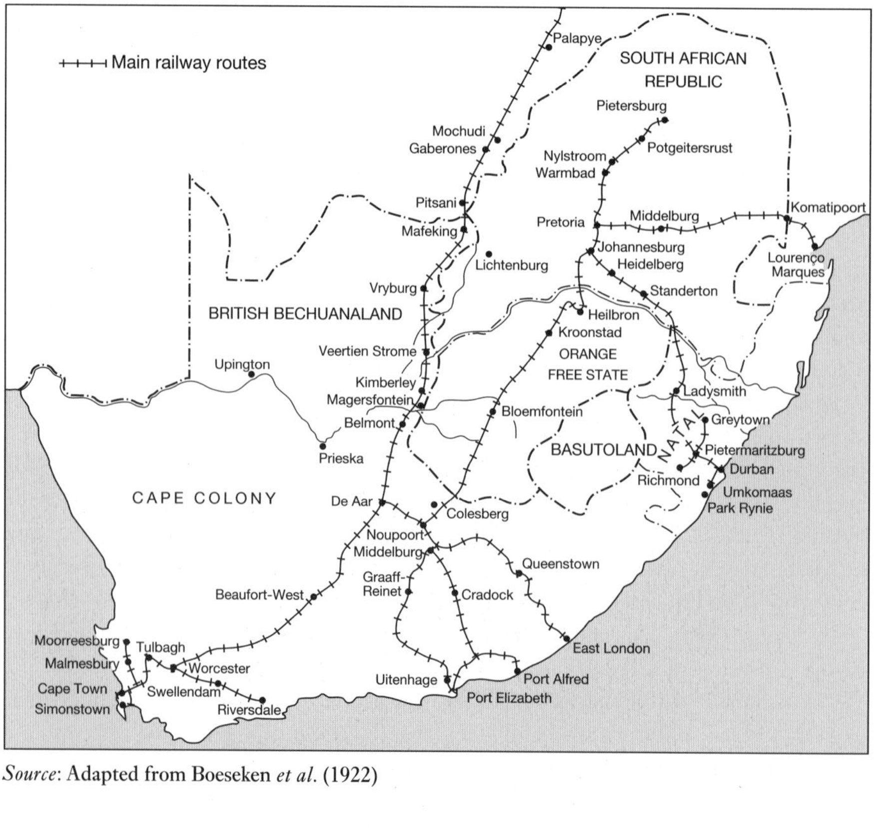

*Source*: Adapted from Boeseken *et al.* (1922)

is more highly concentrated in South Africa today than in most other developing and industrialized countries (Fine and Rustomjee 1996).

Soon after the 1868 discoveries, British officials became determined to do more than just retain control over the diamond fields. The unprecedented prospect of mineral-based industrialization inspired much grander schemes for the subcontinent as a whole. Plans for a confederation of white-administered territories in the region had earlier been devised by the Cape's Governor Grey, but the expense which they would entail had always precluded metropolitan approval. In 1868, however, the British government under Disraeli showed new signs of willingness to intervene in the region in the short term, as long as such intervention generated long-term economies (Marks 1985a). During that year, Britain finally annexed Moshoeshoe's territory, so as to prevent further costly warfare generated by settler land-grabbing and security concerns on the Cape's borders. By 1874, Natal settler capitalists and the governor of the diamond fields were convinced that British rule should be extended over Africans further to the north and up the east coast. Shepstone was confident that the inexpensive techniques which he used to govern Natal's Africans would be similarly effective across this wider region. The expansion of British authority would enable the better administration of labour flows from African chiefdoms, not only to the diamond mines and other anticipated mineral enterprises, but also to Natal's sugar plantations (Etherington 1979).

Shepstone and other local propagandists were able to convince the British Colonial Secretary not only that British rule should be extended over currently independent African polities, but also that a confederation of the white-governed territories – the Cape, Natal, Transvaal and OFS – was necessary. This was primarily so that the long-distance migration of the region's industrial labour force, through networks which spanned each of these territories, could be planned more effectively. But

Plate 3.13  The Kimberley diggings in the early days, c.1880
*Source*: Cape Archives, Cape Town

in addition, confederation would provide the administrative coherence needed to attract investment and install a proper infrastructure for mineral extraction and export. Finally, now that African migrant labourers were acquiring guns through labour on farms and mines, and transporting them back to their home polities in large numbers, confederation would enable the co-ordinated subjugation of those polities and the easing of white fears about a general African uprising (Etherington 1979; Marks 1985a).

What was required for the region, British officials believed, was a single, overarching modern state, whose professional bureaucracy would transcend the parochial concerns of the established colonial and republican governments (Atmore and Marks 1974). In retrospect, as Etherington (1979: 253–4) points out, many of the features of modern South Africa can be found in these late-nineteenth-century imperialist plans: an awareness of the 'essentially unitary nature of the developing southern African economy'; the organization of an extensive labour recruitment and migration network; the economic significance of minerals, and the administration of African territory by a centralized state through the kind of indirect techniques advocated by Shepstone. In all, confederation would be the 'precondition for the expansion of capitalist relations in South Africa' far beyond even those instituted thus far by British settlers in the Cape and Natal (Innes, quoted in Marks and Rathbone 1982: 12).

The opportunity to embark on confederation arose in 1877, when a disastrous campaign against the Pedi polity in the north of the Transvaal bankrupted its republican government. Urged on by commercial interests in the Republic and in the Cape, the British government annexed the territory with relatively little opposition. A series of wars of conquest against many of southern Africa's independent African polities was then compressed into the next few years, their justification being found in Africans' resistance to settlers' and their allies' demands on lands and labour. In rapid succession, imperial armies were launched against the Gcaleka Xhosa, so far relatively sheltered from the Cape to the east of the Kei River, the Zulu and the Pedi. The Sotho, who had successfully resisted disarmament under Cape rule, and those Tswana to the north of the Molopo River, whose chiefs negotiated their incorporation, were brought within British protectorates. Together with Swaziland, these protectorates would ultimately escape settler government and be ruled direct from London as High Commission territories (Bonner 1983; Shillington 1985; Eldredge 1993; Fig. 3.13). The Gcaleka Xhosa, Thembu, Mpondomise and those Tswana to the south of the Molopo, were all eventually incorporated in the Cape Colony and Pondoland was annexed in 1881, while the Zulu were initially left on their lands under rival co-opted chiefs, but ultimately absorbed within Natal. The Pedi were brought within the British-occupied Transvaal and granted a location within their formerly more extensive territory (Guy 1979; Peires 1981; Delius 1983; Marks 1985a). Although the imperial advance of the 1870s ended in the failure of confederation as Transvaal Afrikaners defeated the British occupiers at the Battle of Majuba in 1881, it nevertheless prepared the way for renewed imperial conquest, and a more coherent pattern of African spatial confinement once gold was discovered in the Transvaal itself.

Figure 3.13   The British High Commission territories

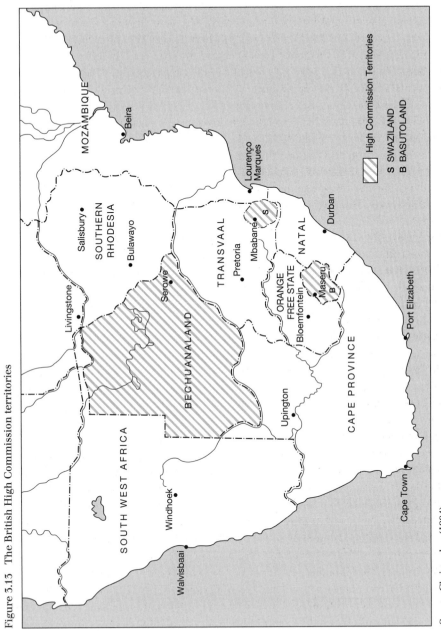

*Source:* Christopher (1994)

The extensive warfare and conquest of the 1870s and 1880s were accompanied by more mundane responses to the opening of the Kimberly diamond fields, on the part of both colonists and Africans. Across southern Africa, the influx of capital from Europe stimulated by diamonds created expanded urban markets for food, especially in Kimberley and the growing port towns which serviced it. 'The massive new opportunities for marketing produce ... altered the fabric of rural life', as speculators, traders, transport-riders and agriculturists seized the chance to accumulate individual wealth (Ross 1982: 90). African peasants who retained access to communal and, in some cases, freehold land, in the Transvaal, OFS, eastern Cape and Natal, were among the first to respond through commercial farming. Organizing the labour of their extended families, homestead patriarchs were able to produce a surplus which in former times may have been yielded as tribute or distributed in long-distance trade. Now it could be transported (often by other African and Coloured entrepreneurs engaging in wagon haulage) and sold for cash in growing urban markets, especially on the diamond fields (Bundy 1988). Not only could the money earned be used to pay colonial taxes and rents; it could also fund the acquisition of fertilizers, seed and equipment to enable further capital accumulation and the acquisition of European consumer commodities.

Despite their territorial struggle, the Sotho had been supplying OFS towns with grain since the 1850s. With the development of Kimberley, many enlarged the scale and the reach of their commercial production (Parsons and Palmer 1977; Marks 1985a; Eldredge 1993). They were joined in this initiative by the subjects of other chiefdoms across the region, and by African sharecroppers located on white farms on the highveld. In each case, as had happened previously among the trekkers, the emergence of a commercially orientated elite was accompanied by increased material stratification. Sharecropping communities, for instance 'began to evolve a distinct identity as an elite class, with a keen interest in education' (Keegan 1991: 43). Within chiefdoms on communal land, certain homestead heads were given a natural advantage in responding to market opportunities as chiefs allocated them the best land. Through commercial production, they were able to pay higher bridewealth, increasing the number of their wives and accumulating greater labour power. Those who owned sufficient livestock were also able to adopt the plough. Although this exacerbated soil degradation, it allowed for a more intensive production than was possible for those who still relied on hand implements (Beinart 1982; Shillington 1982; Crush 1987; Bundy 1988; Lambert 1995).

In some communities, coherent chiefly authority was maintained and applied to the supervision of commercial production, but in others individual accumulation undermined both the chief's command of resources and the established mechanisms of wealth redistribution. Many of the poorer commoners became unprecedentedly impoverished (Shillington 1982, 1985; Bundy 1988; Lambert 1995). Through the latter part of the nineteenth, and the early part of the twentieth centuries, commercialization steadily produced three broad classes, interlocking with 'traditional' strata and displaying differentiated cultural attributes, within most African localities. First came a comfortable and often Christianized petty bourgeoisie with access to land and

capital, and the ability to dominate production; secondly, an insecure middle peasantry, partly Christianized and trying to feed itself and produce a surplus to compete
with the better-off, and thirdly, a poor, 'traditionally' oriented peasantry including
smaller producers and landless proletarians (Bundy 1977; Beinart and Bundy 1987;
Murray 1992; Comaroff and Comaroff 1997).

In the eastern Cape, the Mfengu were the most notable of elite African commercial farmers. As we have seen, they had accepted colonial authority, missionary
patronage and a separate Mfengu identity during the 1834–5 frontier war, when the
colony required their military assistance against the Xhosa. Some had been granted
freehold title, and others their own communal locations as a reward for their co-
operation. With the encouragement of missionaries, a minority landholding Mfengu
elite had adopted Western clothes and houses. Seeking a good mission education for
their children, they were determined to advance socially and economically within
their new colonial environment. By the late nineteenth century, they had ploughs,
wagons, 'respectable' houses and all the accoutrements of a flourishing middle class,
and, alongside white farmers in the western Cape, they were major suppliers of produce to the growing towns of the colony (Bundy 1988). Their consumption of
colonial goods and their status as propertied voters were strong motivations for white
liberal merchants and politicians to maintain a non-racial tradition in the Cape,
resisting settler demands for their subjugation despite the general waning of metropolitan humanitarian influence (Trapido 1980a). In 1891, for instance, one
missionary estimated that each Christian African man required £20 worth of
imported manufactured goods per annum, compared to a typical £2 per annum con-
sumed by a 'traditionally' oriented African (Welsh 1971: 188).

In Natal, elite mission-educated Africans were known as Kholwa. Like the prosperous Mfengu, they saw themselves as 'the heirs to the future, "progressive and
civilised"' (Marks 1985a: 401). Most of them were clustered on mission-owned land,
but some, including a small minority of women inheritors, gained freehold possession (Etherington 1978; Meintjes 1990). Like many Khoikhoi in the late eighteenth
century, groups such as the Mfengu and Kholwa had chosen no longer to defend
their pre-colonial cultural and social integrity against the assaults of colonialism.
Rather, they had selectively adopted and adapted aspects of colonial culture, seeing in
such a strategy their best hopes for social acceptance and advancement. Their conversion to, and reformulation of, Christianity was one of the first signs of this
cultural shift (Elbourne 1995; De Kock 1996; Comaroff and Comaroff 1997).

By the late nineteenth century, Christianity was being adopted by a far wider
range of African groups, and in a variety of distinctively African guises, but the
Mfengu and Kholwa elites were among the first to seize upon its potential significance for their future status (Etherington 1978, 1989a; Bredekamp and Ross 1995;
Elphick 1995; Hodgson 1997). As the Kholwa Johannes Khumalo put it in 1863: 'we
have left the black race and have clung to the white. We cling to them in everything
we can. We feel we are in the midst of a civilised people and that when we became
converts to their faith we belonged to them' (Welsh 1971: 58). Often with the enthu-
siastic support of their husbands and fathers, women in Christianized African

communities, especially on mission stations, found themselves being trained for confinement to the homely, rather than the public sphere of life, just as their British counterparts had been conditioned from the late eighteenth century (Cock 1990; Hughes 1990; Meintjes 1990; Walker 1990a; Hall 1992). Nevertheless, for some, the European form of patriarchy attempted on the colonial missions was relatively liberating. Even while their advantages in terms of material progress were often more prominent in the thoughts of African men, missions could represent refuges from unwanted marriages and witchcraft accusations for African women (Gaitskell 1990).

Along with middle-class Coloureds and wealthy 'passenger' Indians, the Christianized, commercialized Africans of the Cape and Natal pressed most closely against the boundaries of white privilege during the late nineteenth century. Their mission-educated offspring frequently became petty-bourgeois church ministers, teachers, clerks and translators. They comprised a new black elite, determined to leap over the hurdles barring them from access to citizenship as soon as possible. If missionaries and the colonial state told them that they must become Christian, educated, respectful of authority and propertied before they could be incorporated in the ruling classes, then these would be precisely their goals. This was why they were so threatening to settlers who felt dependent on racial privilege. Their very success in meeting the criteria of 'civilization' was the reason why new barriers had to be raised to their incorporation, and new legitimations found for denying them a share in power (Bonner 1982; Marks and Rathbone 1982; Murray 1992). As we will see below, a discourse of biological determinism or scientific racism provided the 'knowledge' out of which those legitimations would be constructed. In the meantime, their rejection by whites and their simultaneous self-distancing from 'traditional' culture gave these representatives of the African elite a particularly ambivalent status (Marks 1986). The insecurities of their position prompted them to form their own societies, clubs, churches and ultimately political organizations during the late nineteenth and early twentieth centuries, in which a more coherent identity of their own could be fashioned (Willan 1982).

## Migrants and compounds

While the commercialization brought about by diamond mining, as well as colonial taxation, encouraged Africans to engage in commercial production, they also stimulated other kinds of response. Some Africans were able to engage directly in diamond mining as claimholders, until the depth of exploitation and the costs of mining increased, and ownership became more concentrated in white hands (Turrell 1982). But for a far greater number of African households, and especially poorer peasants, migrant labour either co-existed with limited commercial food production, or constituted the main response to commercialization (Cooper 1981; Beinart 1982).

The long-distance migration of African men to work in the colonies was by no means a new phenomenon. Given women's control of cultivation in non-commercial homesteads, and the progressive decline of hunting during the nineteenth century, young African men could be spared from the polity for extended periods in order to acquire income further afield (Bozzoli 1983/1995; Marks 1985b; Walker 1990b). Pedi and Sotho men were migrating in large groups to the Cape, Transvaal and OFS from the 1840s, sent by their chiefs to earn guns with which to defend the polity, while even farms in the south-western Cape employed large numbers of migrant workers from inland (Parsons and Palmer 1977; Delius 1980, 1983; Marks and Rathbone 1982; Keegan 1991). When the diamond fields opened, the established traditions of migrancy came to ensure that 50,000 African men would arrive seeking work there each year (Marks 1985a). As far as the mining companies were concerned, if Africans could not be extracted from the rural locations and fully proletarianized, the use of migrant labour supplies emanating from those locations was the best alternative (Cooper 1981; Marks 1985b). Nevertheless, the mine-owners were able to find considerable consolation in an expanded and refashioned migrant labour system.

Labourers who had a household base in the countryside could be made to accept lower wages than those which would have been paid to an urban proletariat. What they earned in the mines was seen merely as a supplement to their rural subsistence production, and it was within the rural economy that their families would have to find the resources to sustain themselves (Wolpe 1972). With this, and an insatiable demand for cheap labour in mind, recruiting agents were being sent by the mining companies to entice migrant workers from across the entire subcontinent by the 1880s. At the insistence of the early white claimholders, once they were admitted into Kimberley, they were required to carry a pass stating their entitlement to be there, whether or not they had completed their contracts, and whether they were allowed to leave the town. Although such pass laws would later be used to regulate the entire urban African workforce, they were initially implemented here in order to prevent poorly paid and overworked labourers deserting their employers to find work elsewhere in town (Worger 1987).

As South Africa's first industrial town, Kimberley set precedents not only for an industrial pass system, but for the housing and treatment of the migrant labourers subjected to that system. By 1885 the major mining houses had constructed 'compounds' to house their migrant workers. The innovation was later extended to the gold mines and factories of the Transvaal and the mines of Northern and Southern Rhodesia. The compound became the standard southern African means of controlling a large migrant workforce (Jeeves 1985; Marks 1985a). Compounds (later known as hostels) were enclosed, prison-like complexes, containing barracks for the workers. They were based on the techniques devised by the De Beers mining company to supervise convicts sentenced to hard labour (Turrell 1984; Mabin 1986). With the backing of Kimberley's merchants, who supplied white workers housed in the town, white workers were able to resist subjection to the compounds. But, extended to African migrant labour, compounds assisted the mining companies in the control of diamond theft, enabled the cheaper provisioning of the workforce through economies of scale, and helped prevent desertion during the labourers' period of contract. They enabled the monopolizing mining industry to cut working costs by 50 per cent between 1889 and 1898 (Turrell 1982, 1984, 1987; Worger 1987).

Plate 3.14  Migrant workers in De Beers compound, Kimberley, *c*.1900
*Source*: Cape Archives, Cape Town

The compounds lay at the heart of an industrial system in which, as the African Mineworkers' Union later put it, 'Workers … are not free. Their time and activities are strictly regulated to conform to a routine and organisation determined by the employer and based on considerations of economy and efficiency' (Crush 1994: 303). The compounds were designed to allow for vantage points from which 'inmates' could be observed, ensuring the most efficient means of surveillance. It was intended that the workers, accustomed to such surveillance, and afraid of fines and corporal punishments, would eventually condition themselves to behave productively and profitably even when they were not under direct supervision (Foucault 1977; Crush 1994). The regulation of food, credit, alcohol and recreation were additional and more direct techniques through which the mine managers hoped to obtain consent and conformity from their charges (van Onselen 1976). In most cases they never quite succeeded, since communities of compounded migrants drew upon their own varying cultural resources, and devised their own informal rules to force certain compromises upon the management (Moodie with Ndatshe 1994; Harries 1994). But if their resistance became too intense, the closed space of the compounds would always facilitate the crushing of unrest by mine security personnel or, as a last resort, the state police. If they did not always produce docile African workers, the compounds at least ensured that the struggle between migrant autonomy and employer demands was weighted in favour of the employers.

Compounds also ensured the distancing between African migrants and whites who lived permanently in the town. But if a more coherent separation of the workforce was to be achieved, further measures would be required. By the 1890s, although most Africans in Kimberley were still working on short-term contracts in the diamond mines, ranging from two months to a year, some 8,000 were no longer returning to their rural homesteads between periods of employment. Instead, they rented accommodation either in unplanned and racially mixed districts, or in confined urban locations (later known as townships) on the edge of Kimberley. They were joined in these areas by Coloureds from the Cape, many of whom were skilled tradesmen, and Indians, both passengers and those arriving to pursue trade once they had completed their indentures in Natal (Worger 1987). The mining companies ensured that their white employees in Kimberley were insulated from these groups by housing them in the better-off, but still relatively closely regulated suburb of Kenilworth (Turrell 1984; Mabin 1986; Pirie 1991; Fig. 3.14). This was not the first instance of racial segregation in the Cape's towns. Since the early nineteenth century, the British military had concentrated the Xhosa and Mfengu residents of frontier towns in sites where they could be closely supervised, and, as we have seen, freed slaves and Khoikhoi also tended to live apart, in the poorest urban districts (Fox, Nel and Reintges 1991; Crais

Figure 3.14   Kimberley *c.*1895

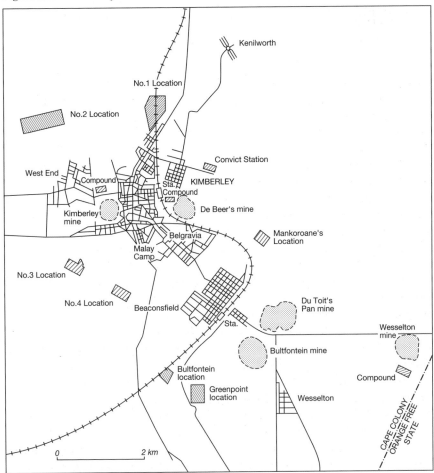

*Source*: Mabin (1986)

1992; Zituta 1997). But practices in Kimberley were significant because they came to mark a more modern form of urban segregation, one accompanying industrial rather than agrarian capitalism.

During the 1890s residential segregation was consolidated in Kimberley and other growing towns fundamentally because their industrialization followed a particular colonial trajectory. It was one driven by capital's attempts to acquire labour power from colonized, yet 'traditionally' oriented, or, as they were called by whites, 'raw' Africans (Mayer 1971). These African workers were unlike the proletariat which had fuelled the industrial revolution in Britain. They had not experienced the almost complete enclosure and dispossession of their land, and they had not had generations to appropriate or capitulate to the time discipline and individualism of a commercialized economy (Hobsbawm 1975; Thompson 1980). They had also not chosen, as the black elite had, to conform as far as possible to the norms of European 'civilization'. Rather, they had only recently experienced a particularly sudden transition in which their political and economic autonomy had been undermined, but their basic social and cultural structure, and their grip on much of the land, had remained intact (Cooper 1981, 1994; Marks and Rathbone 1982). They were not, and neither did they wish to be, completely acculturated to industrial capitalism. Urban segregation, first in Kimberley and later elsewhere, reflected this fact. It manifested a form of economic exploitation in which white employers and workers were riddled with anxiety about the resilient cultures of opposition retained by a very proximate and numerically expanding African workforce.

Such anxieties lay behind a late-nineteenth- and early-twentieth-century colonial discourse of black pollution and contamination. As we have already seen, its genealogy can be traced back to the early British settlers and their anxieties about the consequences of their agrarian capitalist expansion. But in the modern, industrializing era born in the 1880s and 1890s, the practical requirement of this discourse was not simply conquest followed by containment in rural 'locations', but also the more coherent, micro-scale segregation of African workers within confined urban spaces. As economic growth brought more Africans permanently to the towns, and particularly at first to Kimberley, their presence was increasingly associated with disease, and with the moral contaminations of prostitution, drunkenness and violence. White pressure mounted to remove them from mixed-race districts to more closely supervised urban locations on the edge of town. With the spread of bubonic plague in the squalid areas where they were housed, the fear of African 'rawness' was being translated by the early 1900s into what has been called the 'sanitary syndrome'. This in particular drove municipalities to implement urban segregation not just in the Cape, but in Natal and the Transvaal around the turn of the century (Swanson 1977; Parnell 1991).

## Gold and war

In 1884, three years after the British government had signalled the abandonment of confederation with its retreat from the Transvaal, vast quantities of gold were discovered on the Witwatersrand (the reef of gold-bearing rocks, often abbreviated to Rand)

(Fig. 3.15). Again the discoveries came at a time when recession in the Cape and at Kimberley created the conditions for an initial rush of surplus local capital to a new interior outlet even before international finance from Britain poured in (Mabin 1984). The city of Johannesburg, founded on the gold deposits, grew at an astonishing rate. Only three years after the opening up of the gold field, a visitor described 'mile after mile, acre after acre covered with lordly buildings or the humble shanty' (Marks 1985b: 432). Mine-owners and the commercial middle classes built themselves large suburban homes, while single male miners, many of them immigrants, found rooms in the central boarding houses. African migrants from the Cape, Transvaal and Mozambique were housed in compounds owned by the mining companies. Within ten years, the white population stood at 102,000 and within fifteen years, there were a million black workers along the proliferating mining towns of the Witwatersrand extending east and west of Johannesburg itself. The opening of the Johannesburg gold fields 'again [energized] the rural areas with a rising tidal wave of demand for everything from fodder to wood to food crops to dairy produce' (Keegan 1991: 44).

This gold mining 'cauldron of capitalist development' was located in the Transvaal republic, which, having regained its independence in 1881, was led by President Paul Kruger (van Onselen 1982a: xv). However, by the early 1890s, its financial might was centred on the major English-speaking mining magnates, such as Cecil Rhodes,

Figure 3.15   Gold fields in the South African Republic (Transvaal)

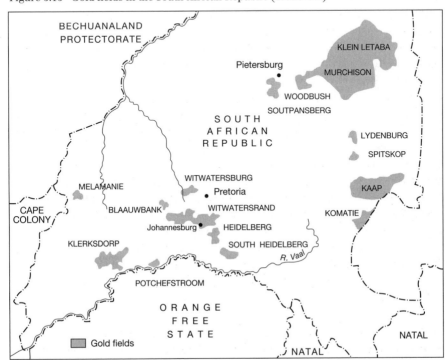

*Source*: Adapted from Reader's Digest (1994)

Plate 3.15  Johannesburg in the 1890s
*Source*: Barnett Collection

J. B. Robinson, Barney Barnato, Julius Wernher and Alfred Beit, known collectively as the Randlords. These were capitalists who were able to tap European financial markets by floating companies on the metropolitan stock exchange and so afford the costs of deep-level mining. These costs could be enormous: the first deep-level mine absorbed £350,000 before it produced its first gold in 1895 (Kubicek 1991). Channelling investments, which amounted to £75 million by 1899, into the mines and the infrastructure needed for their development, the Randlords were able to raise production from £80,000 worth of gold in 1887 to £16 million, one quarter of the world total, by 1898 (van Onselen 1982a; Marks 1985b; Worger 1997). The Randlords were virtually guaranteed continuing investment from Europe because, as the basis of international exchange, there could never be an over-supply of gold (Marks and Rathbone 1982).

Largely due to the gold discoveries, total British investment in southern Africa increased from £34 million in 1884 to £351 million in 1911 (Atmore and Marks 1974). The companies able to attract such vast sums amounted to eight by 1900, but two were dominant. One of these was Cecil Rhodes's Consolidated Gold Fields, created out of diamond profits from De Beers in Kimberley. The other, Wernher, Beit and Eckstein, produced about half of the Rand's gold and by 1900 it alone commanded a budget which exceeded those of both the Natal and OFS governments (Marks and Rathbone 1982). The directors of these companies had a tremendous stake in the future of the gold fields, and thus a particular interest in their governance. In 1889 they formed the Chamber of Mines as a negotiating body to represent their political interests (Marks 1985b).

For Rhodes and his associates, the primarily agrarian Transvaal state, which as recently as 1867 had been so rudimentary that its Treasurer kept no finance books,

Plate 3.16  Cecil Rhodes
*Source*: Readers Digest (1994)

was an anomaly within an industrializing world (Marks and Trapido 1979; Thompson 1982a). Kruger's government was determined to use taxation revenue from gold, which increased government finances by twenty-five times between 1883 and 1895, in order to modernize. But it wished to do so largely for the benefit of the local Afrikaner elite, most of whom were farmers: 'what Kruger … envisaged was the emergence of a state in which industry was the logical outgrowth of, and ultimately dependent on, the output of a domestic agriculture sector' (van Onselen 1982a: 37). For example, an insistence on horse-drawn rather than electrified trams in Johannesburg provided guaranteed markets for producers of fodder, but denied maximum profits to interested industrialists (van Onselen 1982a). Kruger's tolerance of farmers detaining and employing migrant workers heading for the mines was a further thorn in the side of the Randlords (Marks 1985b).

With a fixed gold price on the world market and high costs of extraction, the mine owners considered it essential that the Transvaal state should help not only to supply them more effectively with labour, but also keep the costs of that labour, as well as of other supplies, to a minimum. Skilled white workers, many of them recent arrivals from Europe, amounted to only 11 per cent of the workforce and yet accounted for 28 per cent of total costs. Their high price and strong bargaining position meant that costs would have to be cut in other ways (Marks and Rathbone 1982). A British administration seemed to Wernher, Beit and Eckstein as well as to Rhodes, to offer far more promise for a sustainable and profitable mining enterprise than Kruger's republic. With its greater imperial reach, it could ensure a sufficient and, above all, cheap supply of African workers from across the region. A British-controlled government would also be more likely to reduce the price of dynamite (kept high by a government-recognized monopoly) and cut transport rates in order to assist the mines. Finally, it would be likely to control deserting and drunken labourers more effectively (van Onselen 1982a; Marks 1985b).

At the same time as industrial capitalists were demanding a new form of state capable of meeting their needs in the Transvaal, the British government itself was concerned to maintain some control over shipments of gold from the region. It was the concentration of the gold market in London which enabled Britain's currency to dominate the world economy, but just when other national currencies were coming to be based on gold, Britain's own reserves were becoming limited (Marks and Trapido 1979; Ally 1994). Perhaps more importantly, the British government was also anxious that British finance should be guaranteed access to one of the world's most rapidly growing markets during a period of heightened European competition (Atmore and Marks 1974). Having dominated global industrial production, and the global economy as a whole for much of the nineteenth century, by the 1870s, Britain faced challenges from the industrialization of European and US competitors. By the 1890s, Germany and the USA were both producing more coal and steel than Britain. They were also utilizing the new technologies of rail and steamship to exploit resources and markets in peripheral areas of the global economy, potentially undermining Britain's imperial hegemony (Hobsbawm 1968; Mabin 1984). British politicians feared that their paramountcy in southern Africa would be threatened by a nascent alliance between the Transvaal and Germany. Certainly, Britain's main locus of influence in the region, the Cape Colony, was being overshadowed in terms of the value of exports by the Transvaal. Taking all these considerations into account, British officials were predisposed to agree with local capitalists that the Transvaal should be governed, if not by Britain directly, then at least by commercially minded interests sympathetic to British capital (Porter 1980; Cain and Hopkins 1993; Smith 1996).

With this coincidence of local Randlord and metropolitan government interest established, Rhodes gained tacit acceptance from the British Colonial Secretary, Joseph Chamberlain, for a scheme to overthrow Kruger's republic. A pretext was provided by the disenfranchisement of 'uitlanders' – non-Afrikaners who were largely English-speaking – within the republic. Their claim to citizenship was constructed as Britain's imperial cause in the region. In 1895, Rhodes' friend Leander Starr Jameson launched an attempted coup known as the 'Jameson Raid'. The design was to launch a force of some 500 men from across the Transvaal border, which would link up with an uitlander rising in Johannesburg to overthrow Kruger's government. The plan's dismal failure, marked by the swift capture of Jameson and his men, ended Rhodes's premiership in the Cape, but it by no means alleviated the British pressure on Kruger. Despite a series of concessions from Kruger's government, including the more effective regulation of African migrants through pass laws, the re-examination of the dynamite monopoly and the expressed intention to address uitlander grievances, that pressure finally came to a head in 1899. It was applied mainly by Sir Alfred Milner, who had been appointed as British High Commissioner in the southern African region in 1897. It was Milner who insisted that British force would be required to modernize the Transvaal state, in order that Britain could extend its 'civilizing' sway across modern South Africa (Smith 1996).

The South African War, the bloodiest and most far-reaching in South African history, began in October 1899 when Kruger, gaining the support of the OFS, launched

Plate 3.17  Alfred Milner
*Source*: Cape Archives, Cape Town

a pre-emptive strike against British troops in Natal (see summary box 4, pages 126–8). The war proved a major challenge to British imperial power. In Britain, it generated concern about the poor health and fitness of working-class recruits, raising the spectre of a degenerating imperial 'race'. Mass propaganda was necessary within the country to legitimate this war against colonial whites, in the face of widespread criticism from rival European imperial powers.

In South Africa itself, heavy losses on the British side prompted the revision of military tactics and produced further imperial anxieties. The war ended only after three years of conventional, and then effective Boer (Afrikaner republican) guerrilla warfare. In order to win it, British forces targeted civilians as well as combatants. They launched a 'scorched earth' policy against Afrikaner farms, rounding up their women and children occupants, together with their black servants, and confining them to separate 'concentration camps', so that they would be unable to provision the Boer commandos. Some 28,000 Afrikaner women and children and at least 18,000 Africans (and possibly as many as 28,000) died through disease and neglect in these racially segregated camps, leading some British observers, including the prominent activist Emily Hobhouse, to express public outrage. British forces also enclosed vast areas of the veld within fences punctuated by block houses, in an attempt to trap Boer commandos through grand army 'sweeps' across the country, and employed Afrikaners who had surrendered, many of them landless bywoners, in the struggle against their recent allies (Packenham 1979; Smith 1996). Both sides also armed thousands of Africans to assist in their struggle, something which was immediately regretted once the war was finished (Warwick 1983; Smith 1996).

Commandos from both republics, led by men who were already or were to become prominent political figures, such as Louis Botha, Jan Smuts, M. T. Steyn and J. B. M. Hertzog, as well as Afrikaner 'rebels' from the Cape, finally surrendered to British forces in 1902. In the wake of the war, many Transvaal and OFS Afrikaners were left deeply embittered. Not only had the British authorities been responsible for the deaths of Afrikaner women and children in concentration camps, amounting to a tenth of the republican population; in some districts, African combatants had reclaimed white farmland from which chiefdoms had previously been expelled. Afrikaner survivors returning to their farms were now presented with the problem of removing militant 'squatters' (Warwick 1983; Krikler 1993). In addition, in an attempt to anglicize the inhabitants of newly conquered territory during and immediately after the war, Milner adopted aggressive policies to suppress the Afrikaans language and replace Afrikaner farmers as far as possible with Britons (Murray 1992; Krikler 1993).

Nevertheless, the Afrikaner elite in each former republic was remarkably quickly reconciled to the realities of British domination after the war. The mutual British and Afrikaner determination to protect white political and material privilege from black threats, and an awareness of greater vulnerability to those threats after years of warfare and after the active encouragement given to blacks to fight against whites, were important considerations in this reconciliation (Krikler 1993). But perhaps the greatest concern which British imperial officials shared with the Afrikaner leadership was the imperative for capitalist reconstruction and development. In this respect, the war had brought some 'advances', most notably the development of port facilities on the coast. Within the Transvaal, although Kruger had envisaged a primarily agricultural state serviced by industry, and the British had helped impose the reverse, continued interaction between the industrial and agricultural sectors was in the interests of both English-speaking and Afrikaner capitalists. Together, in the wake of the war, they created what has been called 'the alliance between maize and gold' (Trapido 1971).

Plate 3.18  A Boer commando led by Jan Smuts, future Prime Minister of South Africa (seated middle), 1901
*Source*: Cape Archives, Cape Town

Plate 3.19  A Boer concentration camp
*Source*: Africana Museum Johannesburg

Behind this alliance lay the mutual recognition that the urban and industrial growth wrought by English-speaking, especially mining, capital and British imperialism, provided the greatest market for Afrikaner commercial farmers. As Keegan notes, 'the interior plateau became the vital supplier of life-blood to the industrial economy'(Keegan 1991: 42).

The British government changed as a result of elections in 1905, allowing the easing of Milner's more rigid Anglicization plans. In order to ensure Afrikaner co-operation, representatives from the former republics were allowed to decide their own franchise arrangements. In 1907, self-government was restored to the Transvaal under the former Boer generals Louis Botha and Jan Smuts, and to the Orange River Colony (former OFS) under its own Afrikaner government, later led by M. T. Steyn. As part of the terms of surrender negotiated by Boer forces at the end of the war, these self-governing territories avoided having to introduce the Cape's non-racial franchise and retained their whites-only citizenship. This was despite assurances given by the British government during the war that, in return for their assistance, some blacks would be able to vote in these territories on the same terms as they could in the Cape. By 1910, with the accommodations between British capital and Afrikaner commercial interests in mind, the leaders of the highveld territories had voluntarily agreed to join with the Cape and Natal in a larger South African Union.

The elected Union government was led after 1910 by Botha and then Smuts, at the head of the South Africa Party (SAP). In this first, and in subsequent South African elections, the Transvaal and OFS, by now South Africa's two northern provinces, still maintained their racially exclusive franchise. The formally non-racial constitutions of the Cape and Natal remained intact, but the non-racial franchise was never extended to the Union as a whole (Thompson 1960). Rather, the 1910 Act of Union, in creating the modern state of South Africa, provided the basis for the partial fulfilment of Milner's more exclusive and racially divisive aim: 'a self-governing white community, supported by ... black labour from Cape Town to Zambezi' (Legassick 1995: 46).

By 1910, then, in a process foreshadowing developments after the First and Second World Wars, the mobilization of resources for 'total' warfare had generated a central state within the new Union of South Africa with considerable powers (Lonsdale 1998). This state would intervene extensively to shape social and economic boundaries within its territory. On the one hand, its revenue base was founded on a system of largely English-speaking monopoly capital, concentrated on the diamond and gold fields, deploying some of the most sophisticated industrial techniques in the world, and tied firmly into an industrializing global capitalist economy. On the other hand, this modern industrial dynamo, located at the heart of the Transvaal, was dependent on a mass, unskilled and predominantly migrant labour force – one which had by and large resisted proletarianization and capitalist acculturation (Marks and Rathbone 1982). Only a powerful and interventionist state could marshal such divergent social formations to ensure the continued development of modern capitalism and regulate the restricted privileges which it brought (Yudelman 1984).

# SUMMARY BOXES

## Some key processes and patterns in South Africa's historical geography

In this section, some of the most famous, as well as historically and geographically significant, developments in nineteenth-century South Africa, dealt with in passing in the previous chapter, are examined more closely. The contested ways in which these developments have been interpreted by historians are also outlined. It will become clear that violent conflict between clearly defined ethnic groups has long been the overriding theme of popular interpretations of each of these episodes. Recently, however, historians working within new paradigms have been able to find considerable evidence of the everyday compromises and accommodations, often across ethnic boundaries, which always accompany such instances of violent collision.

---

**Summary box 1**: The eastern Cape frontier wars, 1779–1878*

The frontier between the Cape Colony and Xhosa chiefdoms, located in the region either side of the Fish River (see Fig. 3.5 on page 73) was the first arena in which white colonists partially dispossessed Bantu-speaking Africans. The relations involved are conventionally described as the Hundred Years' War, but they were far more complex than merely a series of armed conflicts between colonists on the one side and Africans on the other. Throughout the extended periods of peace on the frontier, there were everyday exchanges of commodities, labour and even political allegiance between those of different ethnicity, and even during the episodes of warfare, alliances were forged between groups of colonists and Xhosa as well as Khoisan. It was only as colonial imperatives shifted towards agrarian capitalist expansion in the mid-nineteenth century that the balance of power swung against the Xhosa's continued autonomy.

Dutch-speaking colonists occupied the Zuurveld, between the Bushmans and Fish Rivers, from the 1770s. During the first two frontier wars (1779–81 and 1793), these sparsely settled farmers acted in a loose alliance with the Xhosa chief, Ndlambe, who was seeking to establish his authority over the other Zuurveld Xhosa chiefs. In 1795, however, Ndlambe's nephew, Ngqika, rebelled

*(Box continued)*

in order to claim authority in his own right. Ndlambe was able to escape Ngqika's control only by moving west of the Fish River and into the Zuurveld himself. As increasing numbers of both colonists and Xhosa occupied the Zuurveld, differences between colonial and Xhosa modes of authority became critical for the first time. The 'Third Frontier War' (1799) was the result of a British governor deciding to convert a theoretical claim to the Zuurveld into an exclusive occupation. British action was directed particularly against the Gqunukhwebe chiefs, but in the face of a simultaneous uprising of Khoisan servants, the governor was forced to concede the Xhosa's continued presence (Peires 1981).

With a struggle being waged between Ndlambe and Ngqika, and between Ndlambe and the other Zuurveld Xhosa chiefs, it proved impossible for either Xhosa or British colonial authorities to prevent an escalation in mutual cattle raiding during the early years of the nineteenth century. Frustrated in their attempts to administer a zone of political instability and conflict, British officials represented the Xhosa as being imbued with a 'thirst for plunder and other savage passions' (Governor Cradock, quoted Lester 1997: 641). Arising from this cultural construction was the more fervent desire to separate colonial and Xhosa territory. In 1809, small chiefdoms, along with individual Xhosa farm workers, were expelled to the east of the Sundays River, but it was in 1812 that the 'Fourth Frontier War' was launched by British, colonial and Khoi troops. Those Xhosa chiefdoms long established in the Zuurveld, together with the Ndlambe, were forced across the Fish River, a line which had been agreed as the border between the Dutch authorities and the Gwali Xhosa chiefs (but no others) in 1778. The expulsion threw the Ndlambe on top of the rival Ngqika in congested lands just across the frontier. Far from desisting, members of both chiefdoms now intensified their raids against the colony.

Faced with the prospect of 'endless expense without compensatory returns', Governor Somerset attempted to cultivate Ngqika as a colonial ally (Galbraith 1963: 4). At the same time he pressured Ngqika into vacating further land between the Fish and Keiskamma Rivers (Fig. 3.5 on page 73). This 'neutral territory' was to be maintained purely as a buffer for the colony. But it soon became known as the 'ceded territory' and colonists were permitted to move into it. Certain Xhosa chiefdoms were also 'tolerated' within it on the condition of their 'good behaviour'. In 1818, Ndlambe defeated Ngqika's forces and, in order to retain Ngqika as a colonial client, the British authorities went to war again (the 'Fifth Frontier War', 1818–19). A British commando came to Ngqika's aid, but Ndlambe's warriors, spurred on by the prophet Nxele, almost overran the British garrison at Graham's Town. Only the arrival of Khoi troops saved the defenders and allowed colonial forces to effect Ndlambe's surrender.

*(Box continued)*

Between 1809 and 1820, Xhosa grievances against the British colony had steadily mounted. A balanced, if sometimes violent, interaction between isolated colonists and Xhosa chiefdoms had been possible in the late eighteenth century, but British officials' attempts to instil 'order' had brought cumulative land loss, frequent humiliation and greater competition for scarce resources to a number of Xhosa chiefdoms. The establishment of 4,000 British settlers in 1820 was to precipitate a further shift in the overall balance of power. The '1820 settlers' had taken passages to the Cape in order to improve their material situation, not knowing that the colonial authorities required them as a defence against Xhosa 'intrusion'. As we saw in chapter 3, by the 1830s, they were broadly united around a programme of colonial expansion. Alienated by Xhosa raids, and enthused by the profits to be made from wool production, many settlers envisaged the Xhosa's territory becoming sheep farms. Settler spokesmen directed propaganda against the 'irreclaimable' Xhosa in the press and criticized the 'futile' efforts of missionaries to 'civilize' them, while local officials occasionally expelled Xhosa chiefdoms from the 'ceded territory' on the pretext of continuing raids (Lester forthcoming).

Ngqika's son, Maqoma, was especially enraged by arbitrary acts of expulsion and re-admittance. In 1829, his followers were decisively expelled from the Kat River area and a humanitarian 'experiment' was carried out on their land. Khoi and Bastaards were assigned cultivable plots with the aim of generating a loyal and productive peasantry – one which would also act as a buffer against Xhosa raids. 'Hostile' frontier chiefdoms were provoked yet more by colonial commandos which were allowed to 'reclaim' cattle reported stolen from the colony. They often 'retrieved' this lost stock by raiding the first Xhosa kraal to which the tracks were traced. In 1834, it was Maqoma, in alliance with other chiefs, who launched the 'Sixth Frontier War'. Many Xhosa still refer to this war as the 'War of Hintsa', the Gcaleka paramount chief who was first taken captive when entering the British camp to negotiate an end to the war, and then killed and mutilated by colonial troops after trying to escape.

For Maqoma, the 1834–5 war was an attempt to reclaim lost land and reassert Xhosa independence. However, not all the frontier chiefs were behind him. The Gqunukhwebe now assisted the British forces, while Khoi troops, especially those recruited from the Kat River settlement, did most of the colony's fighting. In 1835, British forces also persuaded thousands of 'Mfecane' refugees (see summary box 2, pages 121–3), who had accepted clientship under the Gcaleka Xhosa, to turn against their patrons. Together with Xhosa who chose to desert their chiefs, these new allies became known as the Mfengu. Some were favoured with land grants within and just beyond the colony and others encouraged to take employment with colonists.

*(Box continued)*

During the 1834–5 war, spurred on by settlers, Cape Governor D'Urban declared the annexation of Xhosa territory up to the Kei River, renaming it Queen Adelaide Province. However, the military proved unable to effect the expulsion of the 'hostile' Xhosa from within the new province and in 1836 a humanitarian and economically minded Colonial Office instructed that the Xhosa chiefs' autonomy be restored. Nevertheless, British settlers continued to agitate for colonial expansion. Attempts by the Colonial Office to pursue a policy based on treaties with the chiefs were undermined by exaggerated settler complaints of Xhosa raids and, when a minor incident involving the 'rescue' of a Xhosa prisoner took place in 1846, the next assault on Xhosa independence was launched.

During the 'Seventh Frontier War', or 'War of the Axe' (1846–7), Xhosa territory up to the Kei was annexed again and named British Kaffraria. The Cape government was determined to extend its military power over the Xhosa while allocating much of their land to settler capitalists. Converted to a free trade economic policy and disillusioned with humanitarian policies, the Colonial Office now proved willing to endorse the scheme. By 1850, settler pressure on the land and labour of both the Kat River Khoi and Xhosa chiefdoms were sufficient to create the conditions once more for mass 'rebellion'. Between 1850 and 1852, Xhosa from British Kaffraria united with Khoi from the Kat River Settlement in a final effort to resist subordination within a settler-dominated, capitalist system (the 'Eighth Frontier War' or 'War of Mlanjeni', a Xhosa prophet). After the rebel and Xhosa defeat, British military control was tightened and settler land-grabbing further encouraged within British Kaffraria. When, from 1856 to 1858, the Cattle Killing movement spread through the Xhosa's remaining locations, Governor Grey was able to utilize the ensuing famine and the dramatic decline in British Kaffraria's Xhosa population in order to increase dependency on colonial wages and consolidate settler landholding.

Although the Rharhabe Xhosa were effectively colonized in the wake of the Cattle Killing, the conquest of the Gcaleka to the east of the Kei River was delayed, taking place only in the context of South Africa's mineral revolution (see chapter 3). In 1877, British authorities intervened in a struggle between Gcaleka and some of the colony's Mfengu clients, precipitating the 'Ninth Frontier War'. Despite an uprising in British Kaffraria in support of the Gcaleka, the colonial state was now able to bring the entire Xhosa people within the imperial framework on which South Africa's industrialization would be based.

*This summary box is based on the entry by Alan Lester, 'South Africa: Cape/Xhosa Frontier and the Hundred Years' War, 1779–1878', in K. Shillington (ed.) (forthcoming) *Encyclopedia of African History*, Fitzroy Dearborn, London.

**Summary box 2:** The 'Mfecane': historiographical debates*

The 'Mfecane'/'Difaqane' (*c*.1790–*c*.1830), has conventionally been thought of as a violent episode of restructuring within African societies. This episode is thought to have resulted in the rise of more powerful centralized states such as the Zulu under Shaka and Swazi under Sobhuza on the eastern coastal strip and the Sotho under Moshoeshoe and Ndebele under Mzilikazi on the highveld. In between the heartlands of these states were partially depopulated 'buffer' territories occupied by smaller client polities or refugees. The 'Mfecane' has, however, become the most contested of the nineteenth-century developments to be investigated by South African historians.

The orthodoxy, most comprehensively represented by John Omer-Cooper (1966) and subsequently much adjusted, held that the 'Mfecane' began with a violent revolution in warfare and government among Nguni-speaking Africans on the eastern coastal strip, provoked in particular by Shaka, a Zulu chief (see Fig. 3.1 on page 51). This revolution was stimulated by conflict over increasingly scarce environmental resources or trade or both, but refugee and splinter polities subsequently spread the turmoil on to the highveld, where it became known as the 'Difaqane'. Ultimately, the 'Mfecane' was marked by the colliding migration paths of both Nguni and Sotho-Tswana-speaking peoples across the entire subcontinent. In 1988, however, Julian Cobbing published a radically different interpretation. In general terms, he saw the primary causes of increased conflict among African polities as European-inspired slave-raiding. For Cobbing, the 'Mfecane' was a myth devised by settler apologists to cover the illegal practice of slave-raiding, and to legitimate the trekkers' (see summary box 3, pages 124–6) seizure of land, which had supposedly been depopulated due to conflict among Africans (Cobbing 1988).

The first and most significant of Cobbing's detailed propositions was that European and Griqua slave-raiders, from Delagoa Bay and from across the northern Cape frontier respectively, were the primary agents transforming African states, rather than Shaka's Zulu. Most historians are critical of Cobbing on this point, although some make important qualifications supporting certain of his arguments. Both Norman Etherington and John Wright agree with Cobbing that the idea of a Zulu-centred 'Mfecane' can no longer be sustained, but find his alternative causation lacking in evidence (Etherington 1995; Wright 1995). After defending much of his original account, Omer-Cooper (1995) also now accepts that the role of the Zulu has been exaggerated and suggests that we take more account of the possible slaving activities of the Griqua, but criticizes Cobbing's revisionism for a lack of substantiation. Focusing on the highveld 'Difaqane', Margaret Kinsman and Guy Hartley pursue a parallel line of attack.

*(Box continued)*

They argue in particular that Cobbing's view that missionaries and Griqua colluded in slave-raiding 'gravely distorts the process of historical reconstruction' (Kinsman 1995: 363; Hartley 1995).

Jeff Peires has been the most explicitly scathing of Cobbing (Peires 1993, 1995). He has attempted to demolish one of Cobbing's specific arguments concerning the Ngwane, one of the mobile 'refugee' groups caught up in the conflicts of the highveld. According to Cobbing, the Ngwane were first driven onto the highveld by slave-raiding from Delagoa Bay, then driven from it by Griqua slave-raiding, and finally subjected to a British-led raid for labour at the battle of Mbolompo, which took place beyond the Cape's eastern frontier in 1828 (Cobbing 1988). Despite Peires' extensive use of recorded oral testimony, the evidence surrounding the Ngwane's highveld movements remains scanty, but there is more substantiation for Peires' construction of events at Mbolompo. Cobbing saw the battle as a conscious colonial attempt to seize captive labour. From the available evidence though, it would seem that the Ngwane taken back into the colony from Mbolompo were a 'by-product' of a battle fought to defend the colonial margins from apparently hostile 'intruders', thought by officials at the time to be Zulu.

Cobbing's assertions about Mbolompo raised the question of British settlers' and officials' participation in slave-raiding, and one of Cobbing's former postgraduate students, Alan Webster (1995), has endeavoured to prove that frontier Britons were indeed involved in the capture and exchange of Xhosa labour on a large scale during the 1820s and 1830s. However, his interpretation, too, has been questioned. Peires (1995) and Lester (1998b) accept that some settlers opportunistically seized vulnerable Xhosa and forced them into servitude. But they find it difficult to see how overstretched colonial forces could have retained a large-scale captive labour force along a permeable frontier. Furthermore, this frontier was situated, in many cases, within relatively easy walking distance of the labourers' homes in independent Xhosa territory. In addition, vocal humanitarians who, as we have seen in chapter 3, criticized settlers persistently for their various aggressions against Khoisan and Xhosa, never spoke of a trade in captive labour on the frontier.

Elizabeth Eldredge has questioned another central plank of Cobbing's argument: his assertions about the Portuguese-orchestrated Delagoa Bay slave trade (Eldredge 1994b, 1995). She notes that an expansion of slave-trading activity at the port took place only once the 'Mfecane' was well under way. This would suggest that slave raiders were responding to new opportunities to take captives raised by an increased level of conflict, rather than actually initiating the conflict. But Eldredge (1995) also attempts to generate a new synthesis of the 'Mfecane'. She proposes an hypothesis based upon an initial stratification of

*(Box continued)*

African society exacerbated by expanded trade along the eastern coastal strip. Social inequality was then intensified by the differential impact of drought and famine, leading to the conquest of weaker groups by the stronger. Only subsequently did European and Griqua slaving add further impetus to the turmoil.

Apart from his inferences about European-inspired slave-raiding being the cause of greater defensiveness and aggression among African polities, Cobbing's second main proposition is that the idea of a Shaka-generated 'Mfecane' was used first by British traders as an 'alibi' for slaving and subsequently by historians as legitimation for settler land-grabbing. Christopher Saunders (1995) finds no evidence for such a grand conspiracy among South African historians. Neither does Dan Wylie discover any simple British construction of Shaka as monstrous instigator of the 'Mfecane'. Rather, Shaka's 'portrayal is conditioned by a plethora of Eurocentric prejudices, inherited concepts and narrative conventions' (Wylie 1995: 71). Carolyn Hamilton (1998) argues that pejorative images of the Zulu King were in fact generated first by his own subjects, and that the traders' diverse representations were conditioned by competing interests among themselves.

Rather than engaging directly with Cobbing's ideas, a further set of researchers have concentrated on establishing the wider context within which the whole 'Mfecane' debate should be positioned. Neil Parsons (1995) draws attention to considerable evidence that the 'Mfecane' and the linked highveld 'Difaqane' were not novel phenomena in southern Africa, but episodes in a continuum of violent change revolving around trade, cattle and environmental resources. Simon Hall (1995) and Andrew Manson (1995) provide further material on the 'Difaqane's' historical context, tracing widespread strife in the southern Transvaal and conflicts involving the western Tswana. In each of these studies 'the picture that begins to emerge is that there was a more generalised "time of troubles" sweeping the ... highveld. It was the result of ongoing processes that were not initiated by the impact of European traders and raiders though they may have been hastened by them' (Manson 1995: 360).

*This summary box is based on the review of C. Hamilton (ed.) (1995) by Alan Lester, published in the *Journal of Southern African Studies*, 24, 4, 1998, 839–42.

## Summary box 5: The 'Great Trek'

The term 'Great Trek' was an invention of the late nineteenth century, used to refer to the process by which some 15,000 Dutch-speaking farmers emigrated from the eastern margins of the Cape Colony onto the highveld and into Natal from 1834 to the mid-1840s (Fig. 3.6 on page 75). The term suited a generation of Afrikaner historians who, through their interpretation of the Afrikaners' history, contributed to the modern Afrikaner nationalist movement (see chapter 4). For these historians, the 'Great Trek' represented the escape of Afrikaners from the oppression of the British rulers at the Cape. As such, it was a manifestation of the Afrikaner people's long-standing desire for national autonomy. The trekkers' travails on the highveld and in Natal, confronting and overcoming African 'savagery' in the form of Zulu and Ndebele armies (see below), were tests of their unshakeable Calvinist faith. Their survival, and ultimate triumph, were markers of God's favour. In particular, the battle of Blood River (1838), in which a party of trekkers wreaked revenge on the Zulu for their earlier massacre of Piet Retief and his followers (see below), was seen as being the result of divine intervention (Thompson 1985). These interpretations are etched in stone today in the Voortrekker Monument outside Pretoria, built by an Afrikaner nationalist government of the late 1940s (see chapter 5).

However, both in terms of their motivations for leaving the Cape and of their subsequent activities on the highveld, the trekkers now appear in a very different light to most historians. They certainly do not seem to have been divinely inspired nationalist zealots. Their Calvinist religious principles were first found to be suspect by Andries Du Toit (1983) and their intentions upon leaving the Cape have more recently been investigated by Clifton Crais (1990, 1992, 1994), Norman Etherington (1995) and Timothy Keegan (1988, 1996). It now seems to many historians that the emigrants' goals were largely material.

Many of the trekker parties' leaders were men who had dabbled in commercial ventures in the Cape and who wished to maintain rather than sever their ties with the increasingly commercially oriented Cape society. As we saw in chapter 3, they often felt that preference in such matters was being given to British settlers rather than themselves, and that the accumulation of land on the highveld (especially in trans-Orangia, which was closer to the Cape merchants) or in Natal would enable them to acquire capital more on their own terms (Etherington 1995; Keegan 1996). However, most trekkers probably did not have such exalted aims. For many of them, the acquisition of new land to the north and north-east was a natural development given that the long-established eastern expansion of the colonial margins was coming to a halt as colonists encountered determined Xhosa resistance. They felt that, if left to negotiate their own terms with the African polities of the interior, they could avoid long-running conflicts such as those which plagued the eastern Cape frontier (Keegan 1996).

*(Box continued)*

The emancipation of the slaves in 1834 probably proved decisive for most of the trekker party members. While the colonial state clearly had ambitions to control the colony's newly freed labour force in the ways indicated in chapter 3, it was best able to do so in the settled western part of the colony. The colonial government was by no means so strong in the east, as its constant 'frustrations' in dealing with the frontier Xhosa chiefdoms (summary box 1, pages 117–20) suggest. Many of the trekkers, who came largely from the frontier districts, preferred to maintain their own, more direct control over captive labourers and dependants in the interior, as they had done in the past within the Cape (Crais 1990, 1992).

Once on the highveld, trekkers did not establish the national identity which later Afrikaner historians attributed to them, nor did they abide unswervingly by the principles of strict racial segregation which were later held to have been their particular innovation. Rather, as we saw in chapter 3, they made accommodations with African authorities where they could, and most farmed the land only with the assistance of African sharecroppers or labour tenants, who were able, to a certain extent, to negotiate their own conditions. As Trapido (1980b) and Keegan (1988) have shown, commercially minded elites in each of the republics eventually founded by the trekkers ensured a degree of class stratification which undercut the ethos of an ethnically united, egalitarian community. The trek and its consequences were therefore far more complicated and nuanced affairs than subsequent nationalist historiography would suggest.

Despite this complexity of trekker–African interaction, the most famous episodes of the 'Great Trek' tend to be those which involve conflict rather than accommodation. Foremost among these episodes were the struggles of particular trekker parties with the Ndebele and the Zulu. Having already recently fought off both Griqua and Zulu raiders, Ndebele warriors attacked the trekker party which entered their kingdom's buffer strip in October 1836. The attack initiated a trekker campaign which culminated in the Ndebele's defeat at the hands of a commando led by Hendrik Potgieter in 1837. Following the trekker victory, most Ndebele emigrated to found a new kingdom in what is now Zimbabwe.

After disputes between Potgieter, Gert Maritz and the other trekker party leaders over the ultimate destination and governing principles of their combined parties, Piet Retief led his followers across the Drakensberg, seeking the Zulu King Dingaane's permission to occupy land to the south of the Tugela River. Notoriously, Dingaane had Retief and his followers killed in February 1838, apparently after he had agreed to cede them the territory in question. However, Dingaane had also heard Retief boasting of the trekkers' victory over the Ndebele, and Retief's party had already started to settle on the edges of Zulu lands before formal permission to do so had been granted. With Retief and his immediate entourage dead, Dingaane sent his warriors to massacre the remainder of Retief's followers, mostly white and Coloured servant women and children, who were

*(Box continued)*

encamped around a place subsequently named Weenen ('Weeping') in the Tugela headwaters (Fig. 3.6 on page 75). In December 1838, a commando led by Andries Pretorius avenged Retief when, from a strong defensive position, it fought off a large Zulu army. Some 3,000 warriors were killed at Ncome (Blood River), with not one trekker fatality. A subsequent trekker alliance with Dingaane's half-brother Mpande then allowed the trekkers to found the republic of Natalia with its capital, Pietermaritzburg, named after Piet Retief and Gert Maritz.

As we saw in chapter 3, during the 1840s, Natalia would become the British colony of Natal. After the amalgamation of numerous and fractious governments, by the mid-1850s, the Orange Free State (occupied by Harry Smith's British forces from 1848 to 1854) and the South African Republic (Transvaal) had emerged as the remaining highveld trekker republics. Despite the reliance of many trekker farmers on labour arrangements negotiated with local Africans, these republics both had racially exclusive constitutions, setting them apart in principle, if not so much in practice, from the British colonies of the Cape and Natal.

## Summary box 4: The South African (Boer) War

The war fought between the British empire and the two Boer republics (South African Republic (Transvaal) and Orange Free State (OFS)) in 1899–1902 was conventionally known as the Anglo–Boer War, or simply Boer War. Recently, however, South African and imperial historians have rejected this name in recognition of the thousands of participants in the war who were neither 'Anglo' (British, or more properly English) nor 'Boer'. They are thinking in particular of the black men who fought and assisted on both sides of the war, and the black women and children who died alongside Afrikaners in the 'white' concentration camps, or in racially segregated camps of their own.

As we suggested in chapter 3, this most destructive of all the wars fought in South Africa was engineered out of a number of different British agendas for the Transvaal region. There have been long-running disputes about the relative significance of material, diplomatic and strategic 'factors' which led the British to provoke war with the republic once Johannesburg had developed as its largest and most cosmopolitan city, founded on gold (Marks and Trapido 1979; van Onselen 1982a; Porter 1990, Smith 1996). But the distinctions between such 'factors' are often unhelpful. They were invariably conflated in the minds of protagonists.

While the Randlords (see page 110) who sought a more 'rational' British administration of the gold fields and their hinterland may have had greater profits clearly in sight, some of them, like Cecil Rhodes, were as convinced by the

*(Box continued)*

notion of retaining Britain's imperial supremacy in the southern African region, as were metropolitan officials. For Rhodes in particular, a British administration of the Transvaal, rather than Kruger's 'anachronistic', rurally oriented government made sense because, as the supreme nation on earth, Britain was able to guarantee the most rational economic planning for its territories. Key metropolitan or metropolitan-appointed 'players' such as the Colonial Secretary, Joseph Chamberlain, and the High Commissioner for South Africa, Alfred Milner, may have been consciously motivated to attain hegemony over the Transvaal government (and failing that, seize control of it) due to the strategic considerations of maintaining Britain's imperial supremacy, especially in the face of new threats such as that from Germany. But, along with the British government as a whole, they saw economic prosperity for Britain, and for Britons in the colonies, as being a vital manifestation and index of that supremacy (Porter 1980).

There was thus a complex of notions involving rational planning and profitability for the mines, the need to safeguard the superiority of 'Britishness' and the fear of imperial rivalry, that was shared by powerful interests in South Africa and Britain during the 1890s. It was a complex reproduced with varying emphases and priorities in the representations of each of the British individuals involved in bringing about the war. Ultimately, it was the influence and range of this complex of ideas which meant that Kruger's Transvaal government was forced to fight for its continued independence. It did so by accumulating arms in anticipation of a British assault, by forging an alliance with the Orange Free State (OFS) and, ultimately, by launching a pre-emptive strike against British forces in Natal and the Cape in October 1899.

The Boers fighting against the British army numbered some 88,000 men at the start of the war, including 12,000 'rebels' from the British Cape Colony. British forces amounted to 20,000 in 1899, but grew to 450,000 by the end of the war, including 53,000 colonial South Africans and 31,000 troops from Canada, Australia and New Zealand (Thompson 1990). By the end of 1899, Boer forces had defeated British units in a number of set piece battles and were besieging British garrisons in Ladysmith, Kimberley and Mafeking. In 1900, though, the tide of the war turned as reinforced British troops successively lifted these sieges and occupied Bloemfontein (capital of the OFS), Pretoria (capital of the Transvaal) and Johannesburg. Kruger left for exile in Europe and Britain claimed control of both the Boer republics. However, Boer forces fought on in mobile commandos, often with the assistance of African servants, using guerrilla hit-and-run tactics. As well as attacking British outposts, columns and supply lines, commandos even penetrated deep into the Cape Colony in largely unsuccessful attempts to recruit more Afrikaner 'rebels' (Packenham 1979).

It was due to the frustrations encountered in trying to subdue these commandos that the new British commander, Lord Kitchener, adopted the scorched

*(Box continued)*

earth tactics which had long been used in the Cape's frontier wars (see summary box 1, pages 117–20). Boer farms were burnt and their crops and livestock captured or destroyed. But these colonial tactics were also 'modernized'. The Afrikaner women and children inhabitants of the farms were driven into concentration camps where disease, maladministration and neglect caused the deaths of some 28,000. The numbers of Africans who died in similar conditions, as servants in the white camps, or in segregated camps, is uncertain, but it is thought that the figure is at least 16,000 and maybe as many as 28,000. The same kind of military medical neglect which led to the deaths in the civilian camps also meant that more British soldiers died of disease during the campaign than were killed by the Boers.

By 1902, thousands of poorer, landless Afrikaners were fighting on the British side, along with up to 30,000 Africans. The latter could earn relatively good wages working for the British, and they were led to expect an improvement in their political rights once British control of the Boer republics was secured. In many districts, Africans were regaining control of land on their own initiative, in the absence of Afrikaner farmers and their families (Warwick 1983; Krikler 1993). The gold mines were working again and the British imperial administration, as we will see in the next chapter, was planning their more efficient operation. The remaining, ragged Boer commandos finally agreed to peace terms at the Transvaal town of Vereeniging in May 1902. As we will see in the following chapter, accommodations were still possible in the aftermath of the war between a commercially oriented Afrikaner elite within each republic and the British imperial authorities. It was these accommodations which, above all, would shape modern South African politics.

# Chapter 4

# INDUSTRIALIZING SOUTH AFRICA, *c.*1900–48

## Reconstruction, Union and the reformulation of racial boundaries

By the time that Union was agreed among white politicians and an Afrikaner-led central government had been elected in 1910, the territories comprising South Africa had already been transformed in such a way as to secure the goals of the mining companies and the British imperial government. Decisive metropolitan intervention was invoked and assisted by local commercial interests in the aftermath of the South African War, in a programme known as 'Reconstruction'. This programme was geared in particular towards the modernization of highveld economies and societies that were considered by industrial capital and imperial bureaucrats to be outdated and inefficient. It was implemented by officials appointed by Milner. Most of these were young acolytes recruited among the Oxford University colleges, and they became known as 'Milner's kindergarten'. In liaison with the Randlords and commercial Afrikaner farmers, they laid the foundations for modern South Africa's political economy (Marks and Trapido 1979).

Imperial Reconstruction was achieved between 1902 and 1910, with the Union government continuing many of its policy thrusts in the years immediately following. This programme of transformation, from an apparently 'anachronistic' agrarian-oriented society to a 'modern' industrial-centred one, thus occurred over a far more compressed timescale than the transformation from the slave-based to the liberal capitalist system wrought by British intervention earlier in the nineteenth century. This rapidity was possible partly because of communications and technological improvements, but largely because, in the immediate aftermath of the war, 'Milner's kindergarten' controlled their own well-funded 'conquest state', enjoying an unprecedented degree of power (Marks and Trapido 1979: 72). As one of them wrote, 'one gets things done with a stroke of a pen that in England would entail an Act of Parliament' (van Onselen 1982a: 181). And the members of 'Milner's kindergarten' were highly motivated to 'get things done' by the notions of 'greater state responsibility, national efficiency and planning' which they had acquired in late-Victorian Britain (Marks and Trapido 1979: 55).

## Capitalism and racial discourse

Milner's intentions were manifested first in Johannesburg. Here, many of the Rand-lords' prewar demands were fulfilled. The former monopolies on dynamite and railway transport were broken immediately, in order to reduce mining costs. More problematic was the systematic recruitment of African labour for the mines. Attempts had been made by the Chamber of Mines during the 1890s to form a co-ordinated recruitment body which would prevent mining companies competing for workers, and thus keep general wage costs down, but each attempt had failed as a result of the lack of central regulation (Richardson and van-Helten 1982). Under the patronage of the new state these attempts were revived and made more successful. The result was a labour recruit-ment network, the Witwatersrand Native Labour Association (WNLA), managed by the gold mining companies themselves (Jeeves 1975; Marks and Trapido 1979).

Aside from helping the WNLA to procure labour directly, Milner's state assisted the mining companies through various indirect measures. The most notable was taxation, imposed more extensively throughout the African locations. Not only did this help to finance the central state in the same way that the Natal colonial state had long been sup-ported; it also addressed the mine-owners' complaint that the 'typical' African migrant worker 'cares nothing if industries pine for want of labour when his crops and home-brewed drink are plentiful' (Johnstone 1976: 27). If this 'typical' African man was forced to pay more taxes, then work in the mines, as well as on the state's own infrastructural projects such as railway and harbour construction, would be less easy to avoid.

There was a more urgent problem of recruitment though, which such long-term measures could not address. The mine-owners' difficulties had been compounded during the recent war, when the expansion of other economic opportunities enabled African migrants to find more lucrative and less onerous employment. In the

Figure 4.1   Migrant labour flows to Johannesburg and Kimberley, 1914

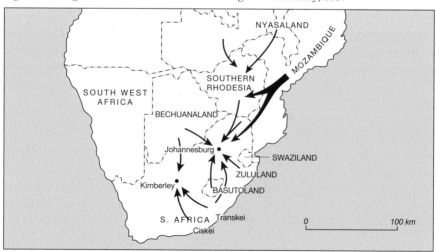

*Source*: Denoon and Nyeko (1984)

immediate aftermath of the war, the Reconstruction state assisted the mining companies by arranging for the indenture and transportation of some 63,000 labourers from China. Although the WNLA was soon able to extend migrant labour recruitment networks into Mozambique and Malawi, thus allowing for the repatriation of most of the Chinese workers, the mining companies had formulated strict regulations in the meantime, establishing a compromise between mining capital and relatively privileged white labour. In the face of white labour unrest, these regulations ensured that Chinese workers did not displace white miners in the performance of certain jobs. Even after most Chinese had been repatriated, this nascent job reservation scheme was perpetuated, allowing white workers to act in protected supervisory roles over Africans (Jeeves 1985). Although, as we will see, the exigencies of the global economy meant that the privileged position of white labour was periodically threatened by cost-cutting exercises, such measures, together with much higher rates of pay than were awarded to blacks, generally allowed an enduring *modus vivendi* between mining capital and skilled white labour (Johnstone 1976).

The new imperial government of the Transvaal ensured that when African labour flows resumed, the migrant workers would be more effectively controlled. Familiar Masters and Servants legislation was enforced by a reconstructed police force and a professional judiciary, and Africans were prevented from being absent from work through drunkenness by a racial prohibition on liquor consumption (Jeeves 1975; van Onselen 1982a). Inheriting the imperial state's imperatives, in 1911, the new Union government co-ordinated labour recruitment measures and pass controls with the Native Labour Regulation Act. This required the registration of all workers recruited for the mines and prescribed the conditions of their employment. The worker's copy of a contract with the employer served as a pass without which he would not be allowed to remain in the urban area (Hindson 1987). In the same year the government implemented a Mines and Works Act which formally reserved skilled and supervisory jobs in the mines for white workers.

While the Reconstruction state assisted mining capital by regulating African migrant labour and helping to secure the co-optation of skilled white workers, mining capital in turn assisted the state. In 1921, the government mining engineer described the gold mines as 'milch cows for State revenue' (Yudelman 1984: 143). The gold mines continued to provide the main channel for foreign investment in South Africa until the 1930s, and taxation of them allowed Milner's state and its successors to pursue their own relatively autonomous goals, including the construction of a more sophisticated national and local administration. The bureaucratic reorganization of, and investment in, the Johannesburg municipality, for instance, allowed housing and sanitation to be provided for working-class whites well away from the city's black proletariat, as had been the case earlier in Kimberley (van Onselen 1982a, 1982b).

But Reconstruction extended further than Johannesburg itself, to embrace the commercial farms of the entire country, and especially the highveld. An imperial loan supplemented gold mining revenue and allowed a start to be made on re-ordering relations between commercial farmers and poorer white and black producers in the interests of productivity (Keegan 1986; Murray 1992). In the immediate aftermath of war, Africans who had seized opportunities to reclaim independent access to the land were brought under state authority and frequently disarmed, while white commercial farmers were replaced securely on their farms (Krikler 1993). Local Afrikaner officials in the Transvaal and Orange River Colony, whose sympathies often lay with the smaller, less commercial farmers, were replaced by magistrates and a police force subordinate to central government and more attuned to the imperatives of commercial agriculture.

Capitalist farmers were given technological assistance and significant financial help through new provincial land banks, experimental farms, agricultural publications, co-operative dairies, the importation of breeding stock, veterinary research centres and a central Department of Agriculture. They were assisted further by the extension of a state-financed railway network to connect the white commercial farming districts with the cities (see Fig. 3.12 on page 98), a more effective co-operative marketing system implemented across the Union after 1910, and the diversion of African mine labour recruiters away from their districts (Keegan 1986). A scheme devised by Milner to bring more commercially oriented British settlers to the Transvaal and Orange River Colony so that they could lead the way in the local capitalization of agriculture failed in its grand ambitions, but nevertheless resulted in the establishment of 972 capitalist settlers on government land, some of which had been cleared by the eviction of poorer Afrikaners (Keegan 1979, 1986; Murray 1992). In 1910, the capitalist transformation of white farming still had a very long way to go, but, as Krikler concludes, 'British imperialism had ... created ... a state in which the capitalization of agriculture could proceed' (Krikler 1993: 225).

Capital growth *per se*, however, was accompanied by another agenda during the Reconstruction period. As has been implied above, Milner's state and its successors deliberately ensured that the beneficiaries of their capitalist programme would be white rather than black. This did not mean, as we have hinted above, and as we will see further below, that *all* whites were beneficiaries, but it did mean that the racial boundaries which had existed in each of the colonies since the era of emancipation were defined and reinforced in order to ensure more watertight black exclusion. By according priority to 'racial' difference rather than to the boundaries of class, region, religion or ethnicity, and by institutionalizing that difference, the Reconstruction state and its successors helped actively to consolidate more robust racial identities (Unterhalter 1995). In order to comprehend the specifically racial design behind imperial intervention, we need to return briefly to the mid-nineteenth-century shift in metropolitan notions of racial difference.

We have already pointed to the resistance which the colonized and the recently emancipated mounted to bourgeois humanitarian prescriptions in the Cape and elsewhere. We also suggested that this resistance allowed settler representations of racial difference to become more influential in the British metropole, where political reform was grinding to a halt. By the 1870s, colonial settlers' assertions of innate racial difference were being complemented by the British bourgeoisie's own notions of inherent class, ethnic and gender differences. Seeking to retain exclusive access to power and privilege, the men of the 'respectable' metropolitan classes found discourses of immutable difference convenient in legitimating the political and economic exclusion of those who challenged their status. Such discourses were being articulated by the late nineteenth century in terms of Darwinian evolutionary theory. Concepts of 'struggle between species' and the 'adaptation and survival of the fittest' entered the language of racial and other kinds of human difference at this time. Men's dominance over women, British rule over the Irish and middle-class privilege over workers, for example, were all commonly portrayed as the quite natural outcomes of evolutionary social progression. In this scheme, those who governed did so because they were best designed, biologically, to govern (Stepan 1982, 1990; McClintock 1995; Perry Curtis Jnr 1997).

Evolutionary discourse applied in particular to differences between groups of people with their origins in different continents. The eighteenth-century Enlightenment notion of a static hierarchy of civilizations, with the European at its apex, was now being modified into an evolutionary progression from the least intelligent and adaptable African female up to the most biologically advanced European male – a notion of progression which underlay all of the European and North American academic sciences and humanities. The progression from African to European was represented by geographers, for example, as being both temporal and spatial, the journey from Europe to Africa comprising a passage backwards through time, with more primitive people being encountered the further south one travelled (Driver 1992; Livingstone 1992).

With such notions of difference and progression constituting Western 'knowledge', it seemed apparent that 'breeding' between groups, whether they were differentiated by race or class, would inevitably result in the dilution of the superior characteristics of the more advanced group – a process known as 'degeneration'. Eugenic schemes for the improvement of the human race by selective 'breeding' became popular in Europe and North America, as well as in the colonies, in an attempt to avoid such racial, ethnic or class degeneration, and they were advocated by philanthropic social reformers as well as more conservative elements (Stepan 1982; Dubow 1995). 'Degeneration' was also thought to occur where the environment and lifestyle of a particular group was unhealthy. For this reason, members of the ruling class were especially concerned about the degeneration of the working class in Britain. Their anxiety seemed to be borne out by the high incidence of sickness among the urban workforce, and the poor health of British recruits during the recent war in South Africa. As Davin (1978) and Bell (1993) show, the emphasis which metropolitan medical experts placed on the biological, maternal 'function' of women became interwoven with such imperial anxieties. Together, the discourses of females' breeding function and racial 'degeneration' led to the promotion of British women's emigration to temperate colonies such as South Africa. There, in a healthy environment, British women would nurture a reinvigorated imperial race, reversing the nation's tendency towards 'degeneration'.

The empire-wide notion of evolutionary difference and the fears of 'degeneration' associated with it had a number of further implications for imperial policy in South Africa. Those of 'mixed race', such as South African Coloureds, as well as 'pure-blooded' Africans, were increasingly seen by 'experts' as being hampered in their struggle to achieve advanced levels of civilization by their black 'blood', even if, in the case of Coloureds, it was mingled with white (Gilman 1985; Goldberg 1990; Young 1995). Only in the context of an entrenched discourse of 'degeneration' can the emotive power of early-twentieth-century racism be conceived. It is apparent, for example, in the relatively liberal South African writer, Olive Schreiner's description of Coloureds, who were living testimony to the practice of inter-racial 'breeding':

> When from under the beetling eyebrows in a dark face something of the White man's eye looks out at us, is not the curious shrinking and aversion we feel somewhat of a consciousness of a national disgrace and sin? (Schreiner 1923: 121).

South African Asians were also not exempt from the paranoia and targeted humiliation of late-nineteenth- and early-twentieth-century racial discourse. In Natal and the Transvaal, many Indian men were operating successfully in business. White fears over their ability to seduce, corrupt and even marry Afrikaner female employees blended neatly with anxiety about the economic competition which they presented, and during the early part of the twentieth century, white politicians were able to gain electoral support by promising to legislate against Indians' 'contamination' of the governing race (Hyslop 1995). In the midst of 'raw' Africans, 'cunning' Indians and the Coloureds who, in Schreiner's words, represented the 'sin' of inter-racial sex, the distinctions to which European racial science drew attention already seemed self-evident to most whites, including Afrikaners, in South Africa (Dubow 1989, 1995). But, in the last years of the nineteenth, and the early years of the twentieth century, British metropolitan racial science provided a new language with which to rationalize and justify political and economic differentiation between racial groups (Dubow 1996).

### Excluding the black elites

As we indicated in the previous chapter, white settlers in late-nineteenth-century South Africa were conscious of the need to protect their privilege during a period of commercialization and the threatened incorporation of a growing black elite. They did so by shoring up strategies of racial exclusion. In the period immediately before and during Reconstruction these strategies were more coherently devised, articulated and legitimated in terms of the current 'scientific knowledge' of racial difference.

Exclusion had different implications for each of the stratified black classes, but its most obvious effect in the short run was the disillusionment of the African, Coloured and Indian elites who had followed humanitarian prescriptions most closely and aspired to full colonial citizenship. Middle-class Coloureds, mostly in the western Cape, and wealthier Indians, mostly in Natal, found themselves in the same predicament. Both groups, because of the taint of their 'black blood' were increasingly classified as 'uncivilized', and placed conceptually alongside 'pure-blooded' Africans. Indians found themselves barred from entering the OFS entirely after 1910, and restricted to specific urban locations in the Transvaal, while in Natal, where the Indian elite competed particularly successfully with white traders, they faced a spate of discriminatory legislation including exclusion from the franchise. It was these measures which prompted Gandhi's first campaign of *satyagraha* (non-violent, passive resistance) in South Africa. Together with the dire conditions for some plantation workers, such discrimination also led the government of India to outlaw further indentured labour flows to Natal in 1911 (Pillay 1976; Brain 1989).

Elite Coloureds, far from achieving their long-held goal of extending the Cape's non-racial franchise to the Transvaal and OFS, found that they too had to mobilize with greater effectiveness merely to defend the rights that they still held in the Cape. The greatest threat facing them was the erosion of the province's non-racial constitution. The incorporation of the Xhosa and Thembu-populated Transkeian territories into the Cape in 1885 had already prompted the exclusion of tribal tenure as a property qualification in 1887 and a raising of the property qualification in 1892, both measures being taken in order to prevent a greater mass of Xhosa peasants from exercising the vote. A similar response to those Coloureds who were currently enfranchised was in prospect. Appeals to the Cape 'tradition' of 'equal rights for all civilised men', formed the agenda of new Coloured organizations, the largest of which, the African Political Organisation (APO) was founded in 1902. The APO's claim that 'there existed "an educated class of Coloured people" ... who could no longer be treated as part of an undifferentiated mass of "uneducated barbarians"' fell, however, on deaf ears (Lewis 1987: 23). Despite the continuing appeals of white political parties to the Cape Coloured vote, the barring of Coloureds from sitting in the new parliament after 1910, and the maintenance of a common pass system for Coloureds and Africans in the Transvaal and OFS, were the most explicit indications that full citizenship was not to be offered to Coloureds in the Union (Marais 1957; Fredrickson 1981).

The broader mass of Coloured workers in the Cape were also affected by reinforced racial distinctions during and immediately after the Reconstruction period. They had already been denied access to state-funded white schools, designed, as the Superintendent-General of Education put it, to 'fit [whites] to maintain their unquestioned superiority and supremacy in this land' (Marais 1957: 271). Now Coloureds were also excluded from welfare services and increasingly discriminated against in favour of whites for craft apprenticeships and employment (Fredrickson 1981; Lewis 1987). While the leadership of the APO sought alliances with the African elite, racial exclusion prompted many Coloureds to emphasize instead the distinctions between themselves and Africans. They placed greater reliance on the Afrikaans language and culture which they shared with many whites, and on their lighter skins. Those who were not pale enough actually to 'pass for white' increasingly asserted that through their 'white blood' they had reached a higher stage in the evolutionary scale than the 'raw' Africans who were being subjected to even greater political exclusion, regulation and control. Through such differentiation, they sought to avoid the same fate. The enhancement of a separate and politicized Coloured ethnicity, transcending established divisions between Coloured classes and between Muslims and Christians, was an outcome of this defensive strategy (Goldin 1987, 1989; Bickford-Smith 1994, 1995).

If the Coloured and Indian elites became partially disillusioned during the years of Reconstruction, mission-educated and petty-bourgeois Africans were similarly alienated (Nasson 1991). In common with the other colonized, black middle classes, both in South Africa and in other colonies, they were told increasingly by government officials that although they might 'mimic', they could 'never exactly reproduce English values'. They must therefore recognize 'the perpetual gap between themselves and the "real thing"' (Loomba 1998: 173). In the early 1900s they found their polite representations to colonial officials ridiculed and themselves accused, as one missionary put it, of being 'puffed up with self-importance' (Comaroff and Comaroff 1997: 239).

The Kholwa in Natal were told by the colony's new Secretary of Native Affairs that it would be defying 'the laws of nature' to distinguish them from other Africans simply because of their superior education. Popular social Darwinism held that all members of the African race stood rather on an equal and inferior footing (Lambert 1995: 68). Accordingly, Kholwa were increasingly prevented from exercising the franchise and only three Africans had the vote in Natal in 1905 (Welsh 1971). Furthermore, a Natal government committee on education reported in 1902 that 'it would be better [for Africans] to be contented with a rather lower standard of attainment' in school (Etherington 1989a: 297). In the Cape, even the famous mission-run educational institutions such as Lovedale accepted that there were formidable obstacles to the vast majority of Africans becoming 'civilized' in the foreseeable future. In 1884 the Principal declared that Africans as a whole were some 2,000 years behind Europeans in the evolutionary scale of civilization. By educating only a tiny African elite, often the offspring of the more successful commercially oriented peasantry, and endeavouring to lower the expectations even of this group, mission educationalists developed a 'working compromise' with those settlers who had always resisted notions of African assimilation (Hughes 1990; de Kock 1996).

One consequence of the missionaries' broken promise of incorporation was the rise among various African classes of separate Ethiopianist and Zionist churches, in which African male leaders, and women followers in particular, could articulate and assert their own, less hypocritical versions of Christianity. Behind their separate spiritual mobilization lay the common grievance expressed in 1895 by the African-edited newspaper, *Inkanyiso*: 'It is neither wise nor christian like to preach the doctrine that all men are created equal before the God of the white man, and then, when the Native has accepted his faith, at the sacrifice of feelings handed down to him from generations, to treat him as a moral and social pariah' (Welsh 1971: 244). Through the early twentieth century the independent African church movements grew to enormous proportions, giving a form of communal security and spiritual uplift to African men and women in the cities as well as the countryside (Comaroff 1985; Gaitskell 1982, 1990; Beinart 1987a; Comaroff and Comaroff 1997; Elphick 1997; Mills 1997). A related response on the part of educated Africans was political organization to bring a new form of pressure to bear on the state.

The South African Native National Congress (SANNC), forerunner of the African National Congress (ANC) was founded by middle-class Africans in 1912 to protest against their increased exclusion from civil rights after the Act of Union. It came in the wake of unsuccessful Coloured, Indian and African elite delegations to London. Its leaders, men like John Tengo Jabavu and Tiyo Soga from the eastern

Plate 4.1  Sol Plaatje, journalist, author
and a founder of the SANNC, later ANC
*Source*: Readers Digest (1994)

Cape, Sol Plaatje from the OFS and Pixley Seme and John Dube from Natal, were often educated in the metropole and influenced by the moderate goals of self-improvement articulated by the African-American Booker T. Washington. They never discarded the Christian teachings of the missionaries. Rather, they clung to them more fervently. In drawing attention to the exclusion of Africans who had learnt, as they had been told, to be sober, dignified, educated Christians, they deployed missionary discourse subversively, highlighting the hypocrisy of their white rulers in order to 'undermine [their] moral authority' (de Kock 1996: 4).

Ultimately, it was such members of the African 'petty bourgeoisie' who first developed a moderate ideology of African nationalism. While still speaking to the white authorities in a tone of deference, they spoke to other Africans of the need to transcend the divisive ethnic boundaries which they had inherited from the pre-colonial era and had reinforced in many ways subsequently. Africans were to consider themselves a single people, as entitled as whites were to political and economic privilege so long as they attained the necessary level of civilization (Marks 1986; Cobley 1990; Willan 1996). Aware of the enduring power of African kings, such as the Zulu ruler, Dinuzulu, over their 'traditionalist' subjects, these African nationalists sought to embrace them within their project, turning them and their 'uneducated' followers away from narrow ethnicist politics. At this stage, however, their message was easiest for the Mfengu and Kholwa to absorb. Even before they joined the missionaries, these groups had been descended from members of different polities and 'in the mission station melting pots, they [had] lost old identities and [already become] simply African Christians' (Etherington 1989a: 296).

## *Marginalized Afrikaners*

While Africans were being barred from citizenship, many whites, long entitled to full political citizenship, felt themselves to be similarly marginalized by policies designed to foster capitalist development dating from the Reconstruction period. As well as provoking the formation of defensive black political organizations such as the APO and the SANNC, British imperial-led modernization prompted the creation of workerist and Afrikaner nationalist movements representing these marginalized white groups. While distinct political parties such as the Labour Party (LP) emerged after 1904 to represent the relatively small numbers of skilled English-speaking workers, the movements intended to secure a place for specifically Afrikaans-speaking interests within modernizing South Africa carried more political clout. As was the case with the SANNC, it was not the most impoverished classes who took the leading political role, but a petty bourgeoisie which was afraid of increasing peripheralization.

In the late 1870s Afrikaans farmers, professionals and church ministers in the Cape first formed exclusively Afrikaans-speaking organizations to further their material and political interests against those of increasingly assertive English-speaking merchants, and to resist the growing secularization and Anglicization of Cape society. Led by the Dutch Reformed Church (DRC) minister, S. J. du Toit, they founded the Afrikaner Bond in 1880 (Davenport 1966; Giliomee 1989). However, Afrikaners endeavouring to create an embracing, nationalist identity in opposition to British imperialism faced the same problems of more established and parochial subjectivities, and of class divisions, as those encountered by the leaders of the SANNC (van Jaarsveld 1964; Thompson 1985). Far from constituting a natural, primordial identity, an Afrikaner 'nation' had to be actively created and politically mobilized against formidable obstacles.

With the Transvaal industrializing in the late 1880s and 1890s, wealthier Afrikaners in the Cape were more inclined to side with English-speaking merchants, in order to protect the Cape's regional paramountcy, than they were to ally themselves with northern economic rivals, even if they were fellow Afrikaans-speakers. Indeed, during the early 1890s, commercially oriented leaders of the Afrikaner Bond found common cause with Rhodes, the arch British imperialist, in extending shared investments to the north, and even during the South African War, many Cape Afrikaners remained neutral or sided with the British (Davenport 1966; Giliomee 1989). As we have seen, after the war, with all the bitterness that it occasioned for the Afrikaners who fought against the British, the new Afrikaner leaders of the Union, Botha and Smuts, 'were keenly aware of the centrality of the mines for the capitalist development of South Africa and, as progressive farmers, were as heavily committed to the industry which constituted their main market as Milner and his kindergarten had been' (Marks 1985b: 489). The new state itself was therefore founded upon commercial Afrikaans as well as British interests. It was against this mixture of commercial interests, rather than British influence *per se*, that the nascent nationalist movement had to fight.

The National Party (NP) was founded by J. B. M. Hertzog in 1913 to represent Afrikaners whose interests were potentially or actually neglected or contradicted by this governing complex. In its early years, the party represented DRC ministers determined to uphold the Calvinist faith; teachers alienated by the superior status of the English language in schooling; small-scale businessmen and professionals side-lined by the predominance of English in commercial transactions and the law; journalists seeking to empower their Afrikaans language and civil servants who felt their promotional advance to be blocked by competition from English-speakers (Giliomee 1989; Grundlingh and Sapire 1989). As Giliomee puts it, 'this educated stratum had an overriding interest in creating Afrikaners who would refill Afrikaner churches, attend Afrikaner schools and buy Afrikaans books' (Giliomee 1989: 48). Representatives of this 'stratum' in the Transvaal formed the secretive Broederbond (brotherhood) in 1918, in order to help mobilize influential Afrikaners behind the NP.

During the early decades of the twentieth century, the nationalist movement, spearheaded by the Broederbond and the NP, concentrated on drawing support from the two provinces in which it had been founded – the Cape and the Transvaal. In each of these regions, the party endeavoured to forge a distinct ethnic class alliance. In the Cape, the original petty bourgeois interests were supported by the large Afrikaans banks, created to service the export-oriented and relatively prosperous Afrikaner wheat and wine farmers. These banks provided the Cape branch of the NP with finan-cial backing and funded nationalist newspapers which disseminated its propaganda. In the Transvaal, the petty bourgeois interests comprising the Broederbond focused on securing the support of a rather different class of Afrikaans-speakers. This was com-prised of smaller farmers and landless bywoners (see chapter 3), who were being challenged by the state-sponsored commercialization of agriculture (O'Meara 1983). Their difficulties had begun during the 1890s when the security of smaller farms was first undermined by 'a surge of speculative capital' caused by mineral prospecting fees and the growth of urban markets for food (Keegan 1991: 44). As land prices rose, so too did mortgages, rendering increasing numbers of marginal farmers indebted. Drought, disease and the disruptions caused by the South African War exacerbated their economic problems, and in the wake of the war, the Reconstruction state evicted many farmers who were unable to keep up with their rents on government-owned land (Beinart and Delius 1986; Keegan 1986; Murray 1992).

By the 1920s, wealthier Afrikaans- and English-speaking farmers were enlarging and rationalizing their own holdings by buying out less economic farms in the Trans-vaal. Not only did this lead to the landlessness of former smallholders; it also prompted the eviction of bywoners who stood in the way of optimal commercial use of the land (Keegan 1986). Within the tenant bywoner families, women tended to be stringently confined to a domestic role, and children were expected to attend school. This meant that landless whites were prevented from utilizing extended family labour in the way that was customary for African tenants. Being less productive, bywoners were thus among the first to be evicted as the farm was geared up for maxi-mum production (Bozzoli 1983/1995; Keegan 1986). As Beinart and Delius point out, these white former tenants 'were in some sense the victims of their own society's entrenchment of private property in that they had no access to communal tenure land when ejected from farms' (Beinart and Delius 1986: 42). By contrast, much of the African tenantry remained in place, at least for the time being. Indeed, it was only the presence of African sharecroppers which enabled some smaller landholders to survive on the land without the assistance of capital (Keegan 1991).

The growing stratum of landless Afrikaners was predisposed to 'challenge not only imperialism but also the entire capitalist order' which they blamed for their dispossession (Giliomee 1989: 49). But this did not necessarily mean that they fell easily into the Transvaal nationalist political camp. Having been removed from the land, many impoverished white families went first to the smaller highveld towns, where they made a living as craftsmen, transport riders or small-scale entrepreneurs (Bundy 1986). Van Onselen (1982a, 1982b) has produced the most enlightening work on these poor whites' fate as they moved subsequently to the larger cities, and particularly, rapidly industrializing Johannesburg. During the 1890s, most acquired housing in the more squalid, and often unsegregated parts of the city, surviving by manufacturing bricks, driving cabs and operating transport wagons, as well as working in the mines. Increasing capital intensity in each occupation around the turn of the century, however, progressively excluded them from taking part in the economy on their own terms. Brick-makers succumbed to competition from a rival, capitalist-backed enterprise, transport riders were forced out by the railway, and cab drivers were superseded by a tram company in which mine owners had invested.

For many of the newly unemployed among urban Afrikaners, and for those forced by impoverishment to seek work in the mines, an alliance with English-speaking workers, similarly aware of exploitation at the hands of capitalists, was more natural than one with the Afrikaner middle classes within a nationalist movement. Popular white socialism, as we will see below, tended to exclude black workers, who would undercut white wages and were not seen as part of the civilized community anyway. But a poor white socialist orientation simultaneously frustrated for some time the NP's aspirations to separate Afrikaans- from English-speaking workers in the cities.

### *Transforming the peasantry*

As Coloured, Indian and African elites were redoubling their efforts to attain even basic citizenship during the Reconstruction period, poorer African peasants were being converted more effectively into migrants, supplying the essential workforce for the industrializing economy. This was not a new phenomenon. It was something which had begun in each of the separate white-controlled states, but particularly in the Cape and Natal, with the maturation and centralization of the mining industry. By the 1890s, industrialization was the main economic agenda across the region. Even the white liberal merchants and parliamentary representatives who had traded with successful African peasants in the Cape and secured their votes under the province's non-racial constitution, were re-orienting their investments and abandoning their political alliance. Rather than insisting on the 'civilizing' potential of African landownership and commercial enterprise, they were emphasizing instead the value of Africans' wage labour in the industrializing economy (Trapido 1978, 1980a; Cooper 1981). The rhetorical shift accompanying this changing economic agenda is exemplified in the comments of the British writer and traveller Anthony Trollope.

For him, efforts to Christianize and 'civilize' a landed African peasantry were futile. Labouring on the diamond fields was the true path to African self-improvement. He wrote of industrializing Kimberley in 1877:

> Who can doubt that work is the great civiliser of the world ... If there be one who does, he should come here to see how those dusky troops of labourers, who ten years since were living in the wildest state of unalloyed savagery ... have already put themselves on the path towards civilisation ... Civilisation cannot come at once ... But this is the quickest way towards it that has yet been found (Trollope 1878: 368).

While the timing and pace of change varied markedly across the country, and according to local landholding arrangements, as Marks and Atmore suggest, 'capitalist development [and the liberal discourse which accompanied it] first stimulated the growth of a peasantry by providing the demand for foodstuffs and cash crops – and then helped undermine it in its search for vast quantities of cheap labour'(Marks and Atmore 1980: 24).

In the 1890s, Rhodes, as the Cape's Prime Minister, wholeheartedly agreed that Africans' subjection to labour was necessary. He asserted that 'every black man cannot have three acres and a cow ... it must be brought home to them that in future nine-tenths of them will have to spend their lives in daily labour, in physical labour, in manual work' (Marks 1985b: 463). The Glen Grey Act, which applied in certain eastern Cape districts from 1894, was Rhodes's own experiment in creating such a 'favourable' division of opportunity within the African population. While it allowed for local administration by a co-opted African elite and the individual ownership of small freehold plots on which a limited peasantry could survive, it precluded access to such plots for the majority. In addition, owners of the new plots were explicitly barred from exercising the vote by virtue of their property. Although local African intransigence, including legal challenges and refusal to pay tax, meant that the system never quite worked as planned, the measure was intended to enhance economic distinctions between the peasant elite and the impoverished mass within rural African society. It would thus ensure the availability of the excluded majority on the regional industrial and agricultural labour market (Bundy 1977, 1987a).

In Natal, rather than maintaining the balance of power between state and chiefs which had been Shepstone's policy, the settler government established in 1893 pursued a similar strategy to that of Rhodes. By 1904 over 95,000 out of a total male African population of 296,000 were already engaged in full- or part-time employment as agricultural or migrant labourers (Lambert 1995). In 1906 a new £1 poll tax was designed not only to enhance government revenue, but also to extract further labour supplies from African homesteads. However, coming on top of drought, depression, and increasing landlessness as tenants were squeezed off commercializing white farms, the new tax provoked a rebellion. It was led by the minor chief Bambatha, but sustained by homestead heads and chiefs fearing a loss of control over their subordinates as they were absorbed into the commercial economy (Lambert 1995). The decisive crushing of the resistance, involving the deaths of up to 4,000 of the 12,000 Africans who participated, reflected the insecurity of the settler state in

Natal, but it also acted as a warning to Africans elsewhere who were being subjected to similar assaults on their productive autonomy (Marks 1970).

As if the Cape and Natal states' separate endeavours to convert African peasants into labourers were not enough, 'nature' itself seemed to conspire with them as the cattle disease rinderpest spread throughout the region during the late 1890s. Combined with drought, the disease destroyed 90–95 per cent of all cattle south of the Zambezi. Without state support and the finances for inoculation, and distanced from the major transport routes, African peasants found it far more difficult than white farmers to maintain commercial production in the face of the disaster. Many were effectively forced by their new poverty onto the labour market, or into increasingly overcrowded reserves as a result (van Onselen 1972; Guy 1982; Bundy 1988; Lonsdale 1998). Once in the reserves, their ploughs had to be abandoned due to accelerated soil erosion, and their draught oxen, while retaining their symbolic value, often became redundant in terms of production. During the early twentieth century, as the production, distribution and marketing of food became increasingly commercialized on the highveld as well, there were far fewer opportunities for reserve-based Africans with little access to capital to compete with African tenants on white-owned land, or with wealthier white farmers (Keegan 1991).

Despite all the pressure that was brought to bear on African peasants, however, small groups continued to farm productively enough to market a surplus. In parts of the highveld, those descended from the chiefs and headmen who had commercialized most successfully still owned land in their own right and occasionally rented it out to white farmers, some of them as late as the 1940s (Murray 1992; van Onselen 1996). The greatest opportunities for individual African land ownership existed in the Cape, but elsewhere, where Africans were denied mortgages, chiefs would collect contributions from followers to purchase and allocate land (Beinart and Delius 1986; Keegan

Plate 4.2  The new hut tax is introduced to African chiefs in Natal, 1906
*Source*: Local History Museum Collection, Durban

1986). Across the rural Transvaal, OFS and parts of Natal, semi-commercial African producers, mostly descended from the 'middle peasantry' of the late 1870s and 1880s, also persisted on white-owned land. These were the sharecroppers upon whom poorer white farmers on the highveld continued to rely. Many of them acquired prodigious local reputations for their productivity and their wealth, held usually in the form of cattle (Matsetela 1982; Keegan 1986; van Onselen 1996). Other African tenants continued collectively to rent out the entire estates of absentee owners, including mining companies, paying either in crops or cash (Trapido 1986).

Since they had not been confined to the reserves, these African tenants were increasingly described as 'squatters' by capitalist white farmers. Fearing African competition in an increasingly competitive market, many white farmers deployed the discourse of scientific racism to secure protection from the Reconstruction state and its successors. Milner and his officials were persuaded that, if they were to stay on white-owned land, African 'squatters' must eventually be prevented from marketing produce autonomously. Ideally, they would become proletarian wage labourers, but during the transition to full-blown capitalist agriculture, they must remain on white farms solely as labour tenants. This meant that, rather than supplying crops, draught animals and equipment in return for access to land, they must yield the farmer and 'employer' work alone (Keegan 1986).

The assault on African 'squatters' reached a critical point in 1913 when the Land Act was passed. The Act sought to define more clearly the communal reserves (amounting to 7 per cent of South Africa's land area: see Fig. 4.2) within which Africans, defined as the subjects of chiefs, could occupy and farm their own land. It was intended to drive the productive peasantry off white-owned land outside these reserves and prevent white farmers from competing with mining companies for labour within them (Marks and Rathbone 1982). Those remaining on the white farms were no longer to be sharecroppers, but employees subject to Masters and Servants laws, and Africans were barred from the purchase of land outside the demarcated reserves. The Act had its greatest impact in the OFS, where thousands of relatively prosperous African tenants were evicted by capitalizing farmers responding to urban growth. Here, many former sharecroppers were reduced to impoverishment, proletarianization and criminality (Plaatje 1916). Even if they remained sharecroppers, their bargaining position was weakened by the farmer's capacity to enforce the act at will (Keegan 1986; Murray 1992).

But on a wider scale the 1913 Land Act was less than effective. It did not apply at all in the Cape due to its implications for a non-racial constitution in which Africans were entitled to own land so that they could exercise the vote. Nevertheless, the African reserves already demarcated in the province continued to be recognized. In the Transvaal and Natal the Act simply forced many 'squatters' on to more marginal white-owned land, especially in the western Transvaal, where farmers lacking in capital still welcomed sharecroppers with their extended family labour and their livestock. Other former sharecroppers moved onto the absentee-owned estates where rent-paying tenants continued to receive protection from their powerful landlords (Beinart and Delius 1986; Keegan 1986).

Figure 4.2   Areas defined as African reserves under the 1913 and 1936 Land Acts

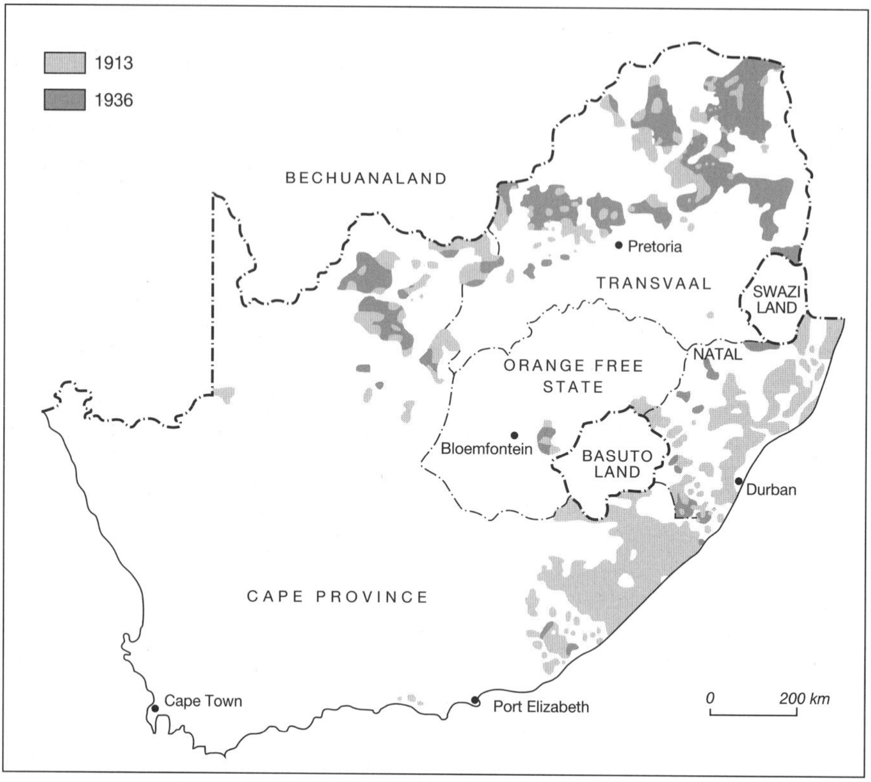

*Source*: Christopher (1994)

Despite government legislation then, the total eradication of African tenants proved impossible (Trapido 1978, 1986; Keegan 1979; van Onselen 1996). As Cooper puts it, 'the problem with capitalist agriculture was that it depended on two of the most coercive and difficult acts possible in agricultural societies: the once-and-for-all alienation of land and the daily struggle to make workers obey' (Cooper 1981: 299). In the first two decades of the twentieth century, neither white farmers nor the relatively new South African state generally had the power or the capital to achieve these transformations. The Reconstruction state's successors continued though the first half of the century to pursue a long-term strategy whereby 'the populations of the reserves [would become] captive labour for the mines while tenants [would become] trapped labour for farmers' (Beinart and Delius 1986: 13). But as we shall see in the following chapter, it was the tractor rather than state-enforced legislation which ultimately put an end to African sharecropping and brought about more modern capitalist relations in the countryside, and that occurred much later in the twentieth century (Keegan 1986).

## 'Native policy'

We have argued that the Reconstruction state wished to lay the foundations for capitalism across South Africa, but simultaneously to exclude blacks, including educated elites and African peasants, from its benefits. If it was to do so, it needed to develop and rationalize a more coherent system of African governance. Millions of African men and women, ranging from the mission-educated elite through the commercialized peasantry, urbanized workers and migrant labourer households, to the chiefs and homestead heads, would have to be administered in such a way that they were discouraged from aspiring to citizenship. But they would also have to be prevented from engaging in outright rebellion by being offered an alternative legal, social and economic framework (Ashforth 1990). This conundrum amounted to what contemporaries universally described as the 'native question' or 'problem'.

If a policy could be devised to settle the 'native question', imperial officials were quite aware that it could serve the additional and vital function of reconciling more Afrikaners to British dominion in the wake of the South African War. The prevalence of racial science allowed Milner's officials to promote the image of themselves and their former Afrikaner enemies as branches of the same white, Teutonic racial group. If they were to manage the common threat posed by the distinctly 'inferior' African population, then they would have to unite. The Reconstruction state's 'native policy' was thus intended as 'a means of persuading white South Africans to bury their internal differences' (Dubow 1997: 78). Only through reconciliation would the new South Africa be able 'to unify, modernize, and survive as a white man's country' (Cell 1982: 211). In the event, as we will see below, Afrikaner nationalist mobilization was continually to upset such plans for a major reconciliation among whites, but a general consensus was nevertheless reached on 'native policy'.

'Native policy' in the Reconstruction period was formulated by the South African Native Affairs Commission (SANAC) set up in 1903 by Milner and headed by Sir Geoffrey Lagden. The commission was comprised of a group of well-educated and, relative to their contemporaries, enlightened imperial and local officials. 'In an intelligent and even scientific manner', they were determined to modernize the country in such a way that brute domination was avoided (Cell 1982: 211). In 1905 their recommendations comprised 'the most far-reaching scheme of social engineering' that the region had ever experienced (Marks and Trapido 1979: 71). These recommendations were drawn from precedents established in each of the colonies, but from Natal in particular, and they included policies to satisfy both commercial white farmers and mining companies (Marks 1978).

Under SANAC's guidance, capitalist white farmers would 'ideally' benefit from the restriction of African land entitlement to more clearly demarcated reserves and the phasing out of sharecropping (as was subsequently attempted under the 1913 Land Act). The mining capitalists would procure migrant labour from the reserves. These would act, in the words of Howard Pim, a mining group accountant, as 'a sanatorium' where workers could be sustained by their 'communal' and family networks whilst recovering their energies (and often their health) in between periods of mine employment (Marks and Trapido 1979: 71). A co-ordinated pass system and separate housing in compounds or urban locations would reassure white townspeople that the 'raw' migrant workers were under control during their spells of employment in the cities. The towns themselves were, as the Cape government put it in a contemporaneous report, 'special places of abode for the white men, who are the governing race' (Comaroff and Comaroff 1997: 203). Through such recommendations, SANAC outlined the system of residence and labour allocation which would characterize segregation and apartheid through most of the twentieth century.

But the fundamental precondition for all the recommendations was a reconstructed version of the 'tribalism' which had helped maintain African distinctiveness during the nineteenth century (Legassick 1995). Provided that they were willing to co-operate, it was suggested, this could benefit African chiefs and homesteads heads themselves. It would thus produce a basis for consensus. A strategy of 'retribalization' seems paradoxical. For much of the preceding century, the Cape, Transvaal and OFS governments had attempted to undermine chiefly authority and 'tribal' structures, rather than resurrect them. However, this had been during a period in which chiefs were still capable of mobilizing effective armed resistance. Even though various other forms of resistance would be encountered, in the wake of the mineral revolution, the formation of an overarching British imperial state and the brutal suppression of the Bambatha revolt, it would be possible to monitor chiefs and their followers more closely. The precedents of Natal and British Kaffraria had demonstrated that this was feasible as long as Africans were confined as far as possible to reserves under the authority of approved chiefs.

Natal had also demonstrated that the laws by which chiefs governed their 'tribes' in the reserves could be codified and monitored by white officials. During the 1890s, the settler government there, liaising with African authorities, had drawn up a particular version of 'native', or 'customary law', based upon a synthesis of 'tribal' customs. As Vaughan (1994) has noted, this process of codification represented one among many instances where colonial officials confronted and responded to pre-colonial African social structures, rather than simply imposing their own will. In the process, they produced new variants of law and governance which were neither wholly European nor wholly African. However 'native law' was negotiated though, Africans' subjection to it served to preclude their participation in a common legal system. They would thus remain outside the framework of citizenship which embraced whites, and this was a particular reason why wider adoption of 'native' law was recommended by SANAC as part of the broader 'retribalization' strategy.

Just as importantly, the codification of 'native law' would serve the purpose of reinforcing gender boundaries within African homesteads. In the context of eugenics (the 'science' of selective race breeding), this was seen by white administrators as being critical to the very survival of white society. If African women freed themselves of the constraints of patriarchal authority in the reserves and migrated to the towns, as many were beginning to do, they were liable to attract white men. Sexual contact between the two groups would then be the surest way to guarantee the 'degeneration' of the white race as a whole. The magistrate of Pietermaritzburg had warned of this as early as 1880. African women in the town, he asserted, were sunk in 'unbridled licence' and 'wildest dissipation ... these women are simply emancipated from all control' (Welsh 1971: 223). African women had accordingly been made the propertyless legal minors of their husbands and fathers under Natal's 'native law', and regulations were devised (although never effectively implemented) to render them unable to leave the homestead or enter into any contract without male consent (Burman 1990; Meintjes 1990; Lambert 1995; Unterhalter 1995). Even in the absence of other economic and social discrimination, this fixing of 'native law', and its subsequent extension with the support of African men, would ensure that African women's experiences remained a world apart from those of white women (Walker 1990a).

'Retribalization' as a whole was the alternative to Africans being offered incorporation. They would be encouraged to find their social role within the structures offered by 'tradition' instead of those which whites enjoyed. The government would henceforth consider the 'authentic African voice' to be that of the male 'tribal' leader, rather than the educated or urbanized claimant of civil rights (Cell 1982: 205). It was only through such a policy that Africans could be prevented from battering at the doors of common citizenship *en masse* during the industrial era. Elite Africans were well aware of the implications. As one Natal Kholwa put it, 'By maintaining and nursing the tribal system and the power of the Chiefs, our Legislators and Rulers are maintaining and nursing the very things which check our progress, which prevent our taking that interest in the works and affairs of our land which we should take' (Welsh 1971: 173). And it was not only educated elite Africans who resented the return to a reformulated and fixed 'traditional' authority. Tension within 'traditionally' oriented homesteads also mounted as young African men, and especially women, protested at the restrictions imposed upon them by a more rigid 'native law' at a time when new opportunities were arising in the towns (Walker 1990b).

## Segregation and modernity in the 1920s and 1930s

### *The global context of modernity*

During the three decades following the First World War, the segregationist principles established by the Reconstruction government were tested by a new round of capital investment and industrial growth centred on manufacturing. The implications of this further capitalist development were systematic, affecting both the countryside and the city. First there was the potential collapse of reserve 'tribal' structures. As white agriculture received support from the state to become progressively more capitalized in an attempt to keep pace with growing urban markets, more African tenants on white farms were evicted. In the reserves they contributed to overcrowding and soil erosion, and they exacerbated younger men and women's existing challenges to chiefly and patriarchal authority. A second implication was partly the direct product of manufacturing growth, but it was also connected with the disintegration of reserve 'tribalism'. This was rapid African urbanization, accompanied by the expansion of urban slums. Finally, in both rural and urban areas, the material strains under which Africans were being placed galvanized them into unprecedented political radicalism.

The new South African state's modern discourse of segregation was, as Dubow puts it, 'elaborated', in order that established elite groups could collectively manage these challenges to an inherited order (Dubow 1987). The discourse continued to be shaped by inherited local circumstances, but it also drew on the latest 'scientific knowledge' about differences between social groups and governmental strategies generated within the global metropoles. As a result of wider imperial trends towards greater autonomy for the settler governments and Afrikaner nationalist pressure, South Africa gained political independence from Britain in 1931, but remained within the Commonwealth of former British imperial territories. The South African state now had its own flag, incorporating both British and Boer republican symbols, and equal recognition for the English and Afrikaans languages, but the country's social boundaries continued to be informed by ideas, as well as capital flows, that circulated across a global terrain. That global terrain itself was undergoing fundamental change. During the 1920s, after the exhaustion of Europe in warfare, a shift in the centre of gravity in the world economy was becoming increasingly apparent. Britain and the other European imperial centres were joined, indeed surpassed, as globally hegemonic powers by the USA. The language of segregation (and later apartheid) in twentieth-century South Africa thus has to be understood in relation to the state strategies adopted not just in the former colonial powers, but in the Western world as a whole, with the USA as its new powerhouse.

As in the global metropoles, nineteenth-century *laissez faire* liberalism was partially abandoned in South Africa during the 1920s. In the wake of the First World War there was a tendency for Western states to continue the wartime expedient of intervening more decisively and extensively in relationships between social groups – a tendency marked by Keynesian state economic interventions (Burk 1982; Constantine, Kirby and Rose 1995). Twentieth-century global 'modernity' was defined above all by such unprecedented state interventions in social life. State planners aimed, as the London Modern Town Planning School put it, to 'ensure happy and ordered development in the place of the chaos which the nineteenth century has left us' (Parnell and Mabin 1995: 54–5). Belief in the state's power to overcome the 'disorder' caused by capitalist industrial and urban growth was almost universal (Brooks and Harrison 1998: 93). The nascent segregationism espoused in South Africa by Milner and Lagden was refined through the late 1920s and 1930s within this broader, transnational context of state-planned modernity.

Under successive SAP, NP–LP Pact, NP and UP (United Party) governments during the 1920s and 1930s, segregation was made into a relatively consistent state project, involving central and local governments, the entire civil bureaucracy and a number of specially commissioned 'experts' (Cell 1982). While this segregationist project was under way, South Africa was seen by the metropolitan powers as sufficiently responsible to administer the colony of South West Africa (now Namibia), which had been confiscated from Germany after the First World War. Although such 'mandated' territories were intended by the League of Nations to be prepared in due course by the victorious wartime allies for independence, South West Africa was to continue under South Africa's sway through most of the century. The fact that the former colony was handed to South Africa was itself recognition from the global powers that South Africa's discourse of segregation certainly did not prevent it being 'one of the club' of economically powerful, 'progressive', Western-oriented nations.

### State intervention and technical solutions: the reserves

Between 1910 and the mid-1920s, most local officials of the government's Native Affairs Department (NAD) saw their role in terms of a liberal 'trusteeship' which was actually pitted against the state interference characteristic of Western 'modernity'. Based in the African reserves, officials would seek to gain personal knowledge of local African 'culture' and politics. Deploying their specialist knowledge, they endeavoured to 'hold a just balance between white and black' by acting largely on their own individual initiative (Dubow 1986; Delius 1996). South African liberals were satisfied that such trusteeship was a just and humane resolution of the 'native question'. For them, it represented an acceptable middle way between two unsavoury alternatives. On the one hand it avoided what was now seen as the naïve nineteenth-century humanitarian notion that differences between Africans and Europeans could quickly be overcome and citizenship shared. Instead, it recognized cultural, if not biological, differences between Africans and whites by endeavouring to keep them politically and territorially apart. On the other hand, it steered clear of the brute domination and exploitation that many capitalists and white workers seemed to favour, and avoided the most extreme notions of social Darwinism. Rather than holding Africans in a permanent state of servitude and 'ignorance', liberals felt that the allocation of 'protected' reserve land and the patronage of sympathetic administrators would enable Africans to continue making progress towards civilization, but at their own pace and without being subjected to outright domination (Rich 1986). The ethnologist G. P. Lestrade described the liberal approach succinctly: 'take out of the Bantu [African] past what was good ... and together with what is good of European culture for the Bantu, build up a Bantu future' (Dubow 1989: 36).

After the mid-1920s, however, trusteeship was subverted by a harsher and more impersonal form of modern state bureaucracy. The shift can be traced partly to internal struggles within the state, as senior NAD officials sought to empower their department at the expense of the rival Department of Justice, which had assumed NAD functions across much of the country (Dubow 1989; Evans 1997). But what ultimately enabled the NAD to take on a more interventionist role was two apparent 'failings' of liberal trusteeship. These were identified in particular by Hertzog's NP government, elected first as part of a pact with the LP in 1924, and subsequently in its own right in 1929. The first of these 'failings' was state impotence in the face of rural unrest, and the second was a failure to halt environmental degradation in the reserves.

African political mobilization in the rural areas continued during the late 1920s to be based on the 'traditional' issues of land and livestock (Beinart and Bundy 1987a). But these issues seemed to government officials to be generating far more threatening responses than they had earlier in the decade. Much of the popular ferment was

Plate 4.3  J. B. M. Hertzog (left) with Jan Smuts, leader of the SAP and Prime
Minister 1919–24; 1939–48 (centre) and Nicolaas Havenga, Minister of Finance
*Source*: Institute for Contemporary History, University of Orange Free State, Bloemfontein

associated with common experiences of impoverishment and confining regulations,
stemming from the progressive, state-sponsored commercialization of white farming.
Modern state support for white farmers dated, as we saw above, from the racial capitalism invoked by Milner's Reconstruction government and its immediate successor,
and particularly from the 1913 Land Act. From the 1920s, white farmers became ever
more reliant on that support to carry them through environmental crises, insulate
them from market fluctuations and lessen their dependence on African sharecroppers
and labour tenants for access to draft animals and labour. Although black sharecroppers would cling to the land and retain their own livestock on white farms in some
parts of the highveld, successive white governments continued the attempt to capitalize white farming, thus tilting the balance of power between white landowners and
African tenants in favour of the former (Keegan 1986; Bradford 1990).

When economic conditions allowed, landlords endeavoured to reduce their African
tenants' autonomy directly. They forced sharecroppers to hand over a greater proportion of their crop, yield more family labour at a time when the younger men and
women of many tenant households were deserting the farms for more lucrative urban
employment, or discard sharecropping altogether in favour of labour tenancy. In the
eastern Cape and the Natal midlands, where labour tenancy was already well established, white landowners would similarly seek to alter conditions further in their own
favour (Bundy 1977; Slater 1980; Beinart and Bundy 1987; Lambert 1995). Landlords

attempted to dictate the numbers of tenants allowed, increase the proportion of their time spent tending the owner's crops and livestock, limit the pay they received for their labour, and restrict the amount of stock or land to which they could have access (Beinart and Delius 1986; Keegan 1986). African tenants who had previously been able to insist on access to land in return for labour service then, were progressively undermined in their resistance to a more complete proletarianization by the combined forces of the state and capitalizing farmers (Morris 1976; Cooper 1981). In the eastern Cape, Natal and the OFS first, and then spreading broadly from the east to the more marginal west of the Transvaal, 'the more independent forms of tenancy were whittled away' during the first half of the twentieth century (Keegan 1991: 52).

For those unable to accept the loss of economic independence, family autonomy and dignity which such pressures entailed, the only other areas in which access to land might be sought were usually the reserves. There, some relatively fortunate families might be allocated land through the patronage of the local chief, but even then, there was generally too little on which to subsist. Between 1929 and 1939 in the Transkeian reserves for instance, the production of maize and sorghum actually fell by 25 per cent, while that of nearby white farmers increased by 40 per cent (Bundy 1977). Increasingly, while older men and women remained in the reserves, younger men migrated to the cities to work. There they earned either supplementary income to boost the resources of the extended family, or more often, the only income to which these families had access (Keegan 1991; van Onselen 1996). By 1936 the death rate in the reserves was 40 per cent higher than it had been in 1921, largely due to 'malnutrition and personal poverty' (Bundy 1988: 226). As greater numbers of Africans from different localities came together in the compounds and townships of the cities, they became more aware of the generality of such experiences of dispossession and impoverishment (Delius 1996).

The rural–urban connections forged by migrants allowed for the spread of new ideologies of resistance to the state and its local officials. Aside from selectively drawing on ideologies of socialism and Christianity, Africans engaging in migrant labour from the Transkei to the western Cape were influenced by the ideas of the West Indian-African Marcus Garvey. They asserted a pride in being black and situated Africans' local struggles within a global contest between white and black races (Bundy 1987c; Hill and Pirio 1987; Fredrickson 1995). However, it was when diverse discourses of popular opposition were formalized in coherent organizations that the government became particularly concerned. Most threatening of all were the activities of the Industrial and Commercial Workers' Union (ICU). Having been created in 1919 after a successful dockworkers' strike in Cape Town, this was a 'fluid, contradictory movement more than a disciplined, uniform organisation' (Bradford 1987: 246). It spread rapidly in Natal, the Transvaal and the OFS during the 1920s, especially among African tenants and workers on white-owned farms. By 1928 its membership consisted of about 150,000 Africans, 15,000 Coloureds and 250 whites, far outstripping that of the ANC (Wickens 1978).

The ICU, with its regionally and locally differentiated structure, was capable of fitting in with rural Africans' local struggles and articulating their aspirations for ' restoration of land, whilst simultaneously expressing a more 'modern' set of demands characteristic of urban workers (Beinart and Delius 1986; Bradford 1987; la Hausse 1989). In parts of the rural Transvaal, for instance, ICU members spread millenarian prophecies of the collapse of the segregationist order whilst simultaneously demanding better wages and written contracts, and organizing strikes and boycotts to achieve them (van Onselen 1997). Ultimately, the ICU disintegrated, due partly to the obstacles which white farmers were able to place in the way of their farmworkers' participation in meetings and strikes, and partly to the personal animosities and corruption among its leadership and the discordant range of its agenda. While it was at its peak during the mid-1920s, however, the movement jolted the South African state into more decisive policies to counter both rural and urban unrest (Beinart and Bundy 1987; Bradford 1987).

Apart from its apparent inadequacies in dealing with mounting black resistance, the second apparent 'failure' of trusteeship concerned conservation of the reserves in the face of environmental degradation. The causes of this again lay in the ongoing restructuring of commercial agriculture and the eviction of former African tenants by white farmers. As increasing numbers of these formerly productive farmers crowded into the reserves in a frequently vain search for land, 'the delicate ecological balance of pre-colonial times', when 'sparse settlement, shifting cultivation and trans-humance had been possible', progressively broke down (Keegan 1991: 53; Beinart and Coates 1995). Not only were whole extended families bringing their livestock with them from the white farms, but some who remained on those farms or lived in the cities also held cattle in the reserves, to see them through their retirement or to pass on to their descendants. Hertzog's government felt that a basic subsistence must be maintained in the reserves or the cities would be inundated with dispossessed and potentially rebellious former peasants. As the liberal historian W. M. Macmillan warned the government, 'we need labour. But our safety demands that labour come in a steady stream, not in a flood that must overwhelm us' (Dubow 1989: 67).

During the late 1920s and 1930s Hertzog's and Smuts's successive governments responded to rural unrest and environmental crisis not only by strengthening the NAD, but by passing major pieces of segregationist legislation. The drive against unrest centred on a more forceful policy of 'retribalization'. Only a return to 'traditionalism' promised to restore rural stability and deflect Africans away from the kind of radicalism manifested by the ICU (Bundy 1977: 210–11). As with so much of the segregationist strategy, Natal had already pointed the way. There, during the mid-1920s the Zulu King, Solomon, whose legitimacy was widely recognized by ᵈditionalist' Zulus, was courted by colonial officials. While Kholwa leaders like ᵔube saw in Solomon a potential ally in their struggle to mobilize Zulu opinion ᵇeir incorporationist demands, colonial officials identified the king as a pos- ᵗ point for an alternative scenario: 'retribalization' based on a distinct ᵗhnicity. The Inkatha movement, named after a symbolic coil repres- ᶜ the Zulu people, was developed with the backing of these colonial ᵛas Solomon himself, but sustaining it were both Kholwa and ᶜ latter, the conservative, 'tribalistic' Inkatha was a means of combating the influence of the ICU within their Zulu-speaking workforce (Bradford 1987; Cope 1993). From having been 'a threat to the colonial order, Zulu history and the Zulu monarchy [thus] became a crucial part of the strategy of social control' (Marks 1986: 112).

Accommodations with 'traditional' African leaders offered a means of pursuing 'retribalization' elsewhere too (Beinart 1982). Across South Africa, segregation gave chiefs more opportunities to invent specific ethnic 'traditions' for their followers, in an effort to buttress 'their position against any challenge from below' and provide a basis for negotiation with the white authorities (Bonner and Lodge 1989: 13). SANAC and then Smuts had already laid the groundwork for resurrecting chiefly authority, but Hertzog believed that the process of 'retribalization' would have to be co-ordinated more decisively by the central state if it was to preclude other forms of African political expression. The 1927 Native Administration Act accordingly marked the erosion of influence for local, liberal NAD officials, and their replacement by a more centralized bureaucracy. Trusteeship gave way to state-planned segregationism. The Governor General (effectively the South African government) was appointed 'Supreme Chief' over all Africans in the Union (except, due to a stronger tradition of local NAD opposition to the undermining of trusteeship, those in the Cape). This allowed central government to pass dictatorial edicts for the reserves. Chiefs would now have their individual powers boosted in mundane and local affairs, but they would have to enforce the government's overarching will.

'Retribalization' was further encouraged by the more coherent and extensive application of 'native law', again drawing on the precedent established in colonial Natal. Under the 1927 Act, and with the support of most chiefs, homestead heads and elders, 'traditional' customs were explicitly incorporated in the governance of Africans throughout the country. Aside from channelling political activity away from potentially radical forms, the codification of patriarchal law was part of the continuing attempt to control the movements of younger African women, now universally defined as legal minors, to the cities (Walker 1990a).

Two pieces of legislation enacted in 1936 – the Representation of Natives Act and the Native Trust and Land Act – imposed a compromise that was essential for the success of the modernized segregationist system and the 'retribalization' upon which it was founded. Africans would gain more reserved land in exchange for forsaking their aspirations to full citizenship. Most African leaders were 'trapped by the acute need for land and struggling to defend traditional resources'. When confronted by Hertzog's alternatives, they 'found themselves forced to trade their claims to political citizenship for a greater share of their territorial birthright' (Dubow 1989: 133). The 1936 Representation of Natives Act thus removed the franchise from Africans in the Cape and replaced it with the representation of three white MPs. Africans elsewhere could qualify to be represented by whites in the upper house of parliament, the Senate. The quid pro quo for the loss of the direct Cape franchise came in the form of the 1936 Native Trust and Land Act. This extended the reserves from the 7 per cent of South Africa's land area which had been set aside under the 1913 Land Act to 13 per cent (Fig. 4.2, page 144), placing the additional land under the direction of a white-administered Native Trust. By 'drawing clear lines of demarcation' between all Africans and whites in terms of territory and citizenship, the two 1936 Acts cut the ground from under the incorporationist African elite and represented the highpoint of pre-apartheid segregationism (Murray 1992: 125). Nevertheless, the fact that a decisive act of political exclusion was tempered by the need to restore some land to Africans, even if it was released only over a period of decades, also signalled that state power would remain dependent on accommodations with certain African groups.

While the Native Trust and Land Act was part of the attempt to resolve the issue of African citizenship, it also provided the crucial mechanism for halting environmental decline in the reserves and the attendant 'social evils' of rural emigration, urbanization and unrest. On the new Trust-administered reserve land, the reorganization of African agriculture, which became known as 'betterment', could be overseen more effectively. Betterment in South Africa manifested the same belief in the capacity of science to solve social and environmental problems that was becoming entrenched within the British and American states after the First World War (Dubow 1989; Evans 1997). In these global metropoles, science was being invoked to guide state intervention in the planning of new towns and the replanning of old ones. It was also guiding the rationalization of agriculture, regional economic development and the provision of housing and welfare to ensure stability among the working classes (Robson 1990; Hennock 1994; Johnson 1994). In South Africa, Africans in the reserves would be socially engineered into conserving what resources they had left through scientific husbandry, irrigation and soil conservation (Beinart 1994).

In the view of experts on 'native' agriculture, only such modern methods could overcome the inherent limitations of 'traditional practices'. Even sympathetic liberals like W. M. Macmillan believed that African agriculture was generally backward. The productivity of the late-nineteenth-century African peasants, who had gained access to land and some capital, was forgotten or ignored and contemporary problems blamed on traits which were seen to be inherent in African farming. Primarily these perceived problems were communal land tenure, the role of women in cultivation and the overstocking of cattle (Bundy 1977; Cell 1982). A 1930 Native Economic Commission blamed environmental decline on the 'grip of superstition and an anti-progressive social system' in the reserves (Delius 1996: 53). Coherent betterment planning, rationally designed to counter such obstacles, included livestock culling, rotational grazing, fencing, dipping, concentrating settlement through relocations, and introducing improved seed and stock. Although their impact remained uneven until after the Second World War, these methods were increasingly imposed by local NAD officials, in liaison with homestead heads and chiefs, in the wake of the 1936 acts (Rich 1996; Evans 1997).

While local chiefs may have been consulted in the implementation of betterment, commoners usually were not, and it was they who had most to lose from restrictions on access to common land and the culling of 'excess' stock. Little had been learned from their refusal to comply, and indeed their blowing up of dipping tanks, earlier in the century when eastern Cape officials had intervened in similar ways to combat the spread of cattle disease (Bundy 1987b). The enmeshing of chiefs within government structures now led to some fine balancing acts between their personal desire to retain power and their obligation to represent their followers (Beinart and Bundy 1987). In the northern Transvaal and elsewhere, the officials attempting to implement betterment schemes in the late 1930s encountered a widespread refusal to register livestock for dipping or culling, the cutting down of fences and protected trees, and the persistence of grazing and ploughing on supposedly fallow land (Delius 1996). As we will see below, when betterment policies were applied more systematically across the country after the Second World War, outbreaks of more direct resistance proliferated. So too did tensions between compliant African chiefs and those grateful for new Trust land on the one hand, and rebellious commoners on the other.

### State intervention and technical solutions: the towns

Enhanced control over the towns was the necessary counterpart to the modernized management of the reserves. Permanent African residence in the towns had been growing since the late nineteenth century, but the First World War stimulated an expansion of import substitution industries as access to European markets was impeded. The partial withdrawal of European capital during the war allowed local South African capitalists to acquire plant and equipment on relatively good terms and promote secondary industrialization (Legassick 1977). It also prompted the further diversification of mining capital into the manufacturing sector, creating more opportunities for non-mining and non-farm employment among Africans (Kubicek 1991; Bloch 1991). With much white labour away participating in the war, unprecedented numbers of skilled and unskilled, male and female African labourers were enticed by relatively high wages into the towns – a process which continued during a post-war boom, when employment in manufacturing rose by 73 per cent (Bloch 1991). It is estimated that the urbanized African population increased from 336,800 in 1904 to 1,146,000 by 1936 (Dubow 1989). Many of the men had previously been migrant workers and even now intended to return to the countryside, but stayed on in the urban areas either because they were no longer assured of independent access to land and livestock in the countryside, or to evade the labour obligations imposed by their elders (Bonner, Delius and Posel 1993; Bonner 1995).

African women often stayed in town to escape the patriarchal constraints of 'native law' and the agricultural burdens imposed by the relative absence of male labour in the reserves. Some came because their marriages had broken down as husbands working in the cities found new lovers and established second families, while others were enticed from the white farms by the prospect of education for themselves and their children (Delius 1989; Bozzoli with Nkotsoe 1991; Bonner 1993, 1995). The timing and pace of women's urbanization depended partly upon the rate of decline in the reserves. Bonner (1990) shows that the early and rapid movement of women from the Basutoland Protectorate to the cities in the 1910s and 1920s was linked to the increased demands put upon them by men in the declining rural economy, while Beinart (1982) and Delius (1996) suggest that women stayed longer in the more viable reserve economies of

Plate 4.4  Slumyard in Johannesburg, 1940s
*Source*: Readers Digest (1994)

Pondoland and Sekhukhuneland. African women's presence also varied by town. While Johannesburg was characterized by a disproportionately large male African population, notably migrant labour engaged on the mines, in other towns and cities such as Port Elizabeth and Durban there was a longer tradition of permanent residence by Africans of both sexes.

The pass laws, implemented by the local authorities since the late nineteenth century, were intended to regulate the residence of these Africans in the urban areas, but in many places they were ineffective. This was so for three main reasons. First, women were generally exempt from such laws. It was widely and incorrectly assumed that the constraints of 'native law' and male domination would be sufficient to restrict them to the reserves, and previous local attempts to impose passes on them, so that they could either be confined to domestic service for whites or expelled from the urban area, had failed in the face of widespread non-compliance. Such resistance could always be expected among urban women who sought both to maintain their households and care for their children on the one hand, and to earn an independent income on the other. These two sets of demands could best be met through informal and flexible work centred on the home, such as laundering and beer brewing. To succumb to the requirements of whites for cheap, live-in domestic service meant the abandonment of their own households and of the care of their own children for extended periods. Given the likelihood of protracted resistance, the central government took the view that the costs of extending pass controls over women would probably outweigh the potential benefits (Wells 1993).

Secondly, the local authorities were torn between the labour demands of different economic sectors and employers. The mining companies and farmers generally demanded that the pass laws be implemented rigidly so as to maintain a relatively cheap migrant workforce for the mines and prevent an exodus of labourers from the farms to better-paying and less onerous jobs in towns. But manufacturers and commercial enterprises often preferred a stabilized urban workforce which could be trained more effectively, hired on a long-term basis and paid at such a rate that it would constitute a viable market for their products. These competing demands meant that local authorities never used the pass laws simply to block African access to the towns. Rather, they were invoked differentially across the country in attempts to balance local sets of employers' demands (Hindson 1987). Thirdly, even had they had been inclined to respond to more coherent labour requirements, local governments had their own modernizing goals. They were often more concerned with the financially sound administration of improved local services than with the costly implementation of influx control (Parnell and Mabin 1995; Robinson 1996).

Given that an urbanized African workforce was unavoidable, the state and urban administrators in South Africa had to find the means of managing it. In the absence of 'traditional' chiefly constraints, the imperative for direct state control over this urban workforce was even greater than that for control over Africans in the reserves (Parnell 1996). Urban blacks in South Africa were also in more intimate contact with whites, raising eugenic fears that white biological superiority itself could be undermined by their uncontrolled presence. In light of the potential for inter-racial sex, the urban 'native woman problem' was usually designated in modern official discourse as a specific, and especially worrying aspect of the 'native problem' as a whole (Manicom 1992). But the further immigration of African men was also eyed warily on 'moral' grounds. Despite the fact that 'mixed marriages' between white women and black men were extremely rare, they were made a central issue in white politics. The 1938 election campaign (ultimately won by the United Party (UP)) hinged around the threat of 'mixed marriages', with the National Party claiming that legislation to outlaw them was essential, and the UP responding that its main rival had offended white women by 'the mere suggestion that they would marry black men' in the absence of such legislation (Hyslop 1995: 59). Periodic 'black peril' scares about the rape of white women by black men highlighted the sexual anxieties which an intimate 'raw African' presence raised among urban whites (Sapire 1989; Cornwell 1996).

The intermeshing of white public and central state concerns about 'miscegenation' on the one hand, and the local governments' own differentiated agendas for urban improvement on the other, produced two parallel techniques for central and local state intervention in the cities. The first was clearer separation between whites and Africans with the more effective surveillance of the Africans. The second was improvements in the standards of housing for both Africans and the poor whites with whom they had mingled in the city's slums (Robinson 1999). Like the restructuring of the reserves, this twin state response both contributed to, and was guided by, modernist notions prevailing in the global metropoles.

In Britain and the USA, as well as in South Africa, urban planning during the inter-war period was concerned above all with 'public health and public administration' (Robinson 1992: 294). The inner-city slums, where impoverished inhabitants were more likely to resort to crime and 'immorality', were seen by the bourgeoisie in each country as especially 'dangerous' places, representing a menace to public health (Robinson 1992: 294; Bonner and Lodge 1989; P. Hall 1996). Large-scale slum clearance schemes began in Britain between the World Wars, and in America during the 1950s. Those Africans evicted in South Africa's equivalent schemes, like African-Americans in the USA, were subsequently relocated in racially discrete areas. In South Africa, white evictees were provided with low-cost, state-subsidized housing, often still near the inner city, while Africans were removed to urban locations (townships) deliberately sited on the city periphery (Parnell 1991; see summary box 5, pages 217–22). South African urban policy was influenced by metropolitan innovations, but it also contributed to them. The British Colonial Office drew upon South African expertise in the effective control of a 'detribalized' African working class in its other colonies, and South Africa's townships became 'model' institutions within the empire (Parnell 1996).

Along with its segregationist outcome, South Africa's discourse of public health manifested the impulse 'towards improved urban living standards' that characterized modernization in the West (Parnell and Mabin 1995: 45). The centrepiece of urban planning during the inter-war period was the 1923 Natives (Urban Areas) Act, introduced by Smuts's SAP. Like Hertzog's 1936 acts, it contained accommodations and 'improvements' which won white liberal as well as African co-operation. The act encouraged municipalities to build new townships on the urban periphery, setting minimum standards of house construction and service provision well above those which had been applied in existing municipal locations. Sixty-four such townships were built between 1924 and 1926, with a further 170 being constructed between 1926 and 1937 (Hindson 1987). The act restricted municipal spending in the townships to the revenue that could be raised from their generally low-paid residents, but even this was an improvement where local authorities had previously collected revenue from urban Africans only to spend it in white suburbs (Marais 1957). Despite the attention that resistance to slum removals has attracted in the literature, many of the Africans removed to new sites under the 1923 act expressed themselves pleased with the improved housing that they received (Evans 1997).

The fears about 'race contact' which drove forward urban segregation in the 1920s and 1930s were closely connected with the drive to stem worker radicalism in the cities. If political militancy among Africans was threatening to accompany economic and social change in the countryside, it was even more of an immediate threat to the stability of social boundaries in the more intimate urban arenas. Immediately after the First World War, African workers launched a series of strikes in Bloemfontein, Johannesburg, Natal, Cape Town and Port Elizabeth, in response to inflationary food prices, housing shortages, high rents and the failure of post-war wages to keep pace. A mineworkers' strike in 1920 was particularly successful in generating higher wages and acted as an example to others (Bonner 1979). Membership of black trade unions soared from 10,500 in 1915 to 132,000 in 1920, in which year there were 66 strikes across the country (Lacey 1981; Simons and Simons 1983; Bradford 1987).

During the 1920s, the increasing popularity of the ICU seemed to promise the disintegration of 'traditional' restraints on African militancy in the cities as well as in the countryside (Bradford 1987). In 1921, the political route which urban African workers might follow in the modern era was spelt out with the formation of the white-led Communist Party of South Africa (CPSA), and after 1924, when the party turned towards mobilizing black workers, communism came to be regarded, in Hertzog's words, as 'the greatest fear that is over-hanging South Africa' (Dubow 1989: 73; Bonner 1982). From the late 1920s, the communist movement in South Africa found its own global metropolitan influence in the Soviet Union-dominated Comintern, which encouraged the CPSA to build a 'Native Republic' in the region. It promised an alliance between poorer white and African workers against the common enemies of capitalist employers and the state, potentially undermining the entire edifice of racially organized capitalism.

However, the communist movement's hope of a non-racial workers' alliance in South Africa appeared increasingly unrealistic after the Rand Revolt of 1922. More than any other specific episode, this demonstrated that white workers preferred to defend their relative racial privileges, rather than jeopardize those privileges by an alliance with their African counterparts. The gold mining companies had been facing a squeeze on profits due to investor withdrawal since the end of the First World War and escalating white wages, spurred on by inflation. To add to their anxiety, by 1921 the gold price was falling due to deflationary policies adopted in Britain. Since it was difficult to cut costs elsewhere, the major mining companies decided to challenge high-priced and well-organized white labour. It was planned to replace some 2,000 white miners with cheaper blacks. In response, white workers formed themselves into armed commandos and rebelled. Joined in many instances by working-class white women, the miners drew upon their recent wartime experience to fight both the police and black miners who broke the strike that they had unilaterally declared. Ultimately, the government brought in troops and even bomber aircraft to crush the revolt and restore the mines' productivity (Johnstone 1976; Simons and Simons 1983; Krikler 1996, 1998).

During the revolt (which paved the way for Hertzog and the NP's election in 1924, as part of the Pact government), white miners had shown that they were determined to maintain, rather than transcend racial boundaries within the working classes. But despite their vehement rejection of an alliance with black workers, there was still the potential for English and Afrikaans-speaking white workers to form socialist oppositional movements. This more confined form of racial socialism had also been manifest in earlier strikes centred in Johannesburg and other urban centres during 1907, 1913 and 1914. In the wake of the Rand Revolt, the new NP–LP Pact government sought to combat such workerist tendencies among whites by bringing the militant working class within the fold of the privileged white population as a whole. Not only would this preclude the threat posed by organized labour to capital and the state; it would also further protect poorer whites from the 'contaminating' influence of blacks, preserving them from 'degeneration' for their own sake as well as the state's. Bywoners and workers on the more isolated farms were never formally protected by the state, but the threat of white worker unrest in the cities had made state intervention on their behalf another fundamental feature of urban segregationist practice.

Segregated public housing schemes and the provision of basic state welfare, funded from gold revenue, were the Pact government's most direct responses, intended to rescue poor whites from poverty, political militancy and 'degeneration'. These policies had their precedents in the incipient welfare state of the British government. But in South Africa they were more selectively targeted on the grounds of race (Parnell 1988). Despite opposition to blatantly racist discrimination from liberal white MPs, who had supported white supremacy through less explicit means, Hertzog's Pact government went further than the mere provision of welfare though. From the mid-1920s it elaborated a 'civilized labour policy'. A series of legislative acts recognized 'civilized labour' as being officially differentiated from 'uncivilized labour'. The former was given preferential treatment in hiring, in the jobs that could be performed, and in pay. The government professed that 'civilized' was a non-racial designation, but in practice it came to mean white, while 'uncivilized' meant Coloured, Indian and particularly, African (Marais 1957; Lewis 1987).

The Pact government developed the means for guaranteeing that 'civilized' labour would be employed on preferential terms. It used revenue acquired through taxing the mining industry to construct South Africa's first heavy industry (as opposed to light import substitution). The Iron and Steel Corporation (ISCOR) was followed by nationalized electricity, transport and other industries. They provided the perfect arena within which the state itself could exercise direct patronage over white workers. Not only did nationalized factories, power stations, railways and harbours, as well as the police force and civil service, become havens of sheltered poor white employment (employing a third of the white workforce by 1960, after the subsequent apartheid government had reinforced this policy), they also acted as a catalyst of further private sector industrial development (Legassick 1977; Lazar 1987).

However, within the private sector the extra costs of protecting more expensive white workers were incurred by employers rather than the state. In the mines, for instance, although the gap between white and black miners' pay narrowed following the mine-owners' victory in the Rand Revolt, it remained substantial, and whites continued to be employed in positions which could have been filled more cheaply by blacks. The excess costs of protected white labour could be tolerated by the mining companies due to the savings made in importing cheap African migrants from beyond South Africa's borders, but many manufacturers would find it increasingly difficult to sustain privileged white rates of pay.

In both the private and public sectors during the 1920s and 1930s, the policy of differentiating more rigidly between white and black labour nevertheless brought those white workers in employment more effectively within the embrace of the dominant classes (Johnstone 1976; Bloch 1991; Jeeves 1991). As a consequence, in the aftermath of the Rand Revolt, 'industrial peace between white workers and their employers ... reigned in South Africa ... on a scale unmatched in any advanced capitalist society' (Legassick 1977: 187). One enduring consequence of the pay and status differential established between 'civilized' white, and 'uncivilized' black labour was the ability of most white families to employ black domestic servants. As Jochelson (1995) points out, it is this continuing ability which relieves most white women of at least some gender-defined burdens, such as cooking, cleaning, ironing and child care, and which thus reinforces the differential experiences of South Africa's white and black women.

## The 'successes' and 'failures' of segregation, *c.*1920–48

By the 1930s segregation had become entrenched because the state and the various interests which it sought to balance found it an acceptable means of exercising a degree of control over resilient African societies, while simultaneously directing them in such a way that modern economic imperatives could be met (Marks and Trapido 1987). However, if we are to understand the reasons why segregation was translated into apartheid after 1948, we need to draw attention not only to the features which made it 'successful' during the 1920s and 1930s, but also to those which made it appear increasingly inadequate to a growing proportion of the white electorate and to the NP itself, during the late 1940s.

As Cell (1982), Marks (1986, 1987) and Dubow (1989) argue, the key to segregation's 'success' during the 1920s and 1930s lay in its flexibility. First, it proved compatible with the competing demands of different sectors of the white economy. By supplying farmers with a tolerably even supply of cheap labour, mining and manufacturing companies with migrant labour and white workers with protection from African competition, it ensured a degree of white consensus. However, segregation could only achieve this at least partial fulfilment of capitalist and white social objectives by providing some basis for active consent among many blacks as well. Through the provision of more land and better urban housing, together with recognition of a degree of chiefly autonomy and patriarchal authority, segregationist policies secured the acceptance of certain powerful African men and women, as well as white liberals.

The fact that segregation never met with coherent, orchestrated African resistance (despite the ICU's potential) was also due to the struggles which continued to take place within African homesteads and former chiefdoms. Conflicts across the boundaries of class, age, gender and locality were often far more immediate and had more apparent social effects for many Africans than any mass struggle against the rather more abstract notion of white supremacy. Even where white–black relations were a major preoccupation, the tensions that arose were not always expressed in broader racial terms. While sharecroppers facing eviction, for instance, might join the ICU to improve their position relative to a particular white landlord, they did not necessarily do so as part of a conscious challenge to the general structures of racial discrimination. Furthermore, because of the autonomy which it allowed to municipal governments, and the central state's desire not to provoke costly resistance, the segregationist order never quite closed off those spaces, both legal and illegal, in which Africans could carve out a degree of autonomy (Bonner and Lodge 1989; la Hausse 1989, 1993).

Even while they continued to be excluded from legal equality with whites, Coloureds and Indians too could find solace during the 1920s and 1930s in their exemption from pass laws and urban segregation, and in their relatively privileged position within the black population. Many enfranchised Coloureds in the Western Cape voted for the stringently segregationist NP rather than the more liberal SAP or UP (Goldin 1987; Lewis 1987). Segregationism thus 'succeeded' to the extent that it did because it left room for a 'complexity of engagement' with government policies, which made any uniform black response unlikely (Cooper 1994: 1534; Beinart and Bundy 1987b). As we will see below, the implementation of apartheid from 1948 was considerably to reduce this 'complexity of engagement' for many, but still by no means all black South Africans.

Finally, segregationism could be sustained during the 1920s and 1930s because it was in tune with the modernizing discourses of the wider, Western-dominated world. As NAD officials frequently pointed out, rural segregation had its counterpart in indirect rule within tropical British colonies. As we have also seen, urban segregation was implemented in accordance with the principles of Western modernist planning. Furthermore, the segregationist system as a whole had its counterpart in the racist Jim Crow laws of the southern USA (Fredrickson 1981; Cell 1982). Accordingly, there was never any significant international pressure brought to bear upon the relatively peripheral South African state to abandon its segregationist practices. In fact, Smuts was applauded when he outlined the principles behind them at Oxford University in 1929 (Dubow 1989).

While the segregationist system had carried white South Africa through the aftermath of the First World War and into the 1940s, however, it was perceived as insufficient to perpetuate white supremacy in the wake of the Second World War. It is this insufficiency which, ultimately, lay behind the implementation of apartheid after 1948. As had happened during the 1914–18 global conflict, overseas economic competition receded during the 1939–45 war, which South Africa entered on the side of Britain. New orders were placed regardless of cost and state economic planning stimulated further industrial development. For the first time, manufacturing became a greater contributor to Gross National Product than any other sector, including mining. Between 1939 and 1944 alone, manufacturing output grew by 82 per cent (Bloch 1991). Once again, white men who were away serving in the armed forces were replaced by both white women and thousands more urbanized Africans. If it was to be 'successful' in the eyes of its white constituents, any postwar South African government would have to demonstrate its ability to manage the social repercussions of this new round of economic growth and urban development (Bonner, Delius and Posel 1993).

Perhaps the most significant of the existing segregationist order's 'failings' in the light of wartime developments was its continuing impotence in the face of African urbanization. Far from holding back this process during the war, Smuts's government encouraged it by temporarily lifting all pass controls in order to deliver labour to crucial and rapidly developing manufacturing enterprises. Between 1936 and 1947, Johannesburg's African population increased by 69 per cent, while the Cape Peninsula's rose by 147 per cent (Evans 1997). Much of this growth was accounted for by natural increase, but immigrants also found better-paid, more highly skilled and more fulfilling jobs than they could aspire to as farmworkers, migrant labourers or small town workers, and many aspired to stay permanently (Bonner 1995). By relocating to the cities, Africans apparently threatened the less skilled white workers once again with competition in the labour market, and potentially drained the countryside of the labour needed by white farmers (Hindson 1987; Jeeves and Crush 1997). Of more immediate concern to local and central governments, urbanizing Africans contributed to a growing housing shortage (although they provided a lucrative source of income for black freeholders and township residents who sublet rooms and backyards to them). By 1947, the demand for family housing in Johannesburg was twelve times the level that it had been in 1939 (Wells 1993).

Neither the NAD, accustomed more to rural than urban African administration, nor the hard-pressed municipalities, proved able to devise a systematic policy in response to this rapid urban African settlement, and divisions within the state, most notably between central and municipal government, contributed to administrative failure (Dubow 1989; Evans 1997). The South African government was not alone in finding it difficult to provide adequate housing and administrative control during the war. Urbanization had posed unprecedented dilemmas for each of the colonial African governments as well (Parnell 1996). But the failure of segregationist responses in South Africa became especially evident when, from 1944 to 1946, a self-appointed Zulu squatter leader in Johannesburg, James Mpanza, initiated the invasion by thousands of homeless Africans, mostly women, of vacant land owned by the mining companies. There the squatters erected new camps, policed and taxed by Mpanza and his followers. Similar land invasions by Africans who were employed but unable to find housing in the city followed, with men summoning their families from the reserves and farmland to join them on the new sites. Soon over 100,000 had housed themselves. Facing a general breakdown of authority, the Johannesburg municipality was forced to provide basic site and service schemes for their inhabitants. The formal township of Soweto emerged as the state's belated response to this squatter initiative, and as an indication of partial official surrender to African urbanization (Stadler 1979; van Tonder 1993; Bonner 1995).

The continued exclusion of African women from pass controls proved particularly damaging for segregationist attempts to police urbanization during the 1940s. Each municipality continued to implement, with varying degrees of rigour, its own regulations pertaining to women's local presence. While some sought to bar them access to the urban area even if they were married to men working there, others actively encouraged them to settle and provide domestic service to white families (Wells 1993). As long as the NAD remained afraid of massive and co-ordinated protest if passes were centrally imposed on women, and as long as it continued to rely largely on African men to confine women within the countryside, they and their children continued to contribute to the urban African population (Eales 1989; Bonner 1990, 1993). While the state could harass those urban African women who tried to earn an independent income whilst maintaining families through beer brewing and prostitution, it lacked the 'bureaucratic capacity and the will' to maintain control over the urban African population as a whole (Jeeves 1991: 109).

Plate 4.5  Pimville, now a part of Soweto, in the 1940s
*Source*: History Workshop Photo Archives, University of Witwatersrand, Johannesburg

Just as it had been in the 1920s, the state's problem of containing urbanization in the 1940s was inextricably tied to its problem of controlling African urban unrest. Not all of this unrest was formally orchestrated. The 1930s had seen the growth of urban African gangs, containing an amalgam of rurally inspired ethnic pride and underclass criminality, and the violence in which they engaged signalled to many whites a more general failure to police urban Africans (Beinart 1992; la Hausse 1993; Delius 1996). But the appearance of a general breakdown of urban strategy was brought into yet sharper relief by formally organized resistance. The CPSA was particularly active in mobilizing such resistance during the 1930s and 1940s, and the contacts which were maintained between permanently urbanized and migrant labourers from the same region allowed its radical ideology to be communicated back to the reserves, particularly during those moments of crisis caused by betterment policies (Delius 1993).

Finally, in 1946 an African mineworkers' strike seemed to be a symptom of the post-war radicalization of urban politics and the malaise of the segregationist order as a whole. The strike had its background in deteriorating rates of pay, dangerous and uncomfortable working conditions and above all, a cut in the supplies of food to miners living in compounds. Each of these grievances was allowed to persist, despite the recommendations of a recent official report, because the mining companies were still able to draw on impoverished labourers from beyond South Africa's borders to make up for shortfalls within the country (Jeeves 1991). The strike was inspired most immediately by the growth of the African Mineworkers' Union and its ability to deploy the Western Allies' wartime rhetoric of a universal struggle for human rights (O'Meara 1976). Workers were mobilized, however, not so much by the union itself, as through networks of friendship established by migrants from the same rural areas working in different mines. This was the best co-ordinated strike that African mineworkers had yet launched. It demonstrated that a finely balanced 'moral economy', reached through uneven compromise between managers and migrant workers over food, pay and punishment in individual mines, had broken down (Moodie 1986). The disintegration of the mines' 'moral economy' was emblematic of a more general breakdown in segregation's ability to secure compliance.

Direct force ultimately had to be deployed to get the African mineworkers back down the mines. In parallel fashion, many of the consensus mechanisms of the segregationist state as a whole were soon to be discarded in favour of more brutal policies. A more coherent system of African containment was considered necessary by many white, and particularly Afrikaner interests, in order to preserve racial boundaries in the aftermath of the Second World War. For such a system to be implemented, far more powerful, inflexible and forcible central state intervention would be essential.

### The consolidation of Afrikaner nationalism in the 1930s and 1940s

The form that this more decisive state intervention would take was ultimately determined by the struggles for hegemony between the major white political identities. During the mid-1930s, Afrikaner nationalists continued their endeavours to mobilize Afrikaners as an ethnic unit, turning their face against the broader white South Africanism espoused by the state. If they were to succeed, nationalists would have to transcend the class divisions within the Afrikaans-speaking population and render ethnic boundaries and an ethnic sense of identity primary among Afrikaans-speaking men and women. Only then could they be sure of securing the votes of the majority of the 60 per cent of the white electorate who spoke Afrikaans as a first language. The global depression of the early 1930s had resulted in the lay-off of many white workers, partially undermining the government's attempts to incorporate all Afrikaners within the boundaries of privilege. It had also led to fusion between Smuts's SAP and Hertzog's governing NP to form the United Party (UP) in 1933. As a result of this fusion, D. F. Malan led a breakaway faction of the former NP into a new, 'purified' NP in 1934. It was this party, backed by the Broederbond, which formed the core of the nationalist political programme during the remainder of the 1930s, and which instituted apartheid policies after 1948.

By 1939 some 300,000 whites, nearly all Afrikaners, were officially estimated to be in 'terrible poverty' and Malan's NP offered them the expansion of the kind of state protection to which more securely employed white workers had become accustomed (Giliomee and Schlemmer 1989). At a series of congresses held during the 1930s, the Broederbond helped Afrikaner capitalists from the Cape on the one hand, and the Transvaal petty bourgeoisie on the other, to agree on the financing and organization of a welfare programme in order to attract Afrikaner farmers and financially insecure workers irrevocably to the nationalist cause. Despite nationalists' frequent condemnation of English-speaking capitalist practice, what lay behind this critical class alliance was a desire not to overthrow the racial capitalism which had been established under British imperial patronage, but to adapt it in such a way that all Afrikaners were included among its beneficiaries (O'Meara 1983; Bloomberg 1990).

Thus, financial institutions were created which would build up specifically Afrikaner participation in commerce and industry and provide employment for poorer Afrikaans-speakers. Poor whites were assisted directly by the *Reddingsdaadbond* (Salvation Deed Bond), which arranged cheap life assurance, loans, training and an employment bureau for its members, as well as Afrikaans cultural events. This material organization 'directed a clear message to all Afrikaners – only as part of the Afrikaner *volk* (nation), only in exclusively ethnic … organisations, would their economic interests be fostered' (O'Meara 1983: 142).

Inextricable from this economic programme was a powerful and emotive discourse of national inclusion, amounting to a 'civil religion' (Moodie 1975). Within this discourse, Afrikaner material improvement was rendered into an explicitly moral project. Whereas liberal segregation had firm, if partial roots in the modern, metropolitan discourses of Britain and the USA, Afrikaner nationalist ideology drew on anti-democratic theological currents from the Netherlands for its religious foundations, and a mystical association between the nation (*volk*), God and the landscape inspired by Nazi Germany, for its political rhetoric.

The Dutch Reformed Church (DRC) theology of Abraham Kuyper (Prime Minister of the Netherlands from 1901 to 1905) placed emphasis on the nation as a special part of God's divine plan, alongside the family and the individual. Invoked by Afrikaner nationalist spokesmen during the 1930s, and particularly by DRC ministers, the theology allowed the boundaries of race, language and culture defining the Afrikaner *volk* to be defined as God's own creation, and therefore as sacrosanct. As well as constituting a powerful argument against class alliances between Afrikaners and their ethnic 'others', the idea of divinely sanctioned ethnic boundaries rendered 'miscegenation' with blacks all the more odious. But nationalist theology in the 1930s did not just sanctify the *volk* – it elevated it above all the other ethnic groups which shared its territory. The neo-Calvinism upon which the DRC was built enabled Afrikaners to be constructed as God's chosen people, elected by divine favour for ultimate redemption: a particularly appealing message to poor whites enduring relative deprivation in a modernizing capitalist society (Moodie 1975; Bloomberg 1990; Kinghorn 1997).

A further component of Afrikaner nationalist discourse concerned the volk's relationship with the South African landscape. Nationalist politicians and writers, some of whom had studied in Germany, insisted that Afrikaners had a special bond with the South African soil. A stream of novels, plays and films produced by the nationalist intelligentsia mirrored the mystical novels of Nazism, in which Aryan heroes were stabbed in the back by Jews and others who did not 'belong' within the nation. Afrikaans plays, poems, novels and films featured the poor white farmer ripped from the land which he had cultivated, and for which he and his forefathers had shed blood, by the cunning designs of the capitalist English-speaker based on the Randlords, who, in a long anti-Semitic tradition, was also often portrayed as being Jewish (Coetzee 1988; Furlong 1991; Shain 1994). It may be partly because of the conjunction of racism and anti-Semitism in Afrikaner nationalist thought that many of the whites who most actively resisted the movement during the later, apartheid period, were Jewish.

Allusions to Afrikaners suffering at the hands of English-speaking and Jewish capitalists and imperialists on the one hand and black competitors in the unskilled labour market on the other, were particularly rife during 1938. While the suffering of Boer women and children in the British concentration camps of the South African War provided a constant reference point in Afrikaner nationalist historical discourse, this year (1938) was designated as the centenary of the 'Great Trek' (the term by now used to describe the settlement of Afrikaner colonists from the eastern Cape frontier in the interior during the 1830s and 1840s: see summary box 3, pages 124–6). The centenary was celebrated with a trek of replica wagons from Cape Town to Pretoria.

The 'Great Trek' had already been made into an icon of the Afrikaner nationalist struggle against British and blacks. It had featured prominently since the late nineteenth century in the histories of South Africa produced by nationalist scholars and disseminated through various media (du Toit 1983; Thompson 1985; Hofmeyr 1991).

Plate 4.6  The centenary reconstruction of the 'Great Trek' in Cape Town, 1938
*Source*: Readers Digest (1994)

At mass meetings along the celebratory trek route in 1938 – meetings at which near-hysteria was generated – nationalist politicians adroitly drew parallels between the 'voortrekkers' (pioneers) of the 1830s and Afrikaners in the contemporary period (Moodie 1975). While the former were interpreted as having escaped British oppression in the Cape only to struggle against African savagery in the interior, the latter had apparently been driven from the security of their farms by English-speaking speculators and commercial farmers, only to face competition with blacks and exploitation by other English-speaking and Jewish capitalists in the cities (Tomaselli 1988; Grundlingh and Sapire 1989).

In particular, the centenary celebrations presented an opportunity for the petty-bourgeois nationalist leadership to mobilize support for the new NP among poor whites. Nico Diederichs, later State President, argued that it was essential to 'create unity so that the poor can identify with us and feel one with us' (Grundlingh and Sapire 1989: 23). During the replica trek, organized by largely NP-supporting nationalist groups, the NP was constructed in speech after speech as the natural rallying ground for *all* Afrikaners, regardless of their class. Those Afrikaners who risked 'degradation' by living among blacks in the inner-city slums, and by sharing casual manual labour or unemployment with them, had to be 'saved for the *volk*', and only the NP, it was suggested, was available to perform such an act of redemption. By generating mass 'cultural euphoria', the 1938 celebrations in themselves went a long way towards persuading many poorer whites that their interests lay with the nationalists' political struggle (Grundlingh and Sapire 1989: 37).

White women had been enfranchised in 1930, primarily in order to 'swamp' black voters in the Cape with an enlarged white electorate. For the NP leadership, therefore, the ideological and political mobilization of Afrikaner women was now as important as that of men (Walker 1990c). During the late 1920s and 1930s, thousands of these women from the rural areas had found low-paid employment in burgeoning secondary industry. Many of them came from the bywoner or small landowner classes being displaced by commercialization in the countryside. While the men within their families were still striving to cling to their holdings, young women who were not yet eligible to marry were free to migrate to the cities in order to earn a supplementary income (Hyslop 1995). On the Rand, they worked in the food, tobacco, clothing, printing and chemical industries (Walker 1990a). By 1935, 73 per cent of the total manufacturing workforce in Johannesburg were white women, mostly Afrikaners (Hart and Parnell 1989). Hyslop argues that a 'situation where young women were playing a growing economic role outside the home was inherently threatening to the kind of patriarchal authority which Afrikaner men had previously exercised' (Hyslop 1995: 63). The NP aimed both to assist Afrikaner men to reimpose their masculine authority over women, and to 'reclaim' these women as part of the nationalist movement. It did so largely through the notion of the *volksmoeder* (mother of the nation), again partly inspired by Nazi thought in Germany.

Reproduced in film, oral and book genres (and especially in the work of Willem Postma), the *volksmoeder* image established 'the Afrikaner woman' as the nurturer and educator of the nation as a whole. The anthem of an Afrikaner women's organization, the Suid-Afrikaanse Vroue Federaise, for instance, declared that 'There's work, my sisters, work for women devout/And free and Strong! To serve her nation, honour her God/O Lord, guide her Yourself, the woman, in service of her nation' (quoted Brink 1990: 286). The accompanying emphasis on the more intimate family promised Afrikaner women's more effective confinement to the patriarchally ordered domestic sphere as well as their service to the *volk* (Brink 1990). National Party propaganda suggested that Afrikaner women had to be returned to the control of their husbands and fathers, and implicitly to the nationalist programme, for fear of their engaging in intercourse with the black men whom they met in their places of employment (Hyslop 1995). In this way, Afrikaner nationalism, like other modern nationalisms, was inflected by gendered discourses. Its 'cultural constructions of womanhood' demanded 'women's submission' at the same time as they reaffirmed Afrikaner men's masculine pride (Stasiulis and Yuval-Davis 1995: 27).

Although it addressed Afrikaner women through the patriarchal discourses upheld by men, the NP also sought directly to gain the loyalty of Afrikaner working-class women. Some of these women had developed workers' organizations, most notably the Garment Workers' Union (GWU), to represent them in the workplace (Berger 1987). For the NP, the socialist GWU, potentially bringing together not only Afrikaans-, but also English-speaking and even black women workers, was one of the 'factors which can estrange [Afrikaner workers] and let them come loose from the ethnic group from which they stem' (*Die Vaderland*, quoted Hyslop 1995: 67). Afrikaner women were therefore encouraged to view themselves not as part of a non-racial workers' alliance against capital, but as specifically 'working for the good of the *volk*' in times of economic difficulty (Bozzoli 1983/1995; Brink 1990; Walker 1995). And as Walker notes, 'at a time of immense social flux', Afrikaner women did indeed increasingly find 'comfort and a sense of identity in both the homemaker ideal and their membership of a strongly defined ethnic group striving to seize control over its future' (Walker 1990a: 22). Despite competition from socialist and other nationalist organizations, the NP was able by the mid-1940s to harness the material plight of many Afrikaners, their anxieties over gendered roles, and the populism of their cultural responses, into a coherent and increasingly powerful political programme. From 1948 it was this programme which generated the policies of apartheid, which are examined in chapter 5.

# Chapter 5

# APARTHEID SOUTH AFRICA, 1948–94

## Apartheid and the fixing of racial boundaries

In a powerful comparative analysis with the southern states of the USA, Cell notes of segregation in South Africa during the 1920s and 1930s that it offered 'a basis for collaboration that … harder, more inflexible forms of white supremacy could never have allowed' (Cell 1982: 19). During the 1940s, the NP elaborated its policies for just such a harder, more inflexible system. It did so because its leadership, along with many other whites, constructed the prevailing segregationist system as having proved impotent against African urbanization and radicalism, and because the segregationist order had failed to ensure the Afrikaner *volk*'s wellbeing during a period of economic difficulty. Known as 'apartheid' (separateness), the NP's alternative system was designed to maintain white 'civilization' and uplift the Afrikaner *volk* in the post-war context of 'industrialization, urbanization and popular struggle' (Bonner, Delius and Posel 1993: 1). Before the critical 1948 election, which brought the NP to power, the governing UP itself was far from reticent in imposing new racially discriminatory legislation where it seemed feasible and politic. It did so particularly against Indians in Natal. There the government was persuaded during the 1940s, not least by rival local white commercial interests, to recognize restrictions on Indian landholding and commercial enterprise. But while the UP set out its own more encompassing state interventionist programme in response to segregationism's weaknesses in the wake of the war, the NP successfully politicized a yet more defensive and virulent racism as part of its alternative programme (Ashforth 1990; Brooks and Harrison 1998).

Under the NP racial discrimination would be converted 'into a biblically derived moral project' (Evans 1997: 5). Nevertheless, there was considerable disagreement over specific policies within the nationalist movement and apartheid was never founded upon a consensual 'master plan'. A particularly significant debate was waged between those 'idealists' who favoured complete racial separation, or 'total apartheid', and pragmatists who emphasized the ways in which white material privilege was tied to the use of a cheap African workforce. The former argued for the reduction of white dependence on African labour, the reversal of black urbanization, the establishment of self-sufficient reserve economies for Africans and, if an African workforce was absolutely essential, the decentralization of white-owned industry towards the reserve areas, so that it could be employed away from the towns. The latter advocated mechanisms by which necessary African labour could be absorbed into white urban and rural enterprises whilst maintaining greater state control of the process (Lazar 1993). The policies which comprised early apartheid were hammered out during the 1940s as a compromise between such contrasting positions and between the competing labour demands of capitalist employers. However, while seg-regationist officials had often been resigned to the ambivalences produced by policies which sought to satisfy conflicting interest groups, under apartheid, 'inconsistencies called forth more laws, more regulation, and ultimately more of an authoritarian state' (Evans 1997: 58).

At the same time as the varying demands of white constituencies were met, the NP insisted that apartheid would help fulfil the requirements of Africans themselves. Enhanced state intervention would allow the better provision of township housing for those who were urbanized, and thus the diminution of their grievances. It would also ensure more effective betterment in the reserves, thus maintaining a viable economy for the low-paid migrant workforce. Aside from meeting Africans' material needs however, the NP leadership addressed itself to their cultural welfare. Under apartheid, Africans would be further encouraged in the development of their 'tribal' identities, only now there would be a greater emphasis on the distinctiveness of each 'tribe', and the erection of more coherent boundaries between them. The party leadership, as we have seen, was convinced by the theological doctrine that race and ethnicity contained distinctive, and even divinely ordained essences. This biblical notion was further substantiated by the 'ethnos' theories of Afrikaner anthropologists such W. W. M. Eiselen – theories which also tended to essentialize ethnicity and bemoan Africans' 'detribalization' in modern, industrial contexts (Kuper 1988). Both ideological strands, religious and secular, combined to produce a discourse which ruled out the concept of ethnicity as a fluid social construct. The blurring of boundaries between African ethnic groups was accordingly viewed with almost as much dread as the erosion of distinctions between broader racial groups.

### Apartheid planning: citizenship, population groups and boundaries

Following its narrow electoral victory in 1948, the first move of the new NP govern-ment was to restructure the civil service, military and police by replacing English-speakers with Afrikaners. Its next major endeavour was to establish a firmer basis by which racial boundaries could be protected. This was begun with the Popu-lation Registration Act of 1950. The act sought to 'fix' well-defined racial groups far more rigorously and systematically than ever before, so that each group could be sub-jected to specific legal restraints on location, education, mobility and occupation. NP bureaucrats found, however, that the fixing of race was not as easy as it might have appeared. Since 1911, when the Cape Supreme Court had decided that Coloureds could legally be excluded from white schools, administrators and judges had encoun-tered great difficulty, and produced considerable inconsistency in deciding exactly who was Coloured or white (Fredrickson 1981). Similar problems of categorization existed at the margins of Coloured and African and Indian and Coloured identity. The entire population's categorization into the major groups of White, Coloured (subdivided into numerous groups such as Malay and Griqua), Asian/Indian and

African (the terms used varied over time, with Africans being called, for instance Native, then Bantu, then Black), thus depended on the arbitrary decisions of government-appointed bureaucrats.

Once made, however, these decisions could be crucial not only in the categorization of people, but in the shaping of their identities and their livelihoods. South African Indians, for instance, had come from, or were the descendants of people from, different regions within the subcontinent, or from the Middle East or East Africa, and they were divided by gender, class, politics, religion, location, possibly caste, and length of residence in South Africa. While it did not remove an awareness of these differences, their uniform classification and treatment as 'Indians' did generate shared experiences, and so encouraged a new construction of racial affinity to be layered over established divides (Unterhalter 1995). Apartheid's distinctions thus not only reflected, but helped to *create* more rigidly differentiated notions of 'race'. Even those who mobilized opposition towards state-imposed racial distinctions were forced to take these distinctions as their starting point.

As far as the state was concerned, simply classifying the population by race would not be enough to ensure that the racial groups remained discrete. 'Miscegenation' would render the legal boundaries between official racial categories meaningless. As the government-appointed De Villiers Commission had suggested in 1939, only government legislation could 'keep the white population intact' (Hyslop 1995: 79). The NP state thus intervened in unprecedented ways to police individual sexual activity. In accordance with long-standing NP objectives, 'mixed' marriages were banned through the first piece of apartheid legislation in 1949. While sex between Africans and whites had already been outlawed as a result of eugenic fears in an Immorality Act of 1927, the Immorality Act of 1950 banned any sexual relationship, and later even 'intimacy' between any white and any non-white.

With genetic boundaries apparently secure, the NP government was free to concentrate on the task of sealing off the remaining avenues for black citizenship. Although the measure was delayed by the party's initially tenuous grip on parliament, the Coloured franchise in the Cape was finally abolished in 1956. Coloureds' partial compensation was a policy giving them preference over Africans in the western Cape labour market. This represented a significant relative privilege, and one which contributed to continued support for the NP among many western Cape Coloureds, but the removal of the Coloured vote nevertheless meant, more explicitly than ever before, that 'full citizenship, community acceptance, and the enjoyment of the fruits of ... social progress' were now limited 'to those who were recognised as white' (Fredrickson 1981: 280).

As black economic and social advance via the franchise was being blocked, so too was that through education. Afrikaner children were taught within a nationalist milieu in separate Afrikaans-language schools, but African children's learning was structured through the Bantu Education Act of 1953 (Thompson 1985). By making education compulsory for African children for the first time, the Act manifested the wider, modernizing Western state impulse to generate universal education for the 'public good' (Fleisch 1995). It was also part of a more specific attempt to enhance the skills of already-urbanized blacks, so that manufacturing and commerce could draw upon them in preference to new African arrivals to the cities (Hyslop 1988). The urban economic sectors would thus be restrained from competing for the mines' and farms' largely migrant workforce (Bonner, Delius and Posel 1993; Hyslop 1993). Like other discriminatory legislation, the Bantu Education Act was supported by liberals on specific grounds. The Act seemed to be a way of educating African youth away from socially disruptive criminal and radical political activity, whilst equipping them with essential skills to pursue their self-improvement (Hyslop 1988).

But as the Bantu Education Act widened African access to primary education, it simultaneously allowed the African syllabus to be prescribed in such a way that apartheid itself was legitimated as the 'public good'. Syllabuses were designed to reinforce 'tribal' identities in African schools, and deter aspirations to professional or managerial occupations. What formal African education there had previously been was effectively taken out of the hands of independent mission schools and provincial departments and placed, along with many other aspects of African administration, under the authority of a restructured NAD, now called the Department of Native Affairs (DNA). By the late 1950s, as the African nationalist Z. K. Matthews explained, 'we had to give back in our examination papers the answers the white man expected' (Matthews 1981: 58–9). Finally, segregated learning was introduced to the university sector, with the legal exclusion of blacks from white universities and the building of new Indian, Coloured and African universities.

### Apartheid planning: bantustans and farmland

The NP, along with an increasing proportion of the white electorate, considered that the segregationist state's attempts to administer the reserves had been insufficiently rigorous in the face of collapsing agriculture, emigration to the cities and 'detribalization'. Fundamental to the new government's response was a considerably enlarged and strengthened bureaucracy, with the DNA assuming primary responsibility. From 1950 to 1958, when he became Prime Minister, the department was headed by Dr Hendrick Verwoerd. Under his leadership, it was restructured to such an extent that it became a 'state within a state', forming the centrepiece of the entire apartheid programme for the African population. As Evans (1997) argues, the fact that many of its initiatives had already been envisaged, even if not completely applied under preceding segregationist governments, enabled the department to gain the support of many 'ordinary' administrators across the country. These men did not necessarily consider themselves affiliated to the NP in particular, but saw themselves simply as the implementers of a 'neutral', modern, more scientific and more efficient system of governance (see summary box 6, pages 223–4).

One of the DNA's first imperatives was to pursue betterment in the reserves, or 'bantustans', as they were now known, far more effectively. One motivation for doing so was the gold mining industry's increasing concern that if the subsistence economy of the reserves collapsed, migrant workers would no longer be able to rely on the low wages paid in the mines, but would desert them for more lucrative employment in other sectors (Jeeves 1991). The government's Tomlinson Commission had followed the 'idealist' line of total apartheid, recommending a massive input of finance to develop separate and self-sufficient bantustan economies. However, Verwoerd, as Prime Minister, rejected most of its suggestions and decided that the state should 'modernize' African agriculture without addressing its fundamental limitation – the lack of capital. During the 1950s, stock culling, enclosure, forced resettlement in planned villages and taxation to pay for them, were imposed in reserves across the country without any other form of assistance being rendered to impoverished bantustan households (Lodge 1983; Murray 1992; Delius 1996).

In most areas these interventions produced evasion and protest, and in some they provoked dramatic, overt resistance of the kind that had always been feared by the segregationist state. Armed rebellions flared in the northern Transvaal, the eastern Cape and rural Natal (Lodge 1983). Migrant labourers returning home were involved in each of them. Based as they were on militant attempts to restore an idealized chiefly order, free from state interference, the revolts manifested a distinctively migrant blend of urban radicalism and African peasant aspiration. Through their experience of both urban and rural areas, their participants were acutely aware of the all-encompassing nature of the apartheid system, of which only selected aspects could be seen at either site in isolation. In the cities, migrants had often been impressed by the activities and rhetoric of local CPSA and ANC organizers, and they formed their own movements, such as 'Sebatakgomo' among Pedi workers, to organize resistance embracing nationalist ideas, witchcraft accusations and visions for the future in the bantustans (Sapire 1993; Delius 1996; Crais 1998). When a rebellion broke out in eastern Pondoland in 1960, again with the involvement of migrant workers, 'there seemed to be the potential for an anti-[state] struggle linking nationalists and peasants, town and countryside' (Beinart 1994: 158; Mbeki 1964). However, with tightened security legislation to deal with 'agitators' and the use of overwhelming force, the NP government demonstrated less regard than its predecessors for the costs of repressing black resistance. The insurrections were crushed, often with the killing of resisters, passive participants were imprisoned or fined, and 'betterment' was relentlessly pursued. The new state's determined display of force in specific bantustan locations during the 1950s and 1960s was intended to be exemplary. It created space for the more mundane imposition of an unwelcome apartheid administration across a far wider territory.

The state's aim of reinvigorating the subsistence bantustan economies through enforced betterment in the 1950s was directly contradicted by another apartheid policy – that of assisting white farmers to become more capital-intensive. During the late 1950s and 1960s, a surge of state-sponsored capital investment took place within the white farming sector. Even poorer farmers who had held out so long against the eradication of African squatting on their land were freed by newly available technology from their accustomed dependence on those tenants. As Keegan puts it, 'it was the tractor which most significantly reshaped productive relationships in the arable districts', since it released white landowners from the need to utilize African-owned and managed draught oxen (Keegan 1991: 55). The gradual process of converting semi-autonomous peasants to labour tenants or rootless refugees was thus all but completed (van Onselen 1996). In many cases, tenants were simply evicted outright, converting them in one swift move to members of a growing bantustan-based proletariat, completely dependent on waged labour.

Even where white farming was already characterized by the employment of waged workers rather than tenants, the availability of capital on credit allowed the partial substitution of these workers by machinery during all but peak periods of activity. Many former agricultural workers joined the hundreds of thousands of impoverished Africans, displaced by betterment schemes, expelled from formerly black-owned 'black spots' in white areas, evicted from white farms and forcibly ejected from the towns under the pass laws, who also had no alternative but to relocate in the bantustans (Marcus 1989). There, 'villages' comprised of tents and corrugated iron shacks were constructed, sometimes by the government authorities and sometimes on the evictees' own initiative, to house those expelled from 'white' South Africa. In all, the proportion of a rapidly growing African population contained in the poverty-stricken bantustans increased from 39 per cent in 1960 to 48 per cent in 1970 and 53 per cent (11 million people) in 1980 (Hirsch 1991). This 'surplus' population exacerbated the symptoms of overcrowding, insufficient capital and competition for limited resources which betterment was supposed to be solving (Platzky and Walker 1985; Wilson and Ramphele 1989).

Despite the generation of this 'surplus' rural population, however, many white farmers' peak seasonal demands for waged labour continued to go unfulfilled, since labour was not readily available in the places and at the times required. The DNA took responsibility, ensuring that Africans who wished to leave farms where they were needed had to have the permission of the farmer entered into their pass books. In 1957–8 alone, the department also handed over some 200,000 convicted pass offenders from the cities to white farmers, so that they could serve their terms as convict labourers (Posel 1991). Officials in town and countryside liaised in order to produce this controllable and directable farm labour supply. Urban bureaucrats organized pass raids and convictions in townships while their rural colleagues channelled the ensuing labour to those farmers across the country who demanded it most vociferously. In taking on this labour allocation role, the DNA (which doubled in size between 1948 and 1960) created a state system that mirrored and reversed the rural–urban networks forged by bantustan migrants. Almost incidentally as far as many local officials were concerned, the system also allowed the extensive abuse of African farmworkers, who were trapped with white employers whether they liked it or not. The district of Bethal in the eastern Transvaal became notorious for the brutal exploitation of labourers by commercial farmers. Here in particular, established patterns of squalid accommodation, long hours, gruelling work, poor diet and frequent beatings were continued, indeed assisted by officials, creating an inordinately high death rate among the workforce (Marcus 1989; Murray 1997).

With all its brutality, apartheid still left space for African involvement within local rural administration. The Bantu Authorities Act of 1951, although it was hesitantly and only gradually implemented in the face of administrative opposition, marked the new government's attempt to pursue 'retribalization' more effectively. It heralded a governmental system in the bantustans based on co-opted chiefs and 'tribal' authorities. As the Act came progressively into play across the country, chiefs confronted the dilemma of either opposing the new authorities and gaining popular support, or participating in them at the expense of local popularity. Many chose the latter option. (By the 1960s this choice was made easier by government intervention to remove non-compliant chiefs and replace them with those selected by the authorities.)

Being a bantustan official within the new structures and helping to implement betterment was at least one way in which African personal advancement could still be secured. In particular, it gave access to 'development' funding from Pretoria, a portion of which could be siphoned off, with the connivance of the white authorities, for personal use (Murray 1992; Delius 1996). Many educated commoners, often descended from the petty-bourgeois landowning classes of the late nineteenth century, realizing that attempts to gain economic and social advancement through common citizenship were now absolutely futile, also channelled their energies within the new structures, becoming salaried bantustan bureaucrats. While overt 'collaboration' and corruption were never the only ways that Africans could advance themselves, corruption in the bantustan administrations quickly became endemic. Indeed, 'apartheid could not have worked nearly as well as it did' without it (Bonner, Delius and Posel 1993: 18). One indication of its prevalence can be seen in the Transkei, where 'tribal levies' imposed by chiefs with the support of the state soared from £438,000 in 1958 to £1,014,000 the following year, with no apparent alteration in services to the ratepayers (Evans 1997).

The presence of a rural African administrative elite, ambivalently positioned between frequently hostile commoners and the apartheid state, and dedicated to containing local militancy, allowed Verwoerd to begin a programme of granting the bantustans 'independence', as we shall see below, when domestic and international conditions made it seem advisable. It also facilitated the state's aim of ethnic exclusivity, since local functionaries could ensure that access to land, jobs, homes and pensions (extended to the bantustans in the 1960s in order to hold back urbanization) was available only to those who 'belonged' within the local 'tribal' community (Murray 1992; Delius 1996). Those who were considered not to belong, because of their 'tribe' or their political stance, could find themselves excluded from relative privilege and threatened by ethnic chauvinism manipulated by chiefs and bantustan bureaucrats.

## Apartheid planning: towns and cities

We have suggested that apartheid was designed to maintain the same balance between urban and rural planning in an industrializing society as had informed segregationist policies, and to address similar but intensified contradictions, only more effectively. The foremost of these contradictions involved the competing labour demands of each white economic sector, and the general white preference for Africans to be barred from the cities. On the one hand, if African worker access to the cities were controlled more rigidly, manufacturing, commercial and domestic employers would face labour shortages and higher wage costs, while the reserves would suffer further overcrowding and economic deterioration. On the other hand, if African urbanization were allowed to proceed unhindered, the state's ability to provide housing and its capacity for urban control might well be overwhelmed, with damaging repercussions for its electoral support (Posel 1991; Bonner, Delius and Posel 1993). As we have indicated, the DNA was enlarged in order to deal with these contradictions by segmenting the African labour force. Urban employers would continue to have access to migrant as well as some urbanized African labour, while 'surplus' urban Africans as well as those barred from the cities would be channelled to white farmers and bantustans. All of this would be achieved through urban and rural labour bureaux, with the DNA designating and managing separate pools of workers for each major group of employers (Hindson 1987). This sophisticated balancing act brought an unprecedented level of state interference in Africans' everyday lives (see summary box 5, pages 217–22).

Some manufacturers continued to be satisfied with migratory labour. But in order to meet other manufacturers' and commercial enterprises' increasing demands for trained, semi-skilled machine operatives, the NP continued to recognize the Natives (Urban Areas) Act, implemented by Smuts's UP government in 1945. Section 10 of this Act permitted certain Africans in long-term urban employment (who themselves became known as Section 10s) to stay in the cities. The apartheid bureaucracy quietly allowed some unemployed Africans to remain as well, acting as a reserve pool of labour to be drawn upon in times of manufacturing expansion. Additionally, the wives and children of Section 10 men were permitted to stay, further enlarging the legal urban population (Posel 1991). The children of Section 10s would gain the skills of numeracy and literacy, required for their own eventual employment in industry, through the Bantu Education system (Hyslop 1988). However, in order to minimize further urban African growth, employers were to draw as far as possible on migrant labour rather than encourage additions to this permitted urban workforce (Hindson 1987; Posel 1991). The greatest indication of the state's willingness to intervene more decisively in order to maintain this segmented labour force came with the progressive extension of passes to women. If African labour in general was to be directed more effectively under apartheid than it had been under segregation, then greater control over African women would be vital.

The idea of central, rather than municipal pass controls for women was first proposed in 1952 and implemented from 1956, but many women refused to accept the passes until they became compulsory in 1963. Some African women would have no difficulty in qualifying under the new pass regulations to remain in town, due to their regular employment in factories, schools, hospitals and white homes, and they often remained passive in the face of the state's initiative. But for others, the extension of pass controls prompted their first overt political resistance. Even 'respectable' African middle-class women, who generally followed the Victorian precept that women should leave public matters to their husbands, were deeply offended by the assault to their dignity and to their person, which police pass raids entailed. In extending the pass laws over women and attempting to direct their labour outside of the home, the state was not only challenging the sanctity of the family, cherished equally in white and African households; it was also breaking one of the tacit accommodations which had helped to sustain the preceding segregationist order (Bozzoli 1983/1995; Walker 1995). With their urban presence rendered potentially illegal, many more African women would be vulnerable to exploitation as isolated domestic workers in white households, or to the control of an African man with Section 10 rights who was able to offer legal accommodation (Cock 1980; Unterhalter 1995). Passes for women were therefore seen by many Africans as 'not only unjust, but immoral, fundamentally wrong and therefore intolerable' (Wells 1993: 128). As Wells notes, the women who organized resistance to pass laws, both previously and in the 1950s, drew eclectically on available notions of resistance to fuel their struggle. 'Whether it was the British suffragettes or Gandhi in 1913, the Communist Party concept of a workers' revolution in 1930 or African liberation and nationalism in the 1950s, these ideas were absorbed, assimilated and then applied to their own life situation' (Wells 1993: 130).

For the new system to be installed in the face of such opposition, the state had to wield systematic coercion. Even before the new pass books became compulsory, those African women without them could be excluded from pensions and transport and their children denied access to education. When protests in Johannesburg were organized by the non-racial Federation of South African Women, backed by the ANC in 1958, mass arrests and the banning of individual leaders followed. However, Wells (1993) notes that it was not state repression alone which prevented women from carrying on with their resistance. African men, including migrant labourers returning home from the towns, initially supported the women in their resistance because they felt that the state was using passes to interfere with their own authority over wives, daughters and sisters. But African men ultimately ended the campaign too, by choosing to pay the arrested women's bail rather than allowing them to continue with their protest. 'By demanding that women be [removed from jail and] restored immediately to their traditional home-making roles, African men dismantled the strongest protest to date against state interference with those very roles' (Wells 1993: 123). The boundaries of gender within township communities had, in this case, proved more durable than combined male and female resistance to the racial pass laws and by the late 1950s, increasing numbers of African women were being forced into domestic service for white families in an attempt to gain the vital pass enabling them to stay within the urban areas.

The preceding segregationist system was perceived to have been weak by NP leaders partly because, in failing to extend passes to women, it had failed to attain the broader goal of influx control. But a further 'weakness' of the previous system was its inadequate response to the threats posed by Africans, especially those living in racially mixed areas, when they were in the cities. This too was addressed by the NP during the 1950s. African political radicalism, in which the South African Communist Party (SACP – successor to the CPSA) was highly significant, was countered by unprecedentedly repressive security legislation including the Suppression of Communism Act of 1950 (Delius 1993; Lambert 1993; Sapire 1993). Together with other laws facilitated by the NP's tighter grip on parliament in the late 1950s, this allowed arrests and imprisonment for a much wider variety of oppositional activities than had ever been contemplated by the segregationist state.

In tandem with the judicial assault on African radicalism came a more determined attempt at control through spatial separation and containment. As had been the case during the 1920s and 1930s, this attempt was enframed by a discourse of urban planning which linked the South African periphery with the contemporary global metropoles (Parnell and Mabin 1995; Robinson 1996; Western 1996). On the one hand, South Africa's urban planners continued to be influenced by their counterparts in Britain and the USA. On the other hand, some municipal planners in the USA borrowed South Africa's more explicitly racist planning techniques to accomplish the more effective informal segregation of African-Americans, while British colonial planners elsewhere in Africa continued to see apartheid measures for dealing with 'detribalized' urban African workers as being particularly sophisticated (Parnell 1996; Robinson 1999). Modern ideas about how to intervene and 'improve' cities then, traversed the globe in both directions, encompassing even the exceptional apartheid system.

Continuing the efforts of the previous government, the NP cleared the African-dominated freehold areas in western Johannesburg, and most infamously, Sophiatown, under a Native Resettlement Bill of 1954, which allowed the central government to override the Johannesburg municipality's authority. (The municipality had proved reluctant to carry out mass removals, professedly on the grounds of more liberal political principles, although the cost of providing replacement housing weighed heavily.) Despite the brutality of the move, in many cases tenants crowded into rented rooms in areas like Sophiatown were able to acquire better-quality accommodation in the rapidly expanding peripheral townships. But the clearance was resented by landlords, by unemployed Africans, and by those engaged in illicit activities. The latter groups in particular had sought to evade the closer police scrutiny which was possible in the townships. With removals and relocation to these new sites, they were more likely to be arrested for pass and other offences, fined, imprisoned, sent to farms as convict labour and ultimately, expelled from the towns. Despite the well-publicized protests of African journalists who lived in Sophiatown, some 40,000 people were moved to the new townships of Soweto in 1957 (Hart and Pirie 1984; Evans 1997). Their accommodation now comprised either self-built shacks in proliferating site-and-service schemes or row upon row of standard township homes, built with female domesticity and nuclear families in mind (Robinson 1996).

Formerly mixed-race areas like Sophiatown (now to be occupied by whites only) became subject, like all other urban areas, to the Group Areas Act of 1950. This was described by Prime Minister Malan as 'the essence of the apartheid policy' (Giliomee and Schlemmer 1989: 87; see summary box 7, pages 225–6). It sought, once and for all, to bring order to the 'racial and physical "chaos"' of the modern industrial city by demarcating those zones in which each group, as defined under the Population Registration Act, was legally permitted to live (Robinson 1996: 1; Western 1996). The local implementation of the Act was decided through contestation between a range of local interest groups, both black and white, the central Group Areas Board and the courts (Pirie 1984; Western 1996). Nevertheless, the general outcome was that African townships were planned in order that they could be policed more effectively and separated from white areas by coherent buffer strips. By confining urban Africans to townships and channelling their daily movement along prescribed routes into the shared space of the city centre, the Group Areas Act became one of the key regulatory devices of apartheid (Robinson 1996; Western 1996). The 'spatial configuration' which it developed was seen to be 'just as effective as – and considerably less costly than – the physical presence of the police in regulating the behaviour of Africans' (Evans 1997: 8). However, the brunt of the Act was felt predominantly not by urban Africans, who, after the 1923 Natives (Urban Areas) Act, were mostly already segregated, but by Coloureds and Indians.

In every town and city millions of people were affected by the Group Areas Act's progressive imposition. Although it was enforced at different times in different areas, the Act ultimately brought the seizure of the property belonging to people declared to be in the wrong area, their forced removal to segregated townships and the frequent demolition of their houses. Though by no means unique, the most notorious cases of the Act's application (in conjunction with other legislative devices) occurred in Cape Town and Durban. In Cape Town, the 'mixed-race' inner-city District Six was cleared of its majority Coloured inhabitants, its lively, if often romanticized, community spirit destroyed, and its inhabitants dispersed into soulless townships across the Cape Flats (Cook 1991; Kruger 1992; Lester 1996). This intervention was the first state attempt to place constraints on the property rights of Coloureds. It also allowed white speculators to make huge profits through the redevelopment of the districts newly zoned for white ownership (Western 1996). In Durban, whose municipality had acted as a model for the Group Areas Act as a whole, the materially progressive, and therefore 'menacing' Indian community was similarly dispersed from valuable retailing and residential sites in the city centre to more distant 'group areas'. The NP's intervention here rounded off the state assault on Indian commerce which had begun under Smuts in the late 1940s (Davies 1991; Maharaj 1992, 1995). The redistribution of Indian-owned land to whites allowed the value of white investments in the city to climb from £6 million to £120 million within a decade (Kuper, Watts and Davies 1958).

In every South African town and city, gradually more effective and more widespread racial zoning provided the basis during the 1950s, 1960s and 1970s for more comprehensive segregation in schooling, health care and public services. 'Petty' apartheid, progressively implemented under the Reservation of Separate Amenities Act, 1953, even entailed separate post office counters, railway waiting rooms, beaches and park benches. In all, this reorganization and segmentation of space according to race assisted, perhaps more than any other specific policy, in ensuring the 'balkanization of civil society' in South Africa (Evans 1997: 297; Lester 1996).

With African labour necessarily tolerated in the cities, but confined as far as possible to controlled townships or migrant hostels, the NP sought to protect its white worker constituency from further competition by restricting the jobs which could be performed by black labour. When the economic growth of the early 1950s was replaced by a slowdown towards the end of the decade, a centralized job reservation scheme was outlined under amendments to the Industrial Conciliation Acts of 1956 and 1959. The Minister of Labour was empowered to reserve any job in any area for a particular race. In this pandering to white labour though, the government risked alienating major employers. They were often reluctant to incur the higher wages of white workers when cheaper black, and especially African labour could perform the same tasks. In practice then, a compromise 'floating' job colour bar was widely introduced. White workers were promoted to perform higher level tasks while their former jobs were performed more cheaply by newly trained black workers. Together with the state's direct provision of employment within the expanding nationalized industrial sector, such practices had the desired effect of improving poorer Afrikaners' job prospects and material standing relative to those of black competitors in the labour market (Lewis 1984; Crankshaw 1996).

### Black identities and fragmented opposition

On the one hand, the exposure of many Africans to common experiences of deprivation and humiliation under apartheid could generate a more universal sense of black identity, one on which the ANC hoped to build its opposition. But on the other hand, the differential position of black people according to gender, legally defined racial group, class, location and status within the administrative structure often led to fragmented experiences and political orientations.

The apartheid authorities unintentionally did more than any other agency to broaden the base of overt African opposition. They did so precisely by generating a more interventionist, more inflexible, and modernized version of segregation. The spatial confinement of Africans from different areas and classes in the townships, for example, generated shared experiences, often for the first time. Previously distinct African groups were mutually alienated by their treatment as mere units of labour, being 'moved about and incarcerated virtually at will' (Evans 1997: 87). Given the reluctance of the state and of employers to pay for the full costs of township housing, and for workers' transportation from township to workplace, they collectively faced higher rents and travel costs. In times of economic recession and inflation, this would continue to provide a unifying grievance (Lodge 1983; Maylam 1990).

A similar function was served by Bantu Education among urban African youth. Due to a minimal state subsidy, this too was tied to the finance which Africans themselves were able to raise through levies. Regardless of their ethnicity or class, African pupils and parents resented not only a prescribed syllabus, but inferior and overcrowded educational facilities and a lack of books and qualified teachers. The boon which Bantu Education had conferred by extending access to education was thus quickly overshadowed by the additional grievances which it generated (Hyslop 1988; 1993). Finally, many sections of the urban African population were alienated by the police's diligent attention to enforcing pass laws and clamping down on illegal beer brewing, rather than combating the high rates of violent crime which made urban African life so insecure.

By 'fixing' race as the primary social boundary in the cities, and exposing all urban Africans to similar experiences then, apartheid actually helped to create a broader African racial identity founded on opposition to the state (Parnell and Mabin 1995). As a result, more than ever before, from the late 1950s, a mutual sense of Africanness cut across the divisions of class, ethnicity and political affiliation (Bonner 1995; Western 1996). For instance, although the workerist SACP remained anathema to many 'respectable' ANC-supporting middle-class Africans, a shared support-base was being forged between the two movements, both of which continued to include or work with white radicals. The nationalist ANC still clung to the liberal ideal of incorporation rather than revolution, while SACP activists found common ground in pursuing such incorporation as the first step towards a more thoroughgoing socialist transformation. Having previously precluded the potential for an alliance between white and black workers by separating their residential areas, the state now found itself confronted by an alliance of township-based urban blacks from a variety of classes. In this sense, the apartheid system generated its own, novel contradictions.

It was not only among Africans that greater potential for alliance was emerging in the 1950s. The common exclusion of Coloureds and Indians from white space and from occupational opportunity was encouraging broader alliances among elite black groups. High-profile, inclusive, ANC-led campaigns were supported by the largely middle-class Indian and Coloured Congress movements (the inheritors of the early-twentieth-century APO and Natal and Transvaal Indian Congress traditions). These movements were known collectively as 'Congress'. Directed against the eradication of the Coloured vote, Group Areas removals, the pass laws, Bantu Education, rural betterment and various other specific or common grievances, Congress campaigns articulated dispersed groups engaged in struggles at distant sites during the 1950s. Aside from numerous strikes and boycotts, the best publicized instances of resistance included the 1952 Defiance Campaign, in which protesters offered themselves for arrest by contravening certain apartheid laws, and the creation of a Freedom Charter in 1955. The latter was a socialistic document produced as a result of extensive consultation across the country and used as the manifesto of the combined African, Indian, Coloured and white radical Congress movement. It included the promise to nationalize the 'commanding heights of the economy', including the mines and banks in order to effect a major redistribution of wealth from whites to blacks, and it resulted ultimately in the long-running Treason Trial of the late 1950s, in which the bulk of the oppositional leadership was tried but acquitted by a then still largely independent judiciary (Lodge 1983).

Accompanying such prominent displays of opposition were the more mundane activities of trade union members, often acting in alliance with members of the SACP. From 1955 the South African Congress of Trade Unions (SACTU) attempted to co-ordinate these activities and connect them with those of the broader Congress movement (Luckhart and Wall 1980). Together, African nationalist, non-racial Congress and workerist campaigns created a nationwide (if somewhat diverse and patchy) network of overt resistance during the 1950s, designed to counter the state's own modernizing apartheid programme.

However, the agenda of the opposition movements was still shaped overwhelmingly by the initiatives of the white state, and the sense of a united black front against apartheid was by no means shared by all blacks. Counterbalancing the emergence of cross-class and cross-race black alliances was the continuing fragmentation of

Plate 5.1  Africans boarding a 'whites only' train carriage as part of the defiance campaign, 1952
*Source*: Drum

experience along these and other lines of social division. In one sense, these lines of social division were exacerbated by the same process of urbanization which had given rise to the state's more draconian interventions. In the cities, freed from 'tribal' authority, for instance, young men converted the age-cohorts of the countryside into urban gangs and often deployed violence against other gangs, older African men, migrants and women in order to demonstrate their masculinity in new guises. The dominant sense of identity generated by these younger, township men was often far removed from the restraint, respect for elders and 'wisdom' expected of men in the countryside (Morrell 1998). During the 1950s, different African urban groups, such as those in full-time employment and housed in municipal townships, rurally oriented migrants in their hostels, urban gang members, and women beer brewers, continued to have 'their own criteria for evaluating the relevance of national organisations and campaigns to their lives and struggles' (Bonner, Delius and Posel 1993: 18). The fact that many Section 10s benefited from the provision of new township houses, and from the security of permission to stay in the city, for instance, made it difficult for the ANC to co-ordinate united opposition to Group Areas removals or to the pass laws (Bonner 1995; Evans 1997). Even at the height of the 1950s Defiance campaigns then, one could not speak of a united African, let alone black, culture of opposition.

## Reformulating apartheid: globalization, decolonization and the apartheid boom

During the 1960s, the ongoing globalization of the Western-dominated world economy allowed for the increasing penetration of South African industry and finance by international, and especially US and British capital. While this investment fuelled an economic boom, it also rendered the South African economy more dependent on external links. The adverse effect of these links, as far as the South African state was concerned, was increased exposure to a new metropolitan-centred discourse of universal humanity – one entailing a non-racial philosophy akin to that upheld by the ANC since its early missionary-inspired years. It was clear that such a discourse had the potential to upset South Africa's economic performance, since it could result in Western boycotts and sanctions.

However, this adverse effect of global dependency was held at bay during the 1960s boom, because of South Africa's value to Western-based transnational capitalist concerns and the South African state's strategic importance to the Western powers in their Cold War struggle against Soviet communism. Thanks to these economic and political ties, the challenge to apartheid posed by a new discourse of universal human rights was, for the time being, muted (Manzo 1992; Beinart 1994). In this section, the contradictory repercussions of South Africa's location within a global network of power and finance during the 1960s are explored.

### Post-Second World War capitalist practice and global discourse

By the early 1960s the American civil rights movement was persuading the US federal state, with its long-standing adherence to a universal declaration of human rights, to intervene more decisively against the explicitly racist southern states. In Western Europe, a discourse of universal humanity had been revived and utilized by the Allies in their struggle against Fascism and Nazism during the Second World War. The notion of equal human rights was purveyed in particular in an attempt to garner black allies from Britain's colonies – a necessary exercise given the 'total' and global nature of the war. With the full horrors of Nazi genocide becoming well known in the wake of the Allied victory, the kind of differentiation between human groups which had characterized scientific racism earlier in the century was being widely and fervently rejected. New understandings within genetic science were similarly undermining eugenic notions of racial or ethnic difference (Dubow 1995).

In the wake of the Second World War, while European states concentrated on reconstruction, the USA, backed by Britain, set about constructing a regulatory system for the global economy (Manzo 1992). This Bretton Woods system allowed for global, but especially Western capital growth until the early 1970s. After 1971, with the USA again leading the way, the system was abandoned in favour of one promoting a more coherent 'liberalization of world trade and globalization of the world economy' (O'Meara 1996: 475). Throughout the post-war period, multinational firms were becoming increasingly capable of investing at various sites around the globe, and in many cases their economic muscle gave them as much localized power as governments themselves. Such trends helped undermine the economic autonomy, and enhance the interdependence of all states within the world economy.

These economic trends also facilitated decolonization. In most African colonies after the Second World War, Western European governments were faced with the rise of urban-based nationalist as well as workerist resistance movements – movements themselves deploying the Western discourse of universal rights. While nationalist organizations led by indigenous elites, labour unions and rural protest movements were generally not powerful enough to overthrow colonial rule, they could certainly disrupt economic growth and stability. They could also discredit colonial administrations by pointing to the inconsistencies between a rhetoric of universal rights and the colonial reality of indigenous exclusion from citizenship. By the 1950s the British and French imperial governments 'were adding up the costs and benefits of colonial rule more carefully than ever before and coming up with negative numbers' (Cooper 1994: 1537). Furthermore, independence could now be conferred in the knowledge that former colonies would continue to be enmeshed in the profitable global network of production and consumption whose hub lay within the Western metropoles. Even if they lacked the same access to metropolitan capital and to an historically constituted political system, their newly independent governments would, it was hoped, pursue the same capitalist agendas, and be subordinate to the same metropolitan-based economic interests, as had the former colonial administrations (Cooper 1996).

Despite their formal political independence in 1931, white South Africans and the state which represented them in the global arena continued to be influenced by metropolitan-initiated cultural and economic trends during the post-war era (Manzo 1992; Beinart 1994). The South African state's diplomatic relations, its economic ties with global financial institutions dominated by the USA, such as the International Monetary Fund (IMF) and the World Bank, and its attractiveness to multinational investors, were all potentially influenced by the correspondence between its own policy base and those formulated in the metropolitan centres of the world economy (Grundy 1991; Price 1991). Generally, whatever the rhetoric that it directed towards internal constituencies, the South African state sought to maintain favourable connections with the global powerhouses, particularly Britain and the dominant USA (Barber and Barratt 1990).

An incident at Sharpeville, a township outside the Transvaal town of Vereeniging, in March 1960, however, highlighted the disjuncture between apartheid's inflexible discourse of racial difference and shifting metropolitan notions. The Pan Africanist Congress (PAC) had broken away from the ANC in 1958, alienated by its continuing accommodations with white liberals and communists, and its alliances with the Coloured and Indian Congress movements. The new African-centred movement marked its birth by organizing a nationwide anti-pass campaign. When police opened fire on the demonstrators who had been mobilized at Sharpeville, killing 69 of them, they unwittingly brought an international media spotlight to bear on the oppressive system of racial exclusion which the NP had been constructing since 1948 (Lodge 1983; Chaskalson 1991). Sharpeville prompted a reformulation of the apartheid programme which had already been envisaged by Verwoerd – one which would be defensively continued, especially in the light of broader black opposition, until the system's demise (Lester 1996).

The government's immediate response to the widespread African protests occasioned by the Sharpeville shootings was to ban the ANC, PAC and other organizations, thus prompting their first turn towards armed resistance and their continued existence only in exile. The ANC, with its longer-established and more coherent organizational structures, went on to build diplomatic relations overseas more successfully than the PAC, attracting particular support from the Soviet Union and the communist eastern bloc (Lodge 1983). The clampdown on opposition in South Africa, however, prompted concerns among multinational companies about future stability and the security of their investments. It also raised more acute anxieties among other governments about South Africa's violation of the discourse of 'human rights'. Two consequences followed: the flight of capital, leading to the plummeting of the South African currency on international money markets, and widespread political condemnation.

In response to economic isolation, the South African government subsidized import substitution industries such as car manufacture, military equipment and computer technology, in order to maintain domestic capital growth (Legassick 1974). In a defensive reaction against political attacks, it left the Commonwealth of former British colonies and became a republic in 1961. In the light of this post-Sharpeville experience, however, the NP leadership saw as imperative some further reformulation of apartheid policy, in order to bring it in line with the post-war global discourse of human rights (Lester 1996). Some such reformulation was also necessary in order to provide a more effective means of dealing with the increasingly potent urban resistance campaigns, and with the internal bantustan revolts of the late 1950s and early 1960s (Lazar 1993).

The government found itself with sufficient breathing space during the remainder of 1960s to effect this reformulation in its own way and at its own pace. While its insistence on preserving explicit racial boundaries made its relationship with the metropolitan governments consistently problematic, the South African state did derive one major benefit from the new global order. This was the prevailing division between two metropolitan blocs divided around the post-war 'superpowers' – the capitalist West dominated by the USA and the communist East dominated by the Soviet Union. Upon their independence, most former colonies in Africa and elsewhere were immediately locked into the Cold War being waged between these blocs. Whether they 'modernized' along Western capitalist or Soviet or Chinese communist lines, they were affected by the 'competing prescriptive formulas' generated by metropolitan struggles (Alden 1998). The West's post-war formula was that of capitalist 'development', and it was elaborated precisely in opposition to the Soviet prescription of socialist revolution. Disseminated by Western economists and the officials of the US-backed global institutions, 'development' became a new discourse guiding 'progress' in the 'Third World' – one similar in many respects to the discourse of 'civilization' which had been designed to encourage 'non-Europeans' in their emulation of Europeans during the early nineteenth century (Crush 1995; Escobar 1995).

As some newly independent African governments chose socialist formulas instead, and were brought within a Soviet or Chinese sphere of influence, the South African government was able to generate solid Western support as the major proponent of capitalist enterprise on the African continent (Manzo 1992). It was also able to supply minerals essential to the West's nuclear arms race with the Soviet Union, especially through its occupation of uranium-rich South West Africa, and, later, to provide satellite tracking station sites for the USA. As long as tacit Western political and economic support was maintained, apartheid governments were given respite from economic sanctions and allowed breathing space to bring their political systems into an apparently closer alignment with non-racial Western norms.

At the same time, a global capitalist boom provided the economic resources for a state-led reorganization. The crisis of investor confidence prompted by Sharpeville proved but a temporary setback to capital accumulation. The post-Sharpeville repression of internal opposition restored the stability which international investors required, and thus contributed to enhanced state revenues. Not only was further overseas investment attracted by the prospect of stable economic growth (the rate of investment doubling between 1963 and 1972, and amounting to an inflow of US$ 3.7 million between 1965 and 1974); so too were white immigrants, especially from Britain (Houghton 1967). The decade became apartheid South Africa's economic 'golden age'. Through continuing, state-fostered import substitution, the textiles, heavy equipment, chemical and consumer durables industries were growing, both in terms of employment and output. Export markets were expanding and new investment was coming in from Japan, Taiwan, Israel, Germany and the USA as well as the old metropolitan power, Britain (Beinart 1994). White incomes increased by 43 per cent. With black incomes rising by 13 per cent (due largely to the gains made by the minority of urban blacks in stable employment), the overall ratio of white to black income increased from 12:1 in 1960 to 15:1 in the early 1970s (Bundy 1992).

Like those in the West, the South African economy was also becoming more centralized as mergers and acquisitions sent large corporations on the road towards monopoly. This tendency was encouraged by the investment of multinationals, but it was also facilitated by the effects of government tariff protection and the high degree of centralization in the local banking sector, which often supported only large-scale business undertakings (Nattrass 1988; Bloch 1991; Clark 1993). The mining sector provided increased foreign exchange and contributed up to 26 per cent of state revenue, while mining companies themselves diversified into manufacturing and finance. Fine and Rustomjee (1996) note that the South African economy is still inordinately influenced by six major conglomerates, each of which has antecedents of some kind in the gold mining industry. By the late 1970s Anglo-American in particular 'covered almost every sphere of mining, industrial production, finance, property and agriculture' (Beinart 1994: 168). In 1982, it controlled 53 per cent of listed shares on the Johannesburg Stock Exchange.

One particular consequence of the mergers, acquisitions and overseas investments of the 1960s was to become increasingly politically important, for reasons which we will explore below, by the late 1970s. This was the merging of Afrikaner and English-speaking capital, allowing for a more homogeneous capitalist 'voice', speaking on behalf of large-scale South African business as a whole. As Afrikaner financial enterprises, and especially the insurance giant Sanlam, diversified into other sectors, their attention to nationalist discourse faded and they became as ruthlessly business-oriented as any English-speaking enterprise (Bloch 1991).

## The 'homelands'

The reformulation of apartheid elaborated during the 1960s boom would also be built upon the foundations laid by the long-term strategy of African 'retribalization'. In the aftermath of Sharpeville, just as the surrounding British High Commission territories of Botswana, Lesotho and Swaziland became independent, so the bantustans would apparently be launched on their own path towards nationhood (see summary box 6, pages 222–4). They would become independent or self-governing 'homelands' (see Fig. 2.9a on page 31). As W. W. M. Eiselen, head of the Department of Native Affairs explained, South Africa would emancipate 'the diverse members of its own little Commonwealth on the lines long followed by Great Britain in relation to countries subject to colonial rule' (Lazar 1993: 385). The Promotion of Bantu Self Government Act of 1959 already provided the bantustans with structures which could lead to self-government. The act was designed, as Prime Minister Verwoerd put it, to take 'into account the tendencies in the world and in Africa', and was implemented by the NP cabinet directly, thus avoiding resistance from the wider party to the idea of African self-government (Price 1991: 22). Although Verwoerd stated that 'homeland independence' 'was not what we would have liked to see', he made it clear that 'in the light of pressures being exerted on South Africa there is ... no doubt that eventually this will have to be done, thereby buying for the white man his freedom and the right to retain domination in his own country' (Barber and Barratt 1990: 94).

Its adoption may have appeared reluctant, but a policy leading to 'homeland independence' had irresistible attractions for Verwoerd's government and its successors. After 1970, the policy would allow even Africans permanently resident in South Africa to be classified as subjects of 'foreign' states and thereby, finally and irrevocably, nullify any claim they might have to citizenship within South Africa. By the early 1980s, African elites, often educated in the new bantustan universities and descended mainly from the chiefly and petty-bourgeois landowning classes, had been persuaded to accept what was in reality closely supervised 'independence' in four of the 'homelands' or bantustans – Transkei, Bophuthatswana, Venda and Ciskei (see Fig. 2.9a on page 31). The other six 'homelands' remained as self-governing units within South Africa. These latter authorities were relatively autonomous with regard to educational, agricultural and urban development policy, but tightly controlled by the NP in any issues to do with security and the mass media (Graaff 1990).

Aside from their role in 'resolving' the long-standing question of African citizenship, 'homeland' governments, whether technically independent or merely self-governing, were seen as being useful by various state departments in other respects. First, they became an integral part of continuing, if not entirely successful attempts to direct the movements and residence of African workers. After 1967 especially, urban employers who relied upon semi-skilled and stabilized African labour were pressurized either to hire more expensive whites rather than blacks in the cities, or to move to industrial decentralization sites initially near, and later inside, the 'homelands' (see summary box 8, pages 227–8). There, they would benefit from government subsidies, tax relief and cheaper African labour. African employees would migrate or commute to the new sites from within the 'homelands' themselves. As an economist close to the government explained, 'the major goal' of industrial decentralization 'was to accomplish a regional distribution of economic activity so that the 9 or 10 million Bantu [Africans] expected to be born during the half century up to the year 2000 should settle in their own areas, rather than in or around the concentrations of white communities in the Republic' (Hirsch 1991: 150).

'Homeland' administrations then, would help to bifurcate the African labour force, keeping urbanized workers to a minimum and providing for employers' needs through the alternative of 'foreign' homeland labour. At the same time, the industrial development brought by decentralization would help to justify the 'homelands' as separate political entities. However, the scheme did not work so well in practice. Although white workers in the towns were often less productive, and certainly less cost-effective than their black counterparts, the decentralization sites, no matter how low the labour costs, were unattractive to most employers due to the lack of transport, infrastructure and marketing facilities. While some labour-intensive enterprises were attracted to the decentralized sites, and while such sites encouraged urban development near bantustan borders, they remained disappointingly few as far as the government was concerned. The rate of joblessness in the bantustans was hardly affected by the policies (Bell 1973; Rogerson 1982; Lemon 1987; Giliomee and Schlemmer 1989).

Secondly, within their 'homeland' governmental structures, African politicians, civil servants, teachers and entrepreneurs could find outlets for their personal ambition and opportunities for material accumulation, which deflected them from mobilizing opposition against the apartheid state. As had been the case even before 'independence', a blind eye was turned by South African officials to these African bureaucrats' misappropriation of 'development' funding (Murray 1992; Delius 1996). The South African government hoped that blame for the failures of 'development', and the antagonism displayed by commoners with no economic prospects and no long-term access to the labour markets of the cities, would be directed onto these corrupt elites, rather than South African government officials.

Thirdly, the 'homeland' elites could benefit South African governments by targeting their political and material patronage in order to promote 'tribal' exclusivity. They would thereby impede broader alliances against the state. Murray for instance, notes the way in which the Bophuthatswana administration under Lucas Mangope 'pursued a policy of strident Tswana nationalism in respect of access to very scarce resources', especially in the fragmented Thaba Nchu portion of the 'homeland' where the majority were ethnic Sotho (Murray 1992: 225). Even if they refused to accept full 'independence', the self-governing 'homelands' could serve similar functions. In KwaZulu, Chief Mangosuthu Buthelezi was responsible for the revival of the Inkatha movement founded during the 1920s by King Solomon. Although Buthelezi first came to prominence during the late 1960s as a critic of apartheid policies, he went on to utilize Inkatha in order to promote an exclusive sense of Zulu nationalism and 'traditionalism', both in the 'homeland' and in neighbouring Natal (Maré and Hamilton 1987; Kentridge 1990; Maré 1993).

In purely international diplomatic terms, the reformulation of bantustan policy as a whole was a failure. Few other governments were prepared to grant diplomatic recognition to the four 'independent homelands' once the United Nations, witnessing their high infant mortality, malnutrition and disease, had declared them too poverty-stricken, fragmented and economically and politically dependent on South Africa to be viable. The South African government's reformulation of apartheid was generally seen by Western governments as an exercise in repackaging, designed to produce the impression rather than the substance of African independence. Nevertheless, as Western states became increasingly concerned about the expansion of Soviet and communist influence in Africa and elsewhere, the South African government was able not only to sustain its policies, but to promote them as a solution to other countries' 'racial tensions'. They suggested that their revised bantustan policy could help resolve such 'racial problems' in the West itself, and especially those in the southern states of the USA (Manzo 1992). Due to its continuing ability to guide South Africa's participation within a restructured global economy on favourable terms (providing US multinationals, for instance, with an average rate of return during the late 1960s of 20 per cent compared to a worldwide average of 11 per cent), the apartheid state remained secure, for the time being, from any direct material effects of Western rhetorical denunciation (Curtis 1991; Grundy 1991).

## The official substitution of class for race, 1970s–1980s

During the 1960s, South Africa's economic boom brought increasing prosperity for those protected by racist legislation, and the opportunity for the state to restructure the bantustan administrations. By the 1970s, however, the greater exposure to global economic cycles brought by overseas investment was producing dire consequences for the apartheid state. Once the 1960s global economic boom had ended, South African producers found their enterprises less internationally competitive during a period of general economic recession. If they were to keep pace with the metropoles of the global economy, they had to be able to use ever more capital-intensive techniques and deploy a more skilled labour force more effectively. They also had to gain access to far larger domestic and export markets in order to be able to afford the costs of capital-intensive production.

Such economic pressures, combined with the mounting costs of repressing black resistance (itself spurred on by general economic deterioration), brought about a critical shift in the alignment of those classes comprising the Afrikaner nationalist movement during the mid-1970s. Established differences between Afrikaner classes were politicized as the NP debated reforms designed to address the new demands of those controlling capital investments. Ultimately, the nationalist class alliance broke apart as Afrikaner capitalists, financiers and the wealthy middle classes advocated partial reform, while workers, small farmers and the bureaucracy defended entrenched and rigid racial privilege (O'Meara 1982). With the fragmented state facing challenges not only from a globalizing economy in recession, but from newly independent governments on its borders, and an upsurge in local workerist and popular resistance, the way was clear by the late 1970s for the ascendancy of a new style of government based on co-operation between state, capitalist and military forces.

### Economic crisis, capitalists and bourgeois imperatives

The South African state had long grappled with contradictions between the labour demands of urban manufacturing and commerce on the one hand, and those of the mining sector and farming on the other. The influx control and Bantu Education systems had been developed in order to resolve these contradictions by supplying both stabilized, semi-skilled urban labour to the former and a cheap, unskilled rural and migrant workforce to the latter. Although many manufacturers had long found it necessary to flout the influx control and job reservation laws in order to obtain stable, skilled African machine operatives at lower cost, they had never felt compelled to challenge the government's restrictions on the use of African labour with any great conviction. With the modernization of production techniques pioneered in the West, however, the labour requirements of the economy as a whole were shifting during the late 1960s and 1970s. The economic boom which expanding world markets had allowed during the 1960s provided local employers not just in manufacturing, but in agriculture and mining too, with the finance gradually to update their practices.

Invariably, this modernization took the form of substituting labour with capital. Imports of machinery and equipment increased and large quantities of cheap labour were replaced in some enterprises by capital-intensive production processes and a smaller and more skilled workforce of machine operatives and supervisors (Lipton 1986). While this restructuring could proceed at a leisurely pace during the 1960s, it soon became a far more pressing concern. The heyday of global economic expansion came to an end early in the 1970s as the Organization of Petroleum Exporting Countries (OPEC) precipitated a global recession by raising the price of oil. Across the capitalist world, production and transportation prices increased and were fed through the economic system, producing inflation. At the same time, the general level of consumption for certain commodities reached a plateau. Given the high degree of interconnectedness between the South African and other economies which had been forged during the preceding boom, these trends impacted upon South African capitalists as much as any others. If they were to remain competitive within a static rather than expanding global economy, they had to adopt modernized techniques more rapidly.

On white-owned farms, greater capital intensity during the 1960s, including the more widespread use of tractors, at first led to the employment of more labour, as extra land was taken into production. But by the late 1970s, with the adoption of combine harvesters and other kinds of sophisticated machinery, a smaller pool of more skilled and better-paid workers, supplemented during peak periods by cheap labour gangs drawn from the 'homelands', was all that the most capitalized white farmers required (Marcus 1989; Keegan 1991; Beinart 1994). There were further consequences of this restructuring of the white farmland, especially on the highveld.

While state support enabled farmers to capitalize, it also generated high levels of farm indebtedness (the aggregate level of which trebled between 1980 and 1985) and encouraged the extension of cultivation onto marginal lands, causing environmental destruction (Murray 1992; Beinart and Coates 1995). A similar restructuring took place from the late 1970s on western Cape farms and vineyards, but there the often brutal patriarchy, which had been consolidated during the nineteenth century, continued to structure relations between white employers and predominantly Coloured employees, even as new technology was introduced (du Toit, A. 1998).

Within the mining sector, producers supplying competitive international markets, such as those for coal and copper, had first been obliged to mechanize more extensively during the late 1960s. The gold mining industry faced more technical difficulties introducing machinery and enjoyed a more sheltered market, but it nevertheless followed suit during the 1970s. Increasing capital intensity in the mines led to a partial move away from traditional long-distance migrant labour recruitment and towards stabilized urban workers (Freund 1991).

In all sectors then, by the late 1970s, capitalists in South Africa had responded to global initiatives and were restructuring. One effect was the redundancy of much of the unskilled African workforce and an increase in the long-term, structural unemployment rate ( a trend which is continuing today, as we will see in chapter 7). Similar outcomes were evident within other political systems in both the global cores and peripheries. In addition, South Africa was not alone in having its general joblessness exacerbated through relatively high population growth rates. But in South Africa, the mass of the unemployed were additionally spatially marginalized as far as possible within 'homelands' and deprived of services and amenities (Beinart 1994). Some 1.13 million African farmworkers, many of them formerly productive sharecroppers or tenants, were evicted from white-owned farms from 1960 to 1982 as their labour and that of their livestock became redundant (Platzky and Walker 1985). Those who migrated permanently to the poverty-stricken squatter camps proliferating on the fringes of all the major urban areas were flouting the influx control laws and faced continual harassment and arrest. Most sought accommodation in expanding peri-urban slum settlements in the 'homelands' (Murray 1987, 1992). Few of these households were able to acquire land in their own right, and many of them relied on family members engaging in migrant labour or long-distance commuting across the 'homeland border'. An increasing proportion of extended families fell back on the pensions of the elderly as their main source of cash income (Murray 1987; Unterhalter 1987; Marcus 1989; Wilson and Ramphele 1989).

The phenomenon of mass unemployment contained its own political threat, since it helped to radicalize the youth in both urban and rural areas. By the late 1980s, 55 per cent of the African population were under the age of 20, and with an unemployment rate of around 40 per cent, few could hope to attain formal employment upon leaving the Bantu Education system. This alone was a major source of disaffection, prompting new and violent expressions on the part of young men especially, who felt increasingly emasculated by their poverty and by their inability to exercise control over their own and others' lives. As we will see below, these men's sense of alienation could readily be mobilized either into militant opposition to the state or into criminal activity (Morrell 1998).

A contemporaneous effect of capital restructuring, however, was the call from major employers for the further reformulation of apartheid. First, enterprises demanded that they be allowed to make greater use of skilled and settled African labour. By the late 1970s, even as a structural surplus of unskilled labour was being generated, most employers were facing a shortage of skilled and supervisory workers, capable of handling sophisticated machinery. With the white population being virtually fully employed, they found it necessary to train black workers to a higher level of competency. However, the need to pacify organized white workers and meet the demands of the 'floating' job colour bar meant that, each time that blacks were promoted, whites generally had to be elevated further and paid increased wages. This was especially the case in the well-unionized gold mines, where the extra productivity of black workers had to be exploited in order to subsidize their white counterparts (Nattrass 1990; Freund 1991). Even then, there was a general shortage of skilled employees. In technical, professional and semi-professional grades of work, it was estimated that there was a demand for some 47,000 extra workers in 1969, and for 99,000 by 1977. While such conditions could be tolerated if the economy was 'cushioned' by boom conditions, by the late 1970s the government was being pressurized by major employers' associations from each sector to liberalize the laws pertaining to black occupational advancement, residential location and education so that 'artificially high' production costs could be lowered (O'Meara 1982; Lipton 1986).

The government was also lobbied to address the fundamental problem of limited consumption. Some urban manufacturers, relying (with state support) on import substitution to supply local markets, had been warning of the potentially restricted size of those markets since the 1940s (Lester 1996). But during the 1970s economic crisis, many more suppliers found that the prosperous white market, supplemented as it was by a small group of wealthy Coloured and Indian consumers, was becoming saturated. Price (1991) for instance, notes that by the mid-1970s the car manufacturing industry centred in Port Elizabeth was capable of producing twice as many cars as it could sell.

The period from 1960 to 1990 did see the growth of a relatively prosperous African middle class, capable of consuming significantly in its own right. The 1960s boom in particular allowed the proliferation of African professionals, traders and businesspeople, not least those who had seized lucrative opportunities to provide transport services to township and 'homeland' commuters. In industries where the job colour bar was allowed to 'float', skilled African manual workers also gained higher rates of pay (Bloch 1991; Crankshaw 1996). The stratification to which these developments contributed continued through the next two decades. Although the mean income of the lowest-earning 40 per cent of African households fell by almost 40 per cent between 1975 and 1991, that of the richest 20 per cent increased by 40 per cent (Marais 1998). In many African townships, the middle class began to carve out identifiably wealthier districts, extending and improving their once-uniform houses and, after 1989, buying property in private developments (see the photographs below). In the 'homelands', 'independence' or self-government provided further opportunities for a prosperous bureaucratic elite (Beinart 1994; Hindson, Byerley and Morris 1994). But even the Coloured, Indian and African middle classes combined still constituted a tiny minority of the black population. As long as the vast majority of black, and particularly African incomes were constrained by such apartheid measures as job colour bars, influx control and poor education, market size would continue to be limited. By the late 1970s, big business was publicly engaging in high-profile politics, suggesting to government not only that the labour market had to be freed, but that greater numbers of blacks had to be allowed to accumulate wealth (Adam and Moodley 1986; Lipton 1986; Lee *et al.* 1991).

Finally, major South African enterprises were well aware that only if domestic markets were enlarged would the production of capital machinery (that required to generate other products) be profitable within the country. Until that time, they remained reliant on importing this machinery (Kahn 1991). They were able to do so relatively cheaply through the foreign exchange earned when the gold price was high, as it was for much of the 1970s global recession. But when the international gold price fell, as happened, for instance in 1976, South African producers had to pay more for the machinery necessary to maintain their production levels. In a knock-on effect, the price of South African manufactured exports rose, making them more vulnerable to international competition. During the late 1970s and early 1980s fluctuations in the gold price thus prompted a series of minor booms and busts contributing to the overall economic slowdown (Clark 1997). Furthermore, as apartheid became increasingly internationally illegitimate, even those domestic companies which supplied formerly secure global markets, such as the western Cape wine and fruit producers, found their access potentially impeded by economic sanctions and anti-apartheid boycotts within the countries to which they exported (Lipton 1986). Only a state-led programme of reform was capable of addressing these structural concerns and reducing the South African economy's vulnerability within the competitive global economy.

Plate 5.2  Middle class African housing in Khayelitsha, outside Cape Town

Plate 5.3  Extended 'matchbox' house, Soweto

## State-led reform and the fragmentation of Afrikaner nationalism

It was not only English-speaking capitalists who were advocating further reform in the interests of capital growth during the 1970s. As we have noted, the interests of major English-speaking and Afrikaner capitalists had converged, along with their stock market holdings, during the 1960s boom. The mining houses, financial capital, the state-run industries and Afrikaner big business by now comprised a capitalist block which had reached a 'broad consensus … about the economic path to follow' (Bloch 1991: 100). As O'Meara (1982) argues, this capitalist block was also being joined by other Afrikaans-speaking, middle-class groups, which comprised key voting constituencies within the nationalist movement.

The Cape branch of the NP had traditionally represented agricultural and finance capital, and quickly became associated with reformist politicking. By the late 1970s however, formerly conservative, bourgeois Transvaal interests had joined them. The Afrikaner bourgeoisie had done well out of the 1960s boom. Afrikaners had moved in unprecedented numbers into urban managerial positions, many of them provided in nationalized state industries, schools, universities and businesses, and the proportion engaged in professional occupations had doubled between 1948 and 1975 (Charney 1984; Grundlingh and Sapire 1989). The swollen Afrikaner middle classes were becoming increasingly secularized, staking their identity on the Western patterns of consumption accessed through TV (after 1976) and other global media, rather than on a narrow and insular Christian nationalism or on the criteria of race or ethnicity (Grundlingh and Sapire 1989; Hyslop 1998). They were therefore less resistant to affiliation with the English-speaking middle classes (many of whom were by now voting NP anyway), and would even tolerate the rise of a wealthy and therefore 'respectable' black middle class, although not yet necessarily within their own group areas. Seeking to ensure continued modernization and economic growth, the representatives of the Afrikaner bourgeoisie within the NP were willing to jettison certain pieces of apartheid legislation if they were thought to be an impediment to capital growth (O'Meara 1982, 1996).

The reformist group within the governing party was labelled '*verligte*', or enlightened. However, this group was increasingly vehemently opposed by those '*verkrampte*', or reactionary Afrikaner classes, who insisted on the continued necessity for rigid, state-protected racial boundaries, including influx control, restrictions on the employment of skilled African labour and the continued banning of African trade unions. These groups included workers who faced competition from Africans in the labour market if the job colour bar was removed; smaller-scale farmers who lacked capital and stood to lose access to African farm labour if influx control was undermined, and state bureaucrats and smaller businesspeople who implemented the plethora of apartheid racial controls and faced competition from the emergent African middle class (O'Meara 1982; Charney 1984). As the class alliance which had brought the NP to power in 1948 fragmented, representatives of the various Afrikaans interests struggled for control of the party. By the late 1970s, as O'Meara has shown, Verwoerd's successor, B. J. Vorster was tentatively leading the state in the direction of reform, having ousted his more conservative opponents and nullified potential opposition from within the Broederbond (O'Meara 1996). Those Afrikaner interests who now found themselves marginalized within the NP turned increasingly towards the more reactionary parties of the right wing.

## The resurgence of overt resistance in the 1970s

The imperative for reform became especially clear to many within the state during the mid-1970s, when a sense of peril was generated by events both inside and outside South Africa's borders. In 1975 the Portuguese government finally succumbed to economic crisis and costly guerrilla independence wars fought by socialist and nationalist movements in its African colonies. African socialist movements gained control of the states in Mozambique and Angola, the former directly bordering South Africa and the latter bordering South West Africa, itself occupied and administered by South Africa. The success of the independence wars in these 'frontline' territories raised morale among those seeking to overturn apartheid, and it caused the South African state deepened anxiety about guerrilla incursions from ANC bases across its borders (see Fig. 5.1) and a heightened sense of isolation in an increasingly postcolonial world. The problem of South Africa's international legitimacy as an explicitly racist state was raised once again, and in the late 1970s only Rhodesia, to South Africa's immediate north, gave the government comfort as another white-ruled minority government. It too, however, succumbed in 1980, despite previous South African military assistance. In the event, the Rhodesian state's eventual collapse may well have been the partial result of a withdrawal of South African support, intended to appease the USA. Rhodesia was renamed Zimbabwe when it fell into the hands of an African government (Hanlon 1986; Price 1991). Surrounded by socialist enemies, the South African state relied increasingly on tacit agreements forged with Western governments in the context of the Cold War.

Figure 5.1   ANC bases in South Africa's 'Frontline states'

*Source*: Christopher (1994)

Within South Africa, a fresh wave of strikes among semi-skilled and skilled African workers had broken out in Durban during 1973, prompted by the OPEC-induced inflation and increased costs of living. The workers' informal organizational structures, including migrant associations, made the state's identification and arrest of the leadership difficult. During the remainder of the 1970s and into the 1980s, other workers capitalized on the relatively successful Durban precedent and began to organize more effective 'grassroots' trades unions. The unions mobilized Africans at exactly the point where they had most power to disrupt capital accumulation – in the workplace. As a result, they exposed the government even more effectively to the grievances of big business. By the early 1980s the revived trade unions were coalescing into politically powerful associations with a combined membership of some 500,000 workers. Their enhanced strike activity would lead to progressive improvements in wages and conditions for relatively skilled black workers during the 1980s, with real wages for manufacturing workers, for instance, rising by 29 per cent on average between 1985 and 1990 (Lodge 1983; Mitchell and Russell 1989; Clark 1997).

Already facing renewed labour unrest, the state's educational policies combined with the unredeemed economic situation to provoke unprecedented student and youth opposition in the mid-1970s. The generation of a politicized black youth was one aspect of the long-term processes which had been transforming relationships within black households during the twentieth century as a whole. In the rural areas, as communities became increasingly dependent on migrant labour, 'older men struggled, and it seems increasingly failed, to maintain access to the bulk of wages earned by their sons'. In turn, 'younger men found it difficult or impossible to gain access to land' and often became alienated from the etiquette and rituals of their elders (Crais 1998: 52; Beinart and Bundy 1987b). Many of them, having had their expectations raised by their first exposure to formal education, found themselves during the 1970s recession, 'consigned … to economic and social limbo', and resorted to petty or violent crime (Delius 1996: 161).

Even such 'wayward' young migrants and rural youth, however, were often regarded by township youth as traditionalist and naïve. During the late 1960s, while the ANC in exile was developing its organizational coherence and gaining international diplomatic recognition, the Black Consciousness Movement was gaining adherents among the urban youth within South Africa. After a decade in which internal resistance had been successfully repressed, the student-led movement picked up on currents within West African and African-American thought, concentrating on encouraging pride in black achievements and culture. While it was refined and disseminated in particular by African schoolteachers, the black consciousness philosophy embraced Coloureds and Indians as well (Pityana *et al.* 1991). It never constituted a political programme in its own right, but black consciousness ideology prepared the way among many of South Africa's youth for more tangible and direct expressions of resistance, many of them connecting ultimately with the agenda of the ANC in exile (Lodge 1983; Marks and Trapido 1991).

The first and most explosive such expression would occur over the issue of education. The government had expended vastly increased sums during the early 1970s enrolling more urban Africans in schools in an attempt to meet employers' demands for semi-skilled labour. However, in 1976, the Bantu Education Department was forced by administrative changes to squeeze greater numbers of primary school leavers into the new secondary schools. On top of the disgruntlement caused by severe overcrowding, a further intervention by *verkrampte* ministers provoked overt youth resistance. In June 1976, the decision to impose tuition in Afrikaans rather than English for various subjects within the curriculum prompted huge student demonstrations in Soweto, bolstered by the confidence that the black consciousness ideological renaissance had instilled. As Hyslop points out, 'the policy provoked such a violent response from students not just because of the symbolic role of Afrikaans as the language of government, but also because the policy cut across the need of students to prepare to sell their labour power on the labour market of urban centres dominated by English-speaking concerns' (Hyslop 1988: 473).

When police opened fire on the protesters in Soweto on 16 June 1976, rioting and attacks on government functionaries and buildings spread rapidly through the townships. Media coverage prompted other Africans, as well as Coloured youth in Cape Town with similar grievances, to emulate Soweto's students. Soon police were deployed against rioters in most urban areas across the country, and the apartheid state was facing the most serious occurrence of black resistance yet (Hirson 1979; Brooks and Brickhill 1980; Lodge 1983). This was no longer a case of particular black interest groups opposing specific government policies within a loosely co-ordinated organizational framework. Although many migrant workers fiercely opposed the militancy of the urban youth, wishing simply 'to make money and return home rather than strike for a better life in town', the fundamental legitimacy of the state as a whole was now being violently rejected, for the first time, by diverse urban black groups across the country (Lonsdale 1998: 300).

In the wake of the Soweto uprising, and the estimated 600 deaths which ensued at various points across the country, the *verligte* programme of retrieving economic growth was pursued with more vigour. This response was spurred not least by the

Plate 5.4  The most famous image of the 1976 Soweto revolt, a photograph taken by Sam Nzima: 13-year-old Hector Petersen is fatally wounded
*Source*: *The Star*

renewed flight of international capital which the rioting had prompted (amounting to $2.5 billion in 1976–7, compared with $334 million in the year after the Sharpeville shootings), and by an upsurge in international condemnation leading to a mandatory United Nations arms embargo (Curtis 1991). In the years following the Soweto crisis, encouraged by an IMF loan, foreign capital resumed its flow into South Africa, but the blow which the riots had dealt to investor confidence was manifested in the form that capital inflow now took: short-term loans which had to be repaid rather than long-term investment in plant, production and services (Clark 1997). Overall, foreign investment plummeted from R1,561 million in 1975–6 to R452 million in 1976–7 (O'Meara 1982). For the first time in the modern period, South Africa experienced a decline of GDP during 1977–8 (O'Meara 1982). The South African economy was clearly suffering from more than just a cyclical downswing. There was the real and mounting prospect that structural crisis could seriously undermine long-term capital accumulation, and thus white as well as black living standards.

In response, the Prime Minister, B. J. Vorster, continued to implement tentative reform. He appointed two commissions headed by Riekert and Wiehahn. Riekert's commission was to advise the government on the training, housing, employment and governance of black labour, and Wiehahn's was to report on industrial relations between the black labour force and white employers. Guided by major capitalists, the two commissions taken together advised the removal of the job colour bar, the revitalization of the industrial decentralization policy, the abolition of restrictions on the mobility of urban Africans and the recognition of black trade unions so that their demands could be met through collective bargaining. Some provision for limited self-governance in the townships, and a gradual equalization of educational spending on white and black students was also recommended. Many of these measures were intended to remove unnecessary restrictions on the use of urban black labour by capitalist enterprises, but each committee still advised that the established distinction between relatively privileged urban and excluded 'homeland' African labour be upheld, and indeed reinforced (Hindson 1987; Murray 1987).

## Capital, the military and the state: Total Strategy in the 1980s

In 1978, Vorster was effectively forced out of power by a scandal in which, as O'Meara (1982) argues, more resolutely reformist business interests, NP politicians and military elements within the state were involved. The issue over which he was removed – the misappropriation and use of funds by a propaganda ministry without parliamentary approval – was hardly a concern of paramount importance to these groups, since they had long overlooked the overriding of parliament. But Vorster's removal paved the way for a preferred successor, bent on restoring political stability and the conditions for capital growth. P. W. Botha inherited the recommendations of the Riekert and Wiehahn commissions and, having himself ironically bypassed parliament by centralizing state power in the hands of a small elite of military and state officials, he incorporated these recommendations within a broader plan to 'guarantee … the system of free enterprise' (O'Meara 1982: 12).

The military leadership, with whom Botha had developed close ties as Minister of Defence, was convinced that South Africa was facing a communist-led 'Total Onslaught', orchestrated both from within and outside South Africa's borders. Botha was determined to respond with an all-embracing counter-revolutionary package of reform and repression, inspired partly by French attempts to deal with insurgency in colonial Algeria, and known as the 'Total Strategy'. This, as a defence white paper put it, would embrace 'interdependent and co-ordinated action in all fields – military, psychological, economic, ideological, cultural etc. (O'Meara 1982: 12). Botha's state was not alone in developing such a co-ordinated strategy during the late Cold War period. As Alden points out, authoritarian 'states as diverse as colonial Malaya, the Philippines, South Vietnam … Chile and El Salvador' also turned to such 'broadly conceived techniques as a panacea for their ills, and specifically for the threat of socialist revolution' (Alden 1998: 5).

Since the continued accumulation of capital was essential not only as a goal in its own right, but as a prop to political stability, Botha saw big business as a vital part of the Total Strategy. Leading capitalists would combine with the military and the state to restore prosperous conditions in which revolutionary elements would be marginalized (Price 1991; Schrire 1992; O'Meara 1996). Recovered prosperity would enable the state to reclaim the consent of strategically vital black groups through economic and political reform. Following policies adopted in Margaret Thatcher's Britain and Ronald Reagan's USA, the state would refrain from shaping (or 'impeding') production and consumption patterns so extensively, and allow 'market forces' to have greater sway. Beinart explains that 'as the free-market ideology of the Thatcher and Reagan era permeated South African politics, reformers found its combination of more liberal economic policy and conservative social philosophy matched well with their developing perception of the country's problems' (Beinart 1994: 227).

Botha's state believed that the progressive privatization of state industry and housing, inspired by Thatcher in particular, would provide opportunities for big business to extend profit-making into new areas, as well as assisting the government to balance its own finances after the economic crisis of the 1970s. Military strategists also advised that privatization would substitute private companies for the state as the target of Africans' political struggles over housing, rates and transport, just as 'homeland' administrations were bearing the brunt of grievances in the bantustans (Heymans 1991). As part of the same strategy of state withdrawal, the former linchpin of apartheid – the Department of Native Affairs (by now called the Bantu Affairs Department) – was deprived of much of its power under Botha's leadership.

The goal of maintaining white bourgeois supremacy would still be defended, despite the freer play of 'the market', due to the inherited concentration of capital. But the transition to liberal policies would nevertheless encourage the additional growth of the black middle class. The government aimed at the progressive incorporation, or at least co-optation of this class. Not only would this fulfil employers' demands for more skilled and dependable labour; it would also raise consumption levels among the black population and produce a significant black constituency with a stake in the current political order. Urban Africans, many of them unionized, skilled and potentially able to disrupt production processes, were promised the electrification of their homes, the provision of recreational facilities and the improvement of their schools through co-operation between the state and the private sector. Informal sector activities, including beer brewing, were tolerated for the first time under apartheid governance and Africans were encouraged to aspire to private home ownership (Beavon and Rogerson 1982; Dewar and Watson 1982).

An easing of spatial constraints (often necessitated by the *de facto* flouting of the law), was particularly suggestive of the government's attempts to 'buy off' middle-class and skilled urban blacks, and not just Africans during the early 1980s. A small number of elite Coloureds, Indians and Africans benefited from less rigid enforcement of Group Areas legislation, largely forced upon government in the first instance by a 1982 Supreme Court ruling that evictees had to have alternative accommodation provided. These elite blacks were subsequently joined by others who acquired homes and ran businesses in white areas. At the same time, African and Coloured employees who lacked state housing were given assistance to upgrade their squatter camps (Cloete 1991; Lemon 1991). Despite some tense meetings with government ministers, big business came effectively to support such initiatives, endeavouring to bring about urban political stability for its own reasons and through its own direct measures. A new private sector organization, the Urban Foundation, was established with backing from Anglo-American and other financial giants. Through its command of capital, the Urban Foundation was able to construct housing while ultimately exerting considerable influence over the nature of state-led urban development planning as a whole (Lea 1982).

Those whose economic role made their consent to the political order less vital – unskilled migrant and 'homeland'-based Africans – would continue to be effectively barred from permanent access to the cities under the Total Strategy, but through less obviously racist, and therefore less domestically and internationally provocative, legislation. In 1985, the influx control laws were scrapped, but superseded by a strategy of 'orderly urbanization' based on 'market-driven' mechanisms. It was intended that the black urban presence would be kept to a necessary minimum by the limited housing supply, with the laws against 'squatting' being tightened to keep those who were not wanted by the state or employers out of the cities. Those who were not required in the cities would thus remain subject to the 'decentralized despotism' of the homelands (Mamdani 1996; Greenberg 1987; Stadler 1987).

Continuing the endeavour to hold further black urbanization at bay, the industrial decentralization programme was revived during the early 1980s. Nine development regions were ultimately identified by state planners, many of them linking 'homeland' labour supply areas with the municipal economies which they serviced. This tacit undermining of the notion of 'homeland' autonomy was a result of a new government conceptualization of the South African economy 'as a single economy, divided in terms of "development needs, development potential, functional relationships and physical characteristics"' (Hirsch 1991: 160). Such a conceptualization fitted in well with the hegemonic Western discourse of 'development' and 'modernization', invoked by organizations such as the World Bank and IMF.

Within the economically 'functional' development regions, selected growth points were identified outside the conurbations, to which labour-intensive industries (some of them foreign-owned, especially Taiwanese) were attracted by unprecedentedly generous grants and the provision of infrastructure. For instance, in the Ciskei 'homeland', investors would have 95 per cent of their wage bill paid back to them by government for a period of seven years after the factory opened. The Development Bank of South Africa (DBSA) was established to help finance this anti-urbanization strategy, but it never stamped out the rampant corruption which accompanied it (Tomlinson 1990; Hirsch 1991; Lester 1996). The 1980s decentralization initiative nevertheless proved more successful than any previous one, creating some 52,000 new jobs in the 'homelands', mostly for lower-paid women, between 1982 and 1986 – a doubling of the previous rate (Hirsch 1991).

With the African population divided into necessary urban 'insiders' and effectively controlled rural 'outsiders', the Indian and Coloured populations would also be offered partial political incorporation. Since Indians were overwhelmingly urbanized, and Coloureds increasingly so, and since the middle classes of each of these 'race groups' performed semi-skilled or skilled and managerial occupations, their co-optation would provide the same advantages of economic rationality and political stability as that of the urban African classes. But their further incorporation within the state *political* framework would generate useful allies in the struggle against any broad African citizenship. The 'trick', as far as Botha's government was concerned, was to gain the benefits of their allegiance without jeopardizing real white political hegemony.

A new constitution was thus implemented in 1984 which gave Indian and Coloured politicians their own houses of parliament and control over the running of 'their own affairs', but within limits prescribed by the dominant white House of Assembly and the Executive (white) President. Africans were still generally regarded as citizens of their respective 'homelands', and were therefore denied any parliamentary representation. Those urban Africans whom the state wished to co-opt would have to remain contented with limited representation and the redistribution of resources, both at municipal level. Even this cautiously constructed concession to the black middle classes, however, was too much for the most conservative elements within the NP. Preliminary plans for the new constitution prompted their breakaway and the formation of the Conservative Party (CP) in 1982. Through the remainder of the apartheid years the CP would represent the Afrikaner unemployed, manual workers, farmers and smaller businessmen actually or potentially being displaced by black competitors, being bypassed by commercial agriculture or losing their grip on state patronage under the reformist NP. These constituencies, more than any other, continued to demand a return to the racial protection which had been afforded under Verwoerd (Charney 1984; Stadler 1987; Grundlingh and Sapire 1989).

The final part of the Total Strategy elaborated in the early 1980s addressed the external threat posed by the newly independent states on South Africa's borders. In order to counter their influence, an aggressive programme of destabilization was implemented. South African troops had been committed in Angola since 1976, with covert US support, in an attempt to replace the Marxist government there with one more amenable to South African and capitalist influence. During the 1980s, South African forces continued to give assistance to a rebel guerrilla army, UNITA, which was still trying to unseat that government. In Mozambique, South African-trained and equipped insurgents known as RENAMO destroyed civilian infrastructure and rendered attempts at economic growth futile. Each of the 'frontline states' as the African-led countries surrounding South Africa came to be called, would be conditioned under the Total Strategy to perpetual economic, and therefore political dependence on South Africa (Hanlon 1986; Price 1991).

By the mid-1980s then, the South African state was engaged in a complex and often contradictory attempt to substitute the most obvious racial boundaries of political and economic privilege with class boundaries which would appear more 'normal' within a global discourse of human rights and free markets. This, it was hoped, would help restore the conditions for capital growth (Stadler 1987). The shift was to be achieved in particular through the partial incorporation of strategically placed black 'insiders'. However, the imperative for Botha's government, and for its security forces, was to manage this transition, made necessary in the final instance by the altered conditions of global capitalism, as conservatively as possible. While the Total Strategy would allow white constituents to retrieve their economic prosperity, and while it would allow co-opted black groups to participate as junior partners in that prosperity, it was intended that political control over the rate and direction of reform would remain overwhelmingly in the hands of the established white electorate's representatives.

## The opening of citizenship

By the late 1980s, it was clear to many within the state that its policies were not only failing to cope with continued crisis; they were actually proving disastrous. The South African government's immediate role as defender of capitalism during the Cold War, together with considerable multinational investments, still persuaded the American and British governments to shelter it from the worst effects of international condemnation. But the fundamental restructuring which, it was hoped, would buy off black urban consent within the country proved impossible to implement.

Although the Total Strategy was designed to allow for the accumulation of capital in the long term, costly improvements in black living standards were necessary to its success in the short-term. But the crucial gold price fell after an early 1980s recovery, and international short term debts, accumulated during the 1970s recession, entailed increasingly onerous repayments. South Africa entered recession, alongside its competitors within the international economy, with GDP growth rates falling well behind that of population growth, inflation and unemployment climbing and, to make matters worse, a severe drought necessitating massive food imports in the early 1980s (Curtis 1991). By the time that the Western economies were recovering from the latest recession in the early and mid-1980s, South Africa was already locked into a cycle of insurrection and repression which deterred international investment and produced new sanctions. As a result, the continued economic growth upon which

Botha's reforms were premised failed to materialize. Domestic white capital, the Afrikaner bourgeoisie and those black urban groups who had been promised improvements all applied varying degrees and kinds of pressure for more fundamental change (Greenberg 1987).

Utilizing the space opened up for organization by the state's reformism during the early 1980s, black-led oppositional movements including youth, women's and church groups, trade unions and township-based civic associations were organized at a variety of scales. As the promised improvement of urban African living standards failed to materialize, and as costs of living actually increased, localized grievances articulated by these organizations and centring on rents, housing, transport and education, became interconnected. Despite its attempts at privatization, the state remained the primary target of their mobilization. Above all, it was these shared material concerns which helped partially to transcend some of the divides of class, ethnicity, gender and location among South African blacks, and which prompted more universal resistance to the state during the 1980s.

In 1983, opposition groups came together under the umbrella of the United Democratic Front (UDF) in order to attempt co-ordination of their diverse, but related struggles. The UDF leadership soon declared its political sympathies when it adopted the ANC in exile's Freedom Charter as its own provisional manifesto. When violent opposition, led by the black youth, flared again during the 1980s, it was the internal UDF leadership which tried, and often failed, to harness its power. But it was the ANC in exile, with the imprisoned initiator of armed resistance, Nelson Mandela, as its figurehead, which became the major beneficiary. By the time that an altered NP leadership decided that negotiations with black leaders able to carry popular support were unavoidable, the ANC was overwhelmingly seen as the legitimate representative of those who had most resolutely resisted apartheid. In this final section of the chapter, we examine the particular dynamics of crisis during the 1980s which finally induced the NP to negotiate the end of apartheid, allowing the assumption of government by the ANC.

### Insurrection

The immediate spur to renewed bouts of violent resistance in the 1980s came from the townships of the industrial Transvaal. There, a series of spectacular rent increases was levied by the new, supposedly representative Black Local Authorities (BLAs) in the urban townships. Charged with carrying out the infrastructural improvements promised to urban blacks, the BLAs found themselves without the capital to do so. The central state's own perilous financial situation ruled out any significant subsidy, so they had to be self-financing. 'Improvement' therefore necessitated the persistent raising of township rents and rates. Having been promised reform, the reality of more onerous exactions provoked widespread anger in the townships. Once it had manifested itself in outbursts in the Transvaal, rioting and attacks on government targets were organized in other townships subject to a similar set of circumstances across the country. Taken together, these localized acts of resistance amounted to an insurrection (Seekings 1988; Price 1991).

Local attacks on security forces and symbols of government authority were generally led by those below the age of thirty. As had been the case in previous episodes of violent resistance, such as the 1922 white workers' Rand Revolt, the intensity and scope of conflict opened spaces for militant young women to engage in public acts of defiance alongside men, even if only ephemerally (Krikler 1996). Radicalized Coloured and Indian students, as well as migrant Africans also sometimes participated alongside African township residents. However, while the insurrectionary black youth is often seen as an homogeneous category, it was in fact differentiated in various ways. The more secure material status of many Indian and Coloured families militated against a broad and enduring alliance of black youth, and many African migrants continued to feel alienated from urban insurrectionary activity. Most of the anti-state violence was carried out by African males permanently resident in the most deprived townships (Marks and Trapido 1991). UDF leaders claimed a degree of control over their activities, but organizational co-ordination of the insurrection was relatively weak and even the civic associations which had first been organized around local issues such as the provision of bus shelters, and which affiliated themselves to the UDF, were often imposed on townships by relatively small groups of militant activists (Seekings 1991).

These activists, generally known as Comrades, helped organize rent and service charge boycotts, and engaged the police directly within each township. They also targeted those identified as 'collaborators'. The fundamental illegitimacy of the state, already revealed in widespread boycotts of Indian and Coloured elections to the new houses of parliament, was more apparent than ever before in most townships, as hundreds of those accused of being even remotely connected with state activities were killed. By 1986, as local civic associations came together to form the South African National Civic Organization (SANCO), 'there were many who believed that they inhabited a "liberated zone" with its own courts and anti-crime campaigns, boycotts and student congresses; that by rendering the townships "ungovernable" they were preparing the way for "people's power"' (Marks and Trapido 1991: 7).

If they were not effectively directed and controlled by the UDF, Comrades were certainly spurred on by the ANC's belated call to make the townships 'ungovernable'. Although the geographical focus of the violence shifted constantly as the state concentrated its resources in 'hot spots', the security forces' brutal responses, including the shooting of participants in the funeral processions of activists previously killed by police, fuelled a general spiral of violence. By the late 1980s the government had declared a succession of states of emergency. Comrades, many of whom were boycotting school, were openly fighting police and troops, including white conscripts drafted into the townships, and rendering certain areas 'no-go zones' for the security forces (Wolpe 1988; Seekings 1991; Lodge 1992; Marais 1998). Some of them received assistance from armed, ANC-trained guerrillas who had infiltrated back into the country. Although these fighters were not generally militarily effective, for many Comrades they were heroes. They did even more than the movement's underground local organizers to raise the prestige of the ANC and establish it as the legitimate representative of the Comrades. One indication of the ANC's resurgent popularity was the way in which banned ANC flags and banners made increasingly prominent appearances, alongside those of the allied SACP, at rallies, marches and politicized funerals.

Police and covert military units responded to the upsurge of open opposition by torturing and assassinating activists. They also sponsored anti-Comrade 'vigilante' forces from within the townships, in an attempt to immobilize direct resistance. Police methods in the rural areas had traditionally centred on the use of local African authorities in the suppression of crime and anti-governmental activity, but during the 1980s urban-based insurrection, the police found it more difficult to identify Africans with political legitimacy, such as local chiefs, with whom to collaborate. Instead, they turned to any Africans who would be willing to resist the 'tyranny' of the militant Comrades (Ellis 1998). Certain squatters, for instance, were armed and set against others who supported the ANC in struggles for the control of scarce resources within the informal settlements. Groups of elders and migrant men, already conscious of being unable to fulfil the patriarchal role of providing for their households, also experienced a marginalization from community decision-making as the militant youth took control. Alienated by the disregard for their authority and the apparent challenge to their own 'honourable' masculinity, many of them too, were induced to help suppress the Comrades (Beinart 1992; Lonsdale 1998). From 1985, the security forces effectively turned the strategies which had been used to destabilize neighbouring states inwards to attack the ANC's constituency within South Africa's own townships (Ellis 1998; Taylor and Shaw 1998).

In KwaZulu and Natal, Buthelezi's Inkatha movement proved a particularly useful ally of the security forces. As the leader of a 'homeland', in order to retain a popular following, Buthelezi had 'either to confront the state seriously or to deliver material benefits to his [Zulu] constituency, through access to state and business patronage' (Marks and Trapido 1991: 9). By the mid-1980s, he seemed to have opted for the latter course, making him the enemy of the ANC-supporting youth. Chiefs and headmen who dispensed patronage locally on Inkatha's behalf were convinced, and in turn helped convince their followers, that the gains in support made by the UDF/ANC in the region would not only undermine Inkatha's political position, but more fundamentally corrupt Zulu identity. What especially alienated many Inkatha-supporting men was the lack of traditional respect for elders, or *hlonipa*, displayed by militant youths. For most Inkatha members, the maintenance of *hlonipa* was essential to the preservation of Zuluness itself. It was certainly observed by members of the substantial Inkatha Youth Brigade in their interactions with the movement's leadership, and in their hierarchical gender roles. When township-based Zulu Comrades forced their elders to comply with boycotts or demonstrations and mocked their 'collaboration' with white authority, they were seen to be challenging this central feature of Zulu masculine identity (Dlamini 1998; Bonnin *et al.* 1998).

Violent conflicts between Comrades and Inkatha sympathizers became especially common in the urban townships and peri-urban squatter camps around Pietermaritzburg and Durban. As a result of these areas' incorporation into the KwaZulu homeland, Comrades there confronted not the South African police directly, but the KwaZulu police, effectively controlled by Inkatha. When a number of black trade unions across South Africa came together to form the powerful, and later UDF-affiliated, Congress of South African Trade Unions (COSATU) in 1985, launching an unprecedented series of strikes, Inkatha formed its own more conservative union, the United Workers' Union of South Africa (UWUSA) in response. With backing from the state, Inkatha members stepped up their physical attacks on COSATU members as well as Comrades in the Natal region during the late 1980s. Altogether some 6,000 were killed in a series of attacks and reprisals, the vast majority of victims being Comrades or supporters of the broad UDF/ANC alliance (Lambert and Webster 1988; Kentridge 1990; Maré 1993).

Despite its resort to unprecedented levels of violence, and its alliance with black anti-Comrade forces, the fragility of the government's control was highlighted in the rural areas during the insurrection as isolated white farmers were targeted by local or externally based guerrillas. Of more immediate concern for the state as a whole, during the late 1980s, was a series of coups in the 'independent homelands'. These coups saw the rise of new leaders such as Bantu Holomisa in the Transkei, who sought legitimacy by proclaiming allegiance to the ANC (Humphries and Shubane 1991). In the self-governing 'homelands' too, returning migrant labourers, as well as students seeking sanctuary from urban schools in turmoil, brought insurrectionary tactics and ideas from the cities to the countryside (Delius 1996). By the early 1990s, only Buthelezi in the self-governing 'homeland' of KwaZulu, Mangope in the 'independent' Bophuthatswana (whose grip had been forcibly upheld against internal opposition by South African troops) and Oupa Gqozo in the 'independent' Ciskei were determined to maintain control in their 'homelands'. As the momentum of political transition gathered during the early 1990s, the latter two leaders did all that they could to prevent their 'homelands' being reincorporated into the rest of South Africa, while Buthelezi had the grander ambition of raising the renamed Inkatha Freedom Party (IFP) to a position of national prominence, rivalling that of the ANC. Each of the other 'homeland' bureaucracies, however, perceived that the insurrection was tilting the balance of power within the country irrevocably towards representatives of the insurrectionary forces.

Taken as a whole, the violence of the 1980s established an unprecedented level of generalized brutality, which is reflected still in high rates of violent crime (see chapter 7). In some 'homelands' and in some white farming districts, Comrades developed a brutal anarchy in which women and elder men were often victimized. Witchcraft accusations mounted in the northern Transvaal and the 'homeland' of Lebowa during the insurrection, with those accused frequently being killed (Delius 1996). While 'poverty, differentiation, and household and generational conflict' are customarily associated 'with the presence of evil, specifically the work of witches', in the 1980s the young Comrades saw the violent eradication of such evil as being a fundamental precondition for an improved post-apartheid social order (Crais 1998: 52; Delius 1996). Those who had accumulated wealth, apparently through 'collaboration' at the expense of others, as well as those who spoke out against the youth's violence, were the most frequently accused, although a variety of more specific and local motivations could also be brought into play (Delius 1996). In the urban areas too, where the direct struggle for citizenship was pursued with most vigour, new levels of violence were reached. Much of the fighting was between youth and migrants, formal township residents and squatters and, especially in the Coloured townships of the Western Cape, between rival gangs. Regardless of its precise and local motivation, the violence was generally facilitated by the breakdown of 'normal' policing (always deeply inadequate in the townships under apartheid) and the general erosion of societal constraints on murder (Beinart 1992; Straker *et al.*1992).

The Comrades' insurrectionary activity, as we will see below, was undoubtedly a major influence in bringing the NP to the negotiating table and thus ushering in the end of apartheid. The 'political' violence deployed by Comrades against the state, however, represented a challenge for other groups within black society as well. As Lonsdale puts it, 'It was the continuing breach between respectable survival and [the Comrades'] disreputable action that ... made the black bourgeoisie ... hesitate, the parents of Soweto fearful, Mandela uncomfortable [and rurally based migrants alienated] and which remains one of South Africa's unresolved problems' (Lonsdale 1998: 299).

## Negotiations

By the end of the 1980s, it was clear that partial reform, by placing some political and administrative organs in the hands of black groups which remained largely noncompliant or ineffective, had only presented greater opportunities for the disruption of the government's Total Strategy. The Coloured house of parliament, for instance, was refusing to co-operate in the passage of legislation, and urban African government had effectively broken down in the face of the immigration of millions of 'squatters' and meaningless elections. State attempts to finance urban African administration through rates and rents had, in many cases, collapsed in the face of widespread boycotts of these charges. By 1985, only a few Black Local Authorities were still functioning, potential councillors in most of the others being too afraid of retribution to stand for election (Price 1991). The collapse of urban African administration in turn encouraged further illegal land invasions and squatter camp construction around and within townships across the country (Hindson, Byerley and Morris 1994). Direct control over the 'homelands' was also slipping from the government's hands, while the security forces were severely overstretched in trying, simultaneously, to destabilize neighbouring states, combat independence fighters in South West Africa, and suppress the internal insurrection.

In addition, the economy was continuing its downward spiral. Unlike other states which had similarly prospered during the 1960s boom years, South Africa had not used the opportunity to restructure the economy towards export-oriented growth. As an exporter of minerals and agricultural products, and an importer of capital equipment and technology, the country remained vulnerable to shifts in the international economic order (O'Meara 1996). From 1986 to 1988 the import bill increased by 60 per cent, while additional costs could not be met by borrowing due to the increasing severity of international sanctions and the wariness of financial lenders. Even the gold mining industry was continuing to fail in its traditional role as guarantor of foreign exchange. A fall in the gold price in 1989 rendered many mines unprofitable, and a mass lay-off of mineworkers began shortly afterwards.

In 1985 the American bank Chase Manhattan, frightened by political and financial instability, caused a financial crisis in South Africa by refusing to roll over its short-term loans. Part of the reason for its nervousness was that during the early 1980s recession, South African banks had increasingly adopted the practice of borrowing from overseas banks only to re-lend the money internally, creating an ever more precarious financial system (Curtis 1991). When other banks followed Chase Manhattan, inflation and a further decline in living standards followed. In response to the ensuing fall of the rand, the government was forced, temporarily, to close the financial and foreign exchange markets. It also declared a moratorium on debt repayment.

Despite state censorship, images of the insurrection and of police brutality were simultaneously being televised across the world, prompting Western anti-apartheid movements to mobilize more effectively against their governments' tacit support for the South African state, and against multinational companies' investment in the country. A sports boycott was particularly painful to many whites in a country where the sporting ethos, and especially that of controlled aggression associated with the masculine sport of rugby, was inculcated within a segregated schooling system. Ultimately, the US Congress overrode a presidential veto and implemented sanctions on new investments, loans, airlinks and South African imports, with the European Economic Community (now European Union) following swiftly. While, in 1984, seven US multinationals had sold off their South African subsidiaries, the following year, 38 followed suit, and in 1986, 45 pulled out (Curtis 1991). (Many of the assets sold off were bought by large South African corporations, thus further increasing the concentration of ownership within the country.) By 1989, the government had been forced to reschedule the payments on its foreign debt, and the effect of sanctions on private sector profitability was becoming apparent in the annual reports of some of South Africa's biggest corporations (Price 1991).

State officials and ministers who were forced to deal with these successive crises were becoming increasingly desperate for a solution which was more than just a short-term response. Without a political accommodation which was widely seen as being legitimate, there was the prospect that 'the state and opposition would become entangled in a death embrace that could destroy South Africa's integrity as a nation-state and a viable zone for capital accumulation – and with it white privilege' (Marais 1998: 64). At a local level, some municipalities ranging in size from Johannesburg to the small eastern Cape town of Stutterheim had already recognized the need to negotiate with insurrectionary civic structures in the townships. Assuming the lead from central government, they realized that this was the only way to bring rent and rates boycotts to an end, restore the revenues necessary to effective urban administration and secure the conditions for renewed white prosperity (Swilling and Shubane 1991; Nel 1994).

In all, by the late 1980s, the economic, social, political and security problems facing the South African state amounted to an 'organic crisis' – one which was far more violent, destabilizing and endemic than the crises which had prompted the consolidation of segregation in the 1920s, or the formulation of apartheid in the 1940s

(Saul and Gelb 1986). Whereas segregation and early apartheid had both been sustained through the acquiescence of key black groups, the late apartheid state was based on far more shaky foundations. The combination of frustrated expectations of reform, deterioration in living standards, awareness of greater international condemnation associated with media coverage and suffering at the hands of brutalized state security forces, had undercut any acquiescence in continued white dominance on the part of the most crucial black groups. These included the urban middle classes, the increasingly organized and mobilized working class, and the newly militant, and frequently unemployed youth. The unprecedentedly vehement denial of consent on the part of these groups necessitated a more dramatic reformulation of ruling group notions of citizenship than had yet been attempted.

Although Botha had recognized that fresh tactics were required, and had engaged in secret negotiations with Mandela, it was left to his successor, F. W. De Klerk, to unify the NP (which had once again become increasingly divided and marginalized in recent years under Botha) around a new grand initiative from 1990. What compelled him was not only South Africa's domestic crisis, but a dramatic shift in the global order. Even as South Africa was plunged into insurrection during the 1980s, its government had continued to rely on the tacit support of the American and British governments. More draconian sanctions, as we have indicated, were frequently resisted by their leaders. However, with the collapse of the Soviet bloc in the late 1980s and the effective ending of the Cold War, the South African state's global significance as a bastion of capitalism and vital Western ally was undercut. A more active and hostile form of international intervention could be expected at any time and, as we have noted, such a form was being increasingly successfully pressed for by Western anti-apartheid movements. As a key state security advisor, Neil Barnard put it, the government had to move 'before our backs were against the wall' (Alden 1998: 267).

At the same time, Soviet backing of the ANC had been stripped away, leaving it more isolated within the international political system, more vulnerable to the demands of the newly confident Western powers and more likely to negotiate on terms favourable to the NP. Like popular resistance movements in El Salvador and the Philippines, in the post-Cold War era, the ANC would be exposed to greater international pressure, directed by Western powers, to reverse its potentially socialist agenda. A precedent of what could be expected to ensue from negotiations within South Africa had recently been set in South West Africa (Namibia). There, the South African government had been induced by military stalemate, and by the victory of military and reformist elements over conservatives within the NP, to withdraw its troops in 1989. Through Western-backed negotiations, the independence of Namibia was secured in 1990 under the South West African People's Organization (SWAPO) – the main guerrilla force which had been resisting the South African occupation. However, the new black-led government was hampered by the need for international funding and support, and it had proved unable radically to transform the existing distribution of wealth (O'Meara 1996).

In South Africa, a similar outcome could be predicted. The Western powers and global institutions which could assist or hinder South Africa's future economic prospects, as well as multinational companies themselves, all insisted on a more inclusive and totally non-racial framework of citizenship and labour utilization, in line with the prevailing global discourse of advanced capitalism. But, in a manner reminiscent of metropolitan prescriptions following the abolition of slavery (see chapter 3), they were also united in demanding that, under any post-apartheid government, a market economy be retained, social stability guaranteed, and radical redistributive programmes ruled out. If ever there was a time for the NP to negotiate the future of South Africa's social and economic boundaries from a position of relative strength, this would be it. By the beginning of 1990, De Klerk and the NP leadership calculated that they could preserve entrenched white privilege better through negotiations with the ANC, than they could through continued repression (Friedman 1993; Marais 1998). The first public sign of De Klerk's new approach was the release from prison of senior ANC figures in 1989.

In February 1990, De Klerk announced the unbanning of the ANC and SACP, as well as other opposition movements, and the imminent release of Nelson Mandela so that negotiations over the form of the future state could begin. The following year, all of the major legislation constituting apartheid was abolished. In 1992, after the defeat of an NP candidate by the right-wing Conservative Party (CP) in a by-election, De Klerk countered the threat of a white backlash against reform by putting the idea of negotiations to the white electorate in a referendum (O'Meara 1996). Confronted by the choice of a return to rigid apartheid under the CP, which would entail further international isolation, violence, insecurity and plummeting living standards, or an initiative to preserve 'essential' white interests through negotiations, over two-thirds opted for the latter.

Negotiations lasted from 1990 to 1994. They involved a number of organizations collected together as the Convention for a Democratic South Africa (CODESA), and they started with the NP holding out for an effective white veto on future government decisions, and the ANC pressing for a more thoroughgoing transfer of state power. More marginal groups such as right-wing Afrikaner organizations, recalcitrant 'homeland' leaders like Mangope, and the IFP under Buthelezi, tried to disrupt any consensus which would result in their exclusion from power. Failing in this, elements among them attempted to divert the negotiations in their favour by threatening further violent instability. At one stage, there was an apparently bizarre common front against both the NP and the ANC, forged between white neo-Nazi racists (the *Afrikaner Weerstandsbeweging* (AWB)), IFP Zulu nationalists and corrupt black 'homeland' leaders. Buthelezi, with a popular base of support within the IFP and among certain white business interests, proved most successful in such tactics, eventually gaining a position in the new Cabinet as Minister for Home Affairs. In Bophuthatswana, Mangope was forcibly overthrown, and in the Ciskei, Gqozo was removed from power by the transitional state after his troops fired on ANC demonstrators.

While negotiations proceeded, security force personnel continued to destabilize the ANC by launching attacks on its leaders or on township supporters, often in alliance with IFP members who had been specifically trained for the task. The degree of collusion between security force elements and the IFP (known collectively as the

'Third Force') was first revealed in investigative journalism and subsequently con-firmed by independent inquiry and the mass of traumatic evidence presented before South Africa's post-apartheid Truth and Reconciliation Commission. Aside from those directly mobilized by the NP-controlled state however, 'ordinary' Zulu migrants in urban hostels on the Witwatersrand also participated in IFP-inspired attacks on surrounding communities, provoking further retaliation. With their rural orientation, their alienation from 'corrosive' urban lifestyles and values and their anxieties over the future of a fragile patriarchal authority, many migrants could be relatively easily persuaded of the need to defend their position through violent assaults (Marks and Trapido 1991).

Among other analysts, Ellis (1998) concludes that even if he did not directly order them, De Klerk certainly tolerated security force and IFP attacks because they were intended to exert further pressure on the ANC to accept compromises for the sake of ending the dangerous transition process (Taylor and Shaw 1998). In 1992, a series of security force-sponsored incidents culminated in the police-supported IFP killing of 43 ANC supporters in the Transvaal township of Boiphatong. Shortly afterwards, in April 1993, the ANC's military commander Chris Hani was assassinated by white right-wing extremists. By then, however, the mass killing was beginning to prove counterproductive for the NP. The party was eventually forced to relinquish its in-sistence on an effective white veto in the negotiations and rein in the security forces when the ANC walked out of negotiations and began organizing a mass defiance cam-paign, threatening a renewed plunge into international isolation and economic disaster (Alden 1998; Marais 1998).

In the run-up to elections, the ANC, influenced by women's as well as trade union and other organizations, articulated a programme of wealth and opportunity redistri-bution between races, classes and sexes (Unterhalter 1995). A Women's National Coalition was formed to press for equal rights under the new constitution – a thorny issue where 'traditionalist' African leaders demanded the preservation or restoration of the patriarchal structures recognized under 'customary law'. At the same time black women began, more vociferously, to point to the ways that white women's priv-ileged racial and class positions have helped to structure women's lives more differentially in South Africa than in most places (Bonnin et al. 1998). They drew attention in particular to the extremely high levels of sexual violence that women suffer in South Africa's male-dominated society and to the 'series of daily frustra-tions, indignities and denials' that low-paid black domestic servants endure in order that privileged white men and women are partially relieved of child care and house-hold burdens (Cock 1980: 8).

The NP, meanwhile, abandoned its racial paradigm, and concentrated on securing a multi-racial constituency around the principles of free-market capitalism, Chris-tianity, civilized 'standards' and law and order (Altbeker and Steinberg 1998). It was most successful in the western Cape, where many older and working-class Coloureds were already afraid of the ANC's potential assumption of power. They sensed not only that the ANC would strip away the relative privilege which Coloureds had enjoyed over African competitors in the western Cape job market, but also that their distinctive Afrikaans-speaking identity would lead to them being marginalized from housing and education under an African-dominated government. When the first democratic elections were held in April 1994, it was mostly these Coloureds who ensured that the NP retained control of the new Western Cape Province. Aside from a dubious IFP victory in KwaZulu-Natal (a compromised outcome reached after fraudulent elections in order to prevent exacerbation of the region's civil war), the rest of the country's new provinces fell overwhelmingly to the ANC. Within the National Parliament, the ANC did not quite attain the two-thirds majority needed to draft a new constitution unilaterally, but it nevertheless claimed 63 per cent of votes cast nationally. The NP emerged as the second party and shared power in the compromised arrangement known as the Government of National Unity.

By the time that the elections were held, the NP had conceded more than it had anticipated. But the ANC had also abandoned much of its programme of wealth redistribution. During the negotiations process, it had been exposed to a variety of reports, emanating from big business and US-dominated global financial institutions, pointing out the threat to investment and prosperity in South Africa contained in any radical socialist policies. Such reports made 'neo-liberal policies synonymous with "realism" ... [constructing them as] politically and ideologically neutral' (Marais 1998: 158). As we will see in the following chapters, like the NP before it, and as the NP leadership had anticipated, the new governing party was forced to recognize the constraints to its autonomy imposed by a Western-dominated, post-Cold War, global economy.

# SUMMARY BOXES

## Some key patterns and processes in South Africa's historical geography

The four summary boxes which follow aim to encapsulate key aspects of modern South Africa's history, that are considered by geographers and development specialists to have left the most enduring spatial imprints and generated some of the greatest development challenges for the future.

---

**Summary box 5:** From the segregated to the apartheid city

From the genesis of urban settlement, informal and later legal, racial segregation was enforced on the urban landscape of South African towns and cities. Starting initially as either immigrant clusters or with the confinement of indigenous people to mission land adjacent to 'white' towns, urban space developed, by and large, on racially exclusive lines. This process received legal sanction at the beginning of the twentieth century. Motivated by pejorative racial discourses, fears of economic competition and often justified on the grounds of the 'sanitation syndrome' – the association of black people with outbreaks of disease – various Native Locations Acts empowered local authorities to segregate most Africans and to set up and administer locations/townships – urban areas set aside for black people, characterized by entrenched poverty, uniform building styles and racial exclusivity (Swanson 1977).

Following increased African urbanization during the First World War (1914–18), the white central government sought to regulate the administration of African urban residents. In 1923 the Natives Urban Areas Act was enacted, which entrenched the principle of segregating urban Africans. Amendments in 1937 and 1945 restricted Africans' rights to property. No national policies segregated Coloured and Indian people prior to 1950, although there were attempts to segregate Indians in the Transvaal and Natal areas. The spatial structure of South Africa's cities assumed the 'typical' form of the Segregation City as identified by Davies (1981) (see Fig. 5A.1). The city model was characterized by distinctly different white suburbs and African townships. Within this pre-apartheid city, many Coloured and Asian people still lived in an integrated

*(Box continued)*

fashion, often with working-class white people, and normally in inner-city areas. Segregation against Africans was not as rigidly enforced as it was later to become and many people lived outside of the areas officially allocated for their racial group.

In 1950, with the passage of the Group Areas Act, the situation was radically transformed in urban areas. Over a period of time, all cities and towns in South Africa were subjected to a number of draconian controls. The Group Areas Board drew up plans for the division of urban space to ensure that racially exclusive areas would be established for each of the major racial groups in the country. The planning dictated that each area should be separated by buffer strips (open land, transport corridors or physical barriers) from adjacent areas/racial groups, such that each area could expand laterally outward without infringing on the space of another group (Lemon 1991). In order to enforce such plans, individual urban residents not conforming to the racial character of the suburb they were living in were forcibly removed to the appropriate Group Area. In areas which were racially integrated, the euphemistically named Department of Community Development systematically expropriated all property, frequently demolished the buildings, and resettled the people in racially exclusive suburbs and townships. In certain cases, owing to the non-availability of alternative housing, people lived for up to 20 years under a permanent eviction order (Nel 1991). In many cases where a city or town was close to an adjacent 'homeland' (see summary box 6: Territorial segregation, pages 223–4), attempts were made forcibly to remove the entire urban African population to newly established homeland cities. In cities such as East London, Pretoria and Durban, this meant the forced removal of hundreds of thousands of people (Western 1996).

The racial re-planning of South African cities caused untold human suffering, led to the destruction of thousands of houses and the forced removal of hundreds of thousands of people purely so that urban areas could conform to the separation ideals of the apartheid city. The transformation of urban space was so complete that by the 1980s the 'apartheid city' (see Fig. 5A.1) had emerged as a unique urban category. The phenomenal extent to which apartheid reshaped urban space is vividly reflected in the photograph below which shows the mixed race suburb of North End in the city of East London. From the 1970s to late 1980s this area was gradually cleared of its 4,000 mixed-race urban residents and redeveloped as a zone of light industry (see Fig. 5A.2). Equally dramatic was the creation of the homeland city of Mdantsane adjacent to East London, but across the imposed homeland border. In 1963 the area had been open farmland, but within 14 years, following the forced removal of tens of thousands of people and the construction of a vast new city, Mdantsane had overtaken East London itself in size (Nel 1991). The dramatic impact of the city's transformation can be seen in Fig. 5A.3.

*(Box continued)*

Figure 5A.1   Segregated and apartheid city models

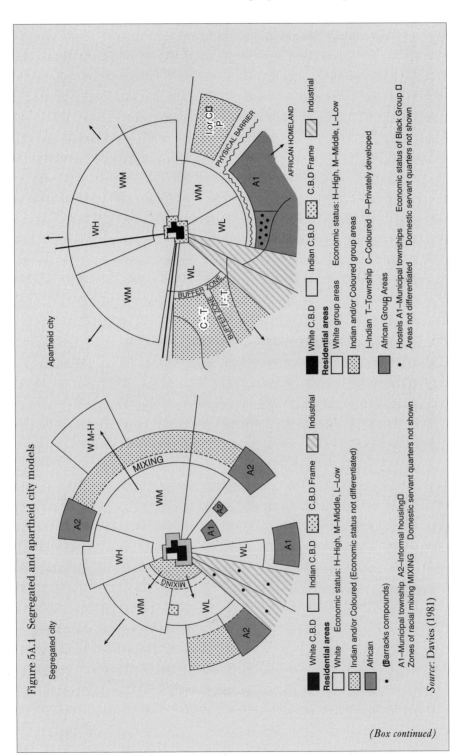

*Source:* Davies (1981)

*(Box continued)*

Plate 5A.1  North End, East London

Plate 5A.2  Red location, Port Elizabeth

Plate 5A.3  Mdantsane, a 'homeland' city created as the result of forced removals

*(Box continued)*

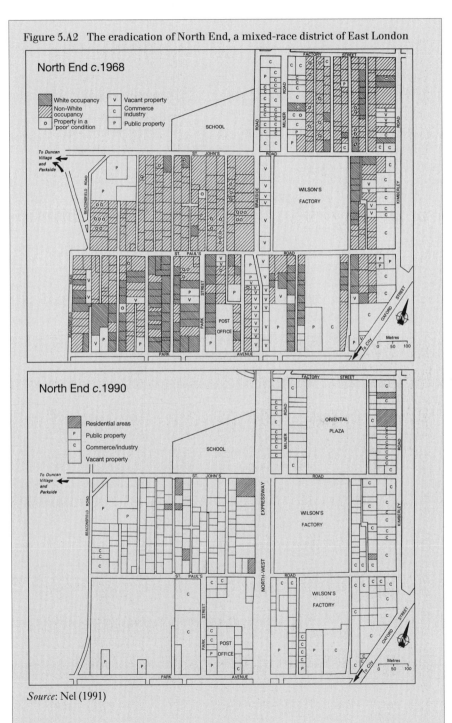

Figure 5.A2   The eradication of North End, a mixed-race district of East London

*Source*: Nel (1991)

*(Box continued)*

Figure 5A.3　Mdantsane and East London

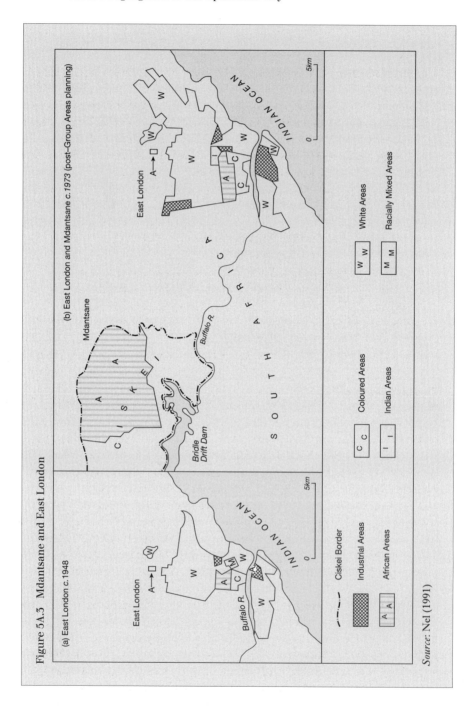

*Source:* Nel (1991)

## Summary box 6: Territorial segregation

One of the most distinctive features of the South African landscape, both spatially and historically, has been the rigid separation of the country into a series of defined regions, based almost exclusively on the grounds of race. In parallel with other settler colonies such as Canada and Australia, distinctive racial reserves were identified on the basis of tribal differences and allocated for indigenous people. As has been described in chapter 3, the first such Crown Tribal Reserves for black people were proclaimed in the area known as British Kaffraria, to the east of the established Cape Colony, in the 1840s and 1850s (see Fig. 3.5 on page 73). In the latter part of the nineteenth century similar proclamations established reserves for the Zulu in Natal Colony and the Xhosa in the Transkei, which was also added to the eastern part of the Cape. A distinctive feature of the reserves was that whereas private tenure, on a freehold basis, was the dominant form of landownership within the colonies as a whole, communal tenure supplemented by various forms of leasehold was the norm within the reserves (Peires 1981; Lester 1997).

The notion of territorial separation was significantly enhanced when in 1913, in terms of the Natives Land Act, 8.9 million hectares of land were identified in the eastern and northern parts of South Africa as Native Reserves for indigenous people (see Fig. 4.2 on page 144). These reserves incorporated earlier proclaimed land and significantly added to them either through the proclamation of land where black people were already living, or through the identification of adjacent farms occupied by whites and 'scheduled' for expropriation. Various restrictive clauses were enacted to prevent black access to land outside of the reserves. In 1936, the Native Trust and Land Act identified an additional 6.2 million hectares for inclusion and took away the rights of blacks to purchase land outside of the reserves (Fig. 4.2 on page 144) (Christopher 1994).

Under apartheid (after 1948) the scattered units making up the reserves began to be consolidated into more unified entities and a gradual attempt was made to separate these new 'black states' from the legal control of 'white' South Africa. In 1959, under the Bantu Self-Government Act, the legal basis to consolidate the Reserves into ten 'independent' 'homelands' was established (see Fig. 2.9a on page 31). Thereafter followed a process of allocation of legal and administrative powers to the 'puppet' governments set up to rule each 'homeland'. In general terms each 'homeland' had a distinctive African tribal/linguistic group living within it. Significant population movements of millions of 'surplus' black people from 'white' South African rural and urban areas to the infamous resettlement camps in the 'homelands' followed (Christopher 1994). Paralleling this redistribution of people, ten unique 'homeland' governments were set up, each with its own bureaucracy, parliament, capital city and all the other trappings of state. They were widely considered to be blatant

*(Box continued)*

dictatorships. Major financial transfers from the South African government were required to prop them up, including the industrial decentralization policy discussed in summary box 8, pages 227–8.

Though technically disestablished in 1994, the former homeland areas remain enclaves of extreme poverty. For decades they were, and often still are, the source areas of migrant labour and they are characterized by severe overcrowding, erosion and the absence of significant private sector investment. As statistics in the following chapter indicate, these areas contain the country's most intense concentrations of poverty and deprivation.

Plate 5A.4　Bisho, capital of the 'independent' Ciskei 'homeland'

Plate 5A.5　Resettlement camp in the environmentally marginal Glenmore area of the Ciskei

## Summary box 7: Key apartheid Acts and their geography

As we saw in the last chapter, apartheid was based on the conscious ordering and control of a society of millions of people on racial grounds, imposing inequalities which favoured white people and discriminated against all others. In order to enforce apartheid a host of laws were introduced. Perhaps the most influential of these from a specifically spatial perspective were the following five:

### The Population Registration Act, 1950

In order to impose racially based segregation on the entire society, the government promulgated this Act, which provided for the compulsory classification of every person in the country into a discrete racial group. Once this had been undertaken it was then possible to compel people identified as belonging to a certain racial group to reside in a specific area and to impose various discriminatory measures against them. The main groups identified were Black (African), White, Asian/Indian and Coloured, with the latter being subdivided into Cape Malay, Chinese, Griqua and Cape Coloured.

### The Group Areas Act, 1950

Having classified people into different groups in terms of the Population Registration Act, it was then possible to compel them to reside in racially homogeneous areas. This Act laid a basis for the subdivision of all towns and cities into defined Group Areas, and the forced removal of people from one area to another to ensure that they conformed to the provisions of the Act (Christopher 1994). The effect of this Act, in conjunction with others aimed at 'slum clearance', such as the Native Resettlement Bill, 1954, is detailed in summary box 5, pages 217–22 and in the discussion of forced removals from Sophiatown in chapter 5 (pages 180–1).

### Natives (Abolition and Co-ordination of Papers) Act, 1952

This Act imposed rigid controls on African people and restricted their rights to stay in 'white' towns and cities. Under the pass system, all Africans had to carry a permit detailing where they could reside and it also reinforced migrant labour whereby workers had to leave their families in the rural areas and work part of the year in distant cities and mines.

*(Box continued)*

## The Reservation of Separate Amenities Act, 1953

This Act provided a legal basis to ensure that all public and many private facilities were for the exclusive use of individual racial groups. Facilities segregated ranged from beaches to train carriages, park benches and toilets. Whites were almost invariably granted exclusive use of the best facilities. In many public buildings, such as post offices, separate doors and counters were built to provide for whites and non-whites.

## The Bantu Self Government Act, 1959

Building on the Bantu Authorities Act, 1950, the territorial division of the country into a predominantly 'white'-owned state and a series of independent 'homelands' for the different African 'tribal' groupings was reinforced through this Act. The Act also provided for measures of self-governance to be granted to the 'homelands' (see summary box 6, pages 223–4) (Christopher 1994).

## Summary box 8: Industrial decentralization strategies

In 1960, South Africa launched an industrial decentralization strategy, which although drawing on international practices of growth point planning and encouragement of economic multipliers, had a very distinctive racial overtone. In the case of South Africa, most growth points were allocated in or adjacent to the 'homeland' areas (see Fig. 5A.3 on page 222) as part of a policy of trying to ensure the economic sustainability of these artificial enclaves. This strategy was one of the key instruments for enforcing the separate development notion implicit in apartheid planning.

In parallel with attempts to encourage industrial expansion in the periphery were strategies to restrict the expansion of the metropolitan areas. Progressively increasing incentives, described as possibly 'the best in the world' were offered to firms establishing in the growth points between 1960 and 1991 (Wellings and Black 1986). Incentives included relocation allowances, loans on land, interest subsidies, labour wage grants, electricity, rail and harbour cost subsidies and

Figure 5A.4   South Africa's industrial development points and development regions, *c.*1960–1991

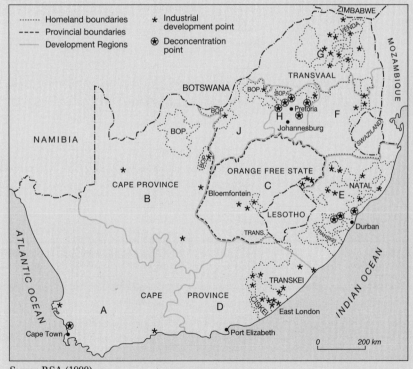

*Source*: RSA (1990)

*(Box continued)*

training grants (Nel 1999). By 1987 some 2,500 firms (14 per cent of the national total) were in receipt of incentives. One of the major flaws of the policy was the fact that firms were often drawn in more by the extremely generous incentives than the inherent locational advantages of the various sites. According to Wellings and Black (1986) 79 per cent of firms chose the growth points because of available incentives, and 51 per cent intimated they would cease operations if the incentives were suspended. The failure to develop local economic multipliers, and the closure of many firms across the country after incentives were suspended in 1991, indicates just how artificial the strategy was. In the case of the former Ciskei 'homeland', industrial growth points lost over 60 per cent of the firms which had established themselves there (Fox and Nel 2000).

Plate 5A.6 Fort Jackson industrial estate in the former 'independent' Ciskei 'homeland'

Plate 5A.7 Dimbaza, near to King William's Town, showing recently closed factory units

# Chapter 6

# INEQUALITY AND DEVELOPMENT:
## a snapshot at the birth of democracy

Having provided a broad account of the contests, alliances and compromises which shaped South African society in the modern period in chapters 3 to 5, we would like to pause in this chapter to consider in more detail the patterns of inequality that characterized the country at a particularly significant point in its recent history. This was the point of transition between the apartheid and post-apartheid political orders in 1994.

There are numerous statistical measures of inequality which could be employed to provide a 'snapshot' of the conditions which the new government inherited at this time. None of them is objective and any assemblage of them is necessarily selective. Nevertheless, by presenting some of them here we hope to achieve three things. Firstly, we aim to convey a general impression of the ways in which relative deprivation and affluence in contemporary South Africa were experienced at the very end of the apartheid era, and, more particularly, of the ways in which these experiences were structured by race, gender and space. Secondly, a closer examination of the patterns of inequality entrenched in 1994 enables us to comprehend something of the scale and implications of the development challenge which is part of apartheid's legacy. This is not to say that South Africa today faces problems of 'development' exclusively because of apartheid policies. As we have emphasized in preceding chapters, and as we will continue to stress in subsequent chapters, the modern South African state, both apartheid and post-apartheid, is unavoidably involved in global networks of power and finance. These networks continue to impose limits on the state's ability materially to assist its favoured constituencies.

Finally, this 'snapshot' chapter provides us with a platform for the following chapters. Essentially, chapters 7 and 8 consider the post-apartheid state's efforts to address the problems which are detailed here. Chapter 7 examines the policies which the new government has adopted, paying particular attention to the ways in which they have been influenced by external and internal business pressure, while chapter 8 sets the current government's endeavours within the context of broader regional, international and global systems of inequality.

## The limitations of political transition

South Africa's democratic transformation in 1994, resulting in the election of Nelson Mandela as President, was clearly identified by the world's media as one of the key events of the century, one which 'has earned its place among the "miracles"' of the era (Marais 1998: 1). In spite of the armed struggle of the ANC and the PAC and the ongoing conflict in the Rand townships and KwaZulu-Natal, the political transformation did not take place as a result of a violent revolution as so many had feared. Instead, a negotiated settlement brought political compromise which in itself proved sufficient to facilitate the ending of apartheid and the attainment of democracy. The remarkable nature of the birth of the 'Rainbow Nation' in 1994 should not be underestimated. According to Patti Waldmeir, what had happened was 'one of the most extraordinary tales of the twentieth century, in which a nation stepped through the looking-glass and emerged as a mirror image of itself' (Waldmeir 1997: front cover). The sense of euphoria which erupted in South Africa over its 'home-grown' solution was matched by awe and admiration from around the world and a phenomenal outpouring of international goodwill and support to the country and its new government. The inauguration ceremony of Nelson Mandela was one of the greatest-ever gatherings of world leaders. They came expressly to witness the victory of 'reason' and endorse what was considered a structural political change (*The Economist* 13 December 1997; Marais 1998).

Whilst not seeking to minimize the profound implications of the political transition, we must nevertheless recognize that the new government has inherited a daunting legacy of economic and social inequality which needs to be addressed. South Africa was, and still is, one of the most unequal societies on earth, as this chapter will demonstrate. Coming to terms with such an inheritance is a formidable challenge and the future stability of South Africa depends on the government's and the society's ability to address social and economic disparities and, simultaneously, to promote sustained growth (Blumenfeld 1997). As Kathryn Manzo wrote in 1992, 'demands for justice, autonomy, and dignity', all of which involve some degree of economic redistribution, 'have been more central to political struggle' in South Africa 'than demands for the franchise. Any negotiated settlement ... that fails to take these factors into account is doomed to fail' (Manzo 1992: 17). The enormity of these challenges was not lost on the outside observers who celebrated South Africa's political transition. Thus, US Vice-President Al Gore noted that the economic challenges might be greater than the political ones. In doing so, he endorsed the view that '[a]fter the political miracle of 1994 ... what is now needed in our country is an economic miracle' (Stutterheim Development Foundation 1995: 1). In the same vein, the World Bank has confirmed that 'reducing poverty is the central challenge confronting the South African economy'(World Bank 1994b: 10).

Inequality in South Africa is structured in a number of ways, not least by the social boundaries of race, class and gender, whose shifting interaction we have been tracing through the last 350 years in previous chapters. Furthermore, as we have argued, these boundaries give rise to, and are themselves shaped by, certain spatial arrangements. The combined forces of racial and gender discrimination, for example, have marginalized black women especially, consigning them, in many instances, to subsistence levels, and to the most impoverished rural territories and urban spaces. In the following sections we examine the manifestation of South Africa's inequalities across the most prominent of its social and spatial divides.

## Race and class

As we have seen in the previous chapter, for much of the twentieth century, race was deliberately made to coincide as far as possible with class in South Africa, notably through the apartheid state's protection of white workers and its educational and occupational constraints on blacks. By 1994, however, a sizeable black middle class was well established and since then, as we will see in the following chapter, new opportunities within the government bureaucracy and arising from government patronage have enabled it to develop further. Indeed, the rise of the black bourgeoisie is one of the most widely noted consequences of the handover of state power. Nevertheless, as Turok argues, the fact that some black people have escaped from relative poverty 'does not diminish the structural character of the [racial] divide' (Turok 1993: 237). The breakdown of development indices according to race and income indicates that a broad coincidence of race and class boundaries is still deeply entrenched.

In comparison with other developing countries, South Africa's aggregate statistics indicate a relatively high level of socio-economic provision. Indeed, the World Bank considers it a middle-income country. But numerous recent studies indicate clear distinctions between two core groupings defined by 'race': whites on the one hand and blacks on the other, the latter including those who would have been classified as Africans, Coloureds and Indians under apartheid (RDP 1995; Crothers 1997; World Bank 1997; du Toit, J. 1998). The 1996 census stated that of the 40,583,573 people in the country, 10.9 per cent were white and of the balance, Africans constituted 76.7 per cent of the population, with smaller percentages of Indians (2.6 per cent) and Coloureds (8.9 per cent). (The category 'unspecified' accounts for the remaining 0.9 per cent.) Average white income in 1995 was R102,857, compared to R70,992 for Indians, R31,835 for Coloureds and R23,228 for Africans (du Toit, J. 1998). It was estimated that by the mid-1990s, 5 per cent of the population, almost all of them white, owned 88 per cent of the nation's wealth, that 50,000 white farmers owned 85 per cent of agricultural land and that the four largest white-owned corporations controlled 81 per cent of South Africa's share capital (Nyerere 1996).

According to the 1995 government Reconstruction and Development Programme study, the poor are identified on the basis of being the 40 per cent of the country's households with the lowest income, whilst the ultra-poor have the lowest 20 per cent of income (RDP 1995). This definition appears to be a logical one given the fact that in 1994, 17 million people (approximately 40 per cent of the total national population) were regarded as surviving below the minimum living level (ANC 1994). The racial dimension of this poverty was overwhelmingly apparent as 64.9 per cent of all African people could be classified as being poor, 32.6 per cent of Coloureds, 2.5 per cent of Indians and only 0.5 per cent of whites (Project for Statistics on Living Standards and Development in 1994, in Crothers 1997). The disparities are even more striking if one examines the distribution of the total number of poor in the population. A racial breakdown of these statistics indicates that 94.7 per cent of South Africa's poor are African, 5 per cent are Coloured, 0.1 per cent are Indian and 0.2 per cent are white (RDP 1995). If one examines human development indices (HDIs), which are internationally applied measures of a country's wealth and its social and economic development (see chapter 8), the country does not perform badly, scoring 0.716 in 1994, compared with a developing world average of 0.576 (UNDP 1997). (Possible scores range from the lowest 0 to the highest 1.) Disaggregation, however, once again reveals stark racial disparities. While the white HDI in 1993 was 0.901, that of Indians was 0.836, Coloureds was 0.663 and that of Africans was only 0.5 (du Toit, J. 1998).

If we move away from an explicit consideration of aggregate racial groupings and examine income levels, we can get some idea of the structuring of inequality by class. The Gini index is a commonly used statistical measure of the degree of socio-economic inequality within a society, based on the nature of the income distribution within that society and the disparities from the norm. The index is derived from the comparison of the percentage of national income earned by key, percentile-based sub-divisions in the society, ranging from the lowest to highest income categories. The higher the statistical score on the test undertaken, the more unequal the spread of wealth that there is in the society. A score of 0 would indicate perfect income equality and 1 indicates perfect income inequality. In 1997, according to the World Bank, South Africa scored a Gini index of 0.58. Out of those countries in the world for which the calculation could be made, this was the second highest score after Brazil which attained 0.63. By contrast, countries in the North average 0.34 (World Bank 1997; du Toit, J. 1998).

The percentage distribution of national income in South Africa amongst the various quintiles (or groupings of 20 per cent) in the population, on which the Gini index is based, is indicated in Table 6.1.

Table 6.1   South Africa's distribution of income

| Societal subdivision (% population) | % share of income |
|---|---|
| lowest 10 % | 1.4 |
| 1st quintile/lowest 20% | 3.3 |
| 2nd quintile | 5.8 |
| 3rd quintile | 9.8 |
| 4th quintile | 17.7 |
| 5th quintile/highest 20% | 63.3 |
| highest 10% | 47.3 |

Source: World Bank (1997)

In themselves, relative inequality levels across overlapping racial and class boundaries demand change. In addition, however, South Africa has 23 per cent of its population living on less than $1 a day. It is these people above all, who experience what Bundy describes as 'the dull ache of deprivation' with its 'intricacies of survival and all its emotions – despair, hope, resentment, apathy, futility and fury' (Bundy 1992: 25). It is thus in absolute as well as relative terms, that poverty needs to be addressed.

Plate 6.1  Up-market, predominantly white suburb: Nahoon, East London

Plate 6.2  Motherwell township, Port Elizabeth

Plate 6.3  Rural poverty in an environmentally marginal area: Thornhill, Eastern Cape

## Gender

Deep racial divisions co-exist with equally profound gender differences. Racial discrimination, confinement to the lowest levels of the job market and poor education, have combined with the marginalized role of women in traditional societies to make black women the double victims of discrimination and inequality. Quite clearly, any attempt to address the legacies of the past has to be particularly sensitive to the socially constructed roles and status of women, and must also acknowledge women's current centrality to household survival in the rural areas and townships.

Any examination of gender disparities reveals a picture which is equally as disturbing as that of racial inequalities. Even though women constituted 51.9 per cent of the population in 1996 (StatsSA 1998), they only comprised 37 per cent of the formal labour force (du Toit 1998). If one also considered the marginal categories of subsistence agriculture, home-orientated activity and the informal sector, the share of the labour market occupied by women increased dramatically to 50 per cent, indicating that a significant percentage of women were eking out a living on the fringes of the economy (DBSA 1994). The overlap between race and gender boundaries is strikingly reflected in the unemployment levels shown in Table 6.2. Whilst 42.5 per cent of Africans were unemployed in 1996, gendered differences mean that this was made up of 34.1 per cent of African men and 52.4 per cent of women. By contrast, in the white population, 4.6 per cent were unemployed, with gender differences being much less marked: 4.2 per cent of men and 5.1 per cent of women (StatsSA 1998). As we indicated in chapters 4 and 5, relatively wealthy white and relatively poor black women in South Africa often seem to live in different worlds. Though less dramatic, the unfavourable position of Coloured and Indian women is also apparent from the table.

Table 6.2  Unemployment rates according to race and gender

| Racial group | Total unem-ployment % | Male % | Female % |
|---|---|---|---|
| African | 42.5 | 34.1 | 52.4 |
| Coloured | 20.9 | 18.3 | 24.1 |
| Indian | 12.2 | 11.1 | 14.0 |
| White | 4.6 | 4.2 | 5.1 |

*Source*: StatsSA (1998)

Low levels of education and their marginalized status have ensured that few black women have top posts in the occupational hierarchy. Whilst 52.7 per cent of African women had low-skilled forms of employment (compared with 26.3 per cent of African men), the corresponding figures for whites were 2.8 per cent and 3.9 per cent (StatsSA 1998). At the other end of employment ladder, 48.3 per cent of white women (and 49.9 per cent of white men) had managerial and professional positions, compared with only 18 per cent of African women and 11 per cent of African men (StatsSA 1998). The last two statistics suggest that educated African women, many from middle-class backgrounds, have been more successful than African men from similar backgrounds in attaining such positions. Post-apartheid efforts to address inequalities in the society will clearly need to give equal attention to the social constructs of both race and gender if greater levels of equality are to be attained and discrimination ended.

## Space

As we have seen, poverty is concentrated most obviously among Africans. This means that it is also concentrated within the geographical spaces designated by apartheid for African occupation, most notably the townships and former homelands (see summary boxes 5–8, pages 217–28). In these areas all socio-economic indicators are a fraction of what they are in neighbouring, predominantly white urban and rural areas. The country's landscape is a conflicting mosaic of all the elements of global inequality in immediate proximity to each other. Nowhere, perhaps, is this more evident than within the cities. Sophisticated urban centres with their established business corporations, which are indistinguishable from those in the Western world, can be viewed from the same spot as the crime-ridden squatter camps, typical of those in South American cities. Indeed, Turok has used the analogy of the 'skyscraper surrounded by a shantytown' to describe the country's entire lopsided socio-economic system (Turok 1993: 237).

Plate 6.4 Johannesburg Central Business District from the air

At a broader spatial scale, in 1994, the four former provinces and the ten home-lands which had existed under apartheid were abolished and replaced by nine new provinces (Fig. 2.8 on page 29), which selectively absorbed or divided between them the pre-existing provinces and homelands. The nine provinces differ fundamentally from each other in a range of criteria as indicated in Table 6.3. Demographically, they vary from less than a million people in the case of the Northern Cape to over 8 million in KwaZulu-Natal. Unemployment levels reveal startling contrasts, varying from under 18 per cent in the Western Cape to nearly 50 per cent in the Eastern Cape. Further noteworthy differences exist between the provinces in terms of their overall contribution to the economy and prevailing poverty levels.

Plate 6.5 Shanty towns outside East London

The economic dominance of the Gauteng province (contributing 37.7 per cent of national GGP, the value of all types of production within a defined geographical area), in which Johannesburg is situated, is immediately apparent from Table 6.3, whilst the more rural provinces, which absorbed several former 'homelands', such as the Northern Province and Eastern Cape, are noted for their low economic contribution, relative to their share of the national population. Generally those provinces having high concentrations of rural people and containing large tracts of former homelands possesses the highest concentrations of poverty. As Table 6.3 indicates, poverty levels in Gauteng stood at 19 per cent of the population, compared with 78 per cent and 77 per cent of the population in the Eastern Cape and Northern Province. Whilst 19.7 per cent of people living in metropolitan areas could be classified as poor in 1994, the figure for smaller centres was 40.5 per cent, whilst that for rural areas was a staggering 73.7 per cent (RDP 1995). Over 43 per cent of rural dwellers are regarded as falling into the category of the ultra-poor. An even more depressing picture emerges if one examines the poverty levels which occurred specifically within the former homelands. As Table 6.4 vividly illustrates, with the exception of the former Kwandebele homeland (see Fig. 2.9a on page 31), all of the

Table 6.3  South Africa's provinces: key indicators

| Province | Population 1996 (million) | % National population | GGP% 1996 | Unemployment % 1996 | Poverty % 1994 |
|---|---|---|---|---|---|
| Gauteng | 7.348 | 18.1 | 37.7 | 28.2 | 19 |
| Eastern Cape | 6.302 | 15.5 | 7.6 | 48.5 | 78 |
| Northern Province | 4.929 | 12.1 | 3.7 | 46.0 | 77 |
| KwaZulu–Natal | 8.417 | 20.7 | 14.9 | 39.1 | 53 |
| Northern Cape | 0.840 | 2.1 | 2.1 | 28.5 | 57 |
| Free State | 2.663 | 8.5 | 6.2 | 30.0 | 66 |
| Mpumulanga | 2.800 | 6.9 | 8.1 | 32.9 | 52 |
| North West | 3.355 | 8.3 | 5.6 | 37.9 | 57 |
| Western Cape | 3.957 | 9.7 | 14.1 | 17.9 | 23 |
| Total | 40.584 | 100 | 100 | 33.9 | – |

*Source*: Crothers (1997); StatsSA (1998)
*Note*: Poverty is defined as the poorest 40 per cent of the population – i.e. those below the minimum living level (the breadline)
GGP: Gross Geographic Product (the value of all types of production within a defined geographical area)

Table 6.4 Poverty levels in the former 'homelands'

| Former 'homeland' | Poverty rate (%) |
| --- | --- |
| KwaZulu | 61 |
| Kangwane | 58 |
| Qwa-Qwa | 88 |
| Gazankulu | 69 |
| Lebowa | 83 |
| Kwandebele | 48 |
| Transkei | 92 |
| Bophuthatswana | 67 |
| Venda | 64 |
| Ciskei | 73 |

*Source*: RDP (1995)

rest had poverty levels either approaching or exceeding two-thirds of the total population. In the case of Transkei the figure is 92 per cent (RDP 1995).

In terms of personal income levels, a similar trend is discernible. In the case of Gauteng the personal income per capita in 1994 was R4,992 compared to only R725

Figure 6.1 Percentage of males absent (−) or present (+) by province

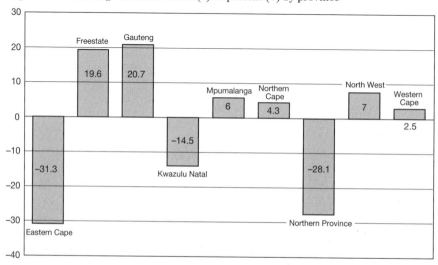

*Source*: Development Bank of South Africa (1994) DBSA

in the Northern Province and a national average of R2,556 (DBSA 1994). The uneven spread of economic opportunities is most poignantly expressed in the significant male absenteeism rate which exists as a result of male migrant workers from former homelands seeking employment in the wealthier provinces such as Gauteng and the Western Cape. While the national average male absenteeism rate was 5 per cent, in the Eastern Cape a figure of 31.3 per cent was recorded in 1994 (Fig. 6.1).

## Service provision

In many key aspects of service provision, South Africa once again does not fare badly relative to the developing world. However, as is the case with poverty and income, very deep differences exist within the society, which averaged statistics obscure. In the case of education for example, 81.4 per cent of the adult population were regarded as literate according to a 1997 survey, which compares favourably with a developing world average of 69.7 per cent (UNDP 1997). However, if one examines the picture more closely, stark differences are evident. Some 24 per cent of schoolchildren do not reach Grade 5 (the fifth year of formal schooling), whilst significant racial differences exist in the pass-rate in the final school-leaving certificate, the 'Matric' examination, which is taken after 12 years of formal schooling. Decades of racially based differences in education expenditure and the deprived environments of the townships and rural areas have played a role. However, perhaps surprisingly, given the historic allocation of government resources, Coloured and Indian pass rates exceeded those for white pupils. In 1994, the African pass rate was 49 per cent, that of whites was 88 per cent, Coloureds 93 per cent and Indians 97 per cent, while the national average was 58 per cent (du Toit, J. 1998). Pupil–teacher ratios also revealed marked racial and spatial disparities and helped to reinforce differences in pass rates. In the Western Cape, where the majority of pupils are white or Coloured, the ratio was one teacher to 23 pupils in 1993, whilst in the predominantly African Eastern Cape the ratio was one to 38 (DBSA 1994; Binns 1998).

Health care in the apartheid era was marked by a focus on first world medicine, a marginalization of primary heath care and the familiar racial inequalities. Whilst health care has been deracialized in the 1990s and the policy focus has shifted towards a greater emphasis on primary health care, clear challenges persist in terms of providing adequate facilities to the entire population. In 1992, the national average number of hospital beds was 5.1 per 1,000 people, but this varied from 6.5 in Gauteng to 2.4 in the largely rural Mpumulanga province (see Figure 2.9a on page 31). In many key areas health scores compare favourably with those in the developing world as a whole. For example, life expectancy in South Africa in 1994 was 63.7 years, compared with a developing world average of 61.8 years (but with considerable spatial variation within the country (Binns 1998)). The daily calorie intake was 2,705 compared with 2,553, while the annual death rate was 8 per 1,000 compared to 15 per 1,000 in sub-Saharan Africa (UNDP 1997; du Toit, J. 1998). Once again, however, these figures obscure stark racial and spatial contrasts.

Plate 6.6  A poorly equipped rural school, Eastern Cape

Some of the greatest inequalities in 1994 were to be found in the provision of water, sanitation and electricity. Even though considerable progress has been made in attempting to address such inequalities since 1994, there is considerably more still to be undertaken. In terms of water supply, 12 million people lacked access to clean drinking water in 1994 and 21 million did not have access to adequate sanitation. In rural areas, the situation was particularly acute, with half of the population lacking access to safe drinking water (ANC 1994). In racial terms the differences are equally striking, with only 27.3 per cent of African households having taps inside their dwellings in 1996. This compares with 72.4 per cent of Coloureds, 96.4 per cent of whites and 97.6 per cent of Indians (StatsSA 1998). Racial disparities were equally

Plate 6.7  Kingswood College, Grahamstown: a private school

stark in terms of electricity connections. In 1994, 64 per cent of households, some three million families, did not have access to electricity supply. Similarly, some 19,000 African schools and 4,000 predominantly rural clinics were not connected to electricity. Household supply varied from nearly 100 per cent in formal, middle- to high-income areas, to an average of only 1–4 per cent in the rural districts of the former homelands (DBSA 1994).

One of the key challenges inherited by the post-apartheid government was an acute housing shortage, primarily in urban areas, and largely for the African residents of the overcrowded townships and squatter camps. In 1994, the housing backlog was estimated to be some three million units, three-quarters of which were in urban areas, while the annual increment in housing demand was estimated to be some 200,000 units. By 1995, there were thought to be 628,000 squatter shacks in the country, the residents of which all desperately require better facilities (ANC 1994; du Toit, J. 1998).

## Employment and the economy

Economic control in the country has always been in the hands of vested white interests, as we have seen in previous chapters of this book. The distribution of economic power, though changing gradually since 1994 in favour of the rising black middle class, still reflects the overwhelming dominance of these white interests. In 1994, for example, 91 per cent of business franchises were owned by whites, compared to only 5 per cent by Africans, 1 per cent by Coloureds and 3 per cent by Indians (du Toit, J. 1998). Despite their increasing prominence, the African middle class constituted less than 7 per cent of the workforce in senior managerial and professional positions compared with over 84 per cent of positions held by whites (du Toit, J. 1998).

Overall employment trends are an even greater cause for concern. From 1960 to 1990, even though the number of job market entrants increased by 2.9 per cent per annum, the economy only generated 1.8 per cent additional jobs per annum (Rogerson 1995). Since 1990, the picture has been a bleak one in terms of the job market. According to recent statistics, between 1990 and 1996, employment opportunities and the overall size of the employment base in most sectors of the economy contracted. The number of mining jobs fell by 27.5 per cent, manufacturing by 8.1 per cent, construction by 21.3 per cent, the wholesale and retail trade by 3.8 per cent and public corporations by 24.7 per cent. The main sectors in which employment growth occurred were private services which grew by 4.3 per cent between 1990 and 1996 and public services which grew by 10.2 per cent (*Mail and Guardian*, 23 October 1998). The extremely high unemployment figures indicated in Table 6.3 (on page 237), which reveal a national average of 33.9 per cent in 1996, are a serious cause for concern in a country which is trying to promote equality of opportunity for all its citizens whilst simultaneously addressing past injustices.

By 1994, as we have seen in chapter 5, South Africa's economy had suffered from years of retarded growth, restricted investment and stagnation. By the early 1990s the economy was contracting by approximately 1.1 per cent per annum (Marais 1998).

Between 1985 and 1995, real Gross National Product per capita had also declined by an average of 1.1 per cent per annum (UNDP 1997). As indicated above, this had a knock-on effect on overall employment levels, which saw the shedding of 400,000 formal sector jobs between 1989 and 1993 (Marais 1998). Although positive economic growth was again achieved in 1994, many deep structural constraints remain. Some of the most serious issues include the fact that 40–45 per cent of the economically active population was outside the formal economy and gross domestic investment, as a percentage of GDP, had fallen from 27 per cent in 1983 to 15 per cent in 1993 (Michie and Padayachee 1997). In parallel, private investment and savings had plummeted, the latter falling to only 3 per cent of disposable income. Capital flight since the mid-1980s had also wrought its toll. By 1993 some US $20 billion had left the country, amounting to the equivalent of 5 per cent of annual GDP. Disinvestment by 40 per cent of transnational companies was one of the key causes of this situation (Marais 1998). Low levels of productivity among the workforce reflected low skills and training levels and volatile industrial relations within most economic sectors. Economic sanctions since the late 1980s have also taken their toll on overall economic growth levels and profitability. In addition, the situation has been compounded by structural considerations. As Marais (1998) argues, and as we suggested in chapter 5, the post-World War Two regime of accumulation in South Africa, based on import-substituting industry, the export of primary products which are subject to price fluctuations and a ruthlessly regimented labour supply, had reached a point of crisis which required the fundamental restructuring of economic, social and labour relations in order to survive in the modern global economy (Marais 1998).

On the positive side, post-apartheid South Africa inherited a reasonably stable macro-economic system. Although the budget deficit had risen steadily to 10.8 per cent of GDP in 1994, tight monetary control in the early 1990s had nearly halved the inflation rate to approximately 8 per cent. Another positive tendency was the reasonably low foreign debt, following debt rescheduling after 1985. By the end of 1993, the debt had been reduced to some US $16.7 billion or 14.8 per cent of GDP, which is a fraction of that in many developing countries, and is less than that in almost all Asian and South American countries (Michie and Padayachee 1997; Marais 1998). Other advantages inherited by the post-apartheid government included a well-developed transport and communication infrastructure system, surplus electricity generating capacity, abundant natural resources and considerable international goodwill, not least associated with the charismatic Nelson Mandela.

### The way forward?

As we have seen, even though democratic transformation and the ending of apartheid were enormous achievements in South Africa, tremendous challenges still exist which must be addressed if the country is to aspire to social and economic equality for all its citizens. The Reconstruction and Development Programme document issued by the ANC in 1994, and later adopted as formal government policy, explicitly recognizes the nature of the inherited inequality: 'Our history has been a bitter one dominated by colonialism, racism, apartheid, sexism and repressive labour policies. The result is that poverty and degradation exist side by side with modern cities and a developed mining, industrial and commercial infrastructure. Our income distribution is racially distorted and ranks as one of the most unequal in the world – lavish wealth and abject poverty characterise our society' (ANC 1994: 2). Segregation has certainly left scars of inequality and economic inefficiency, and the society has been brutalized by apartheid injustice over many decades. Gross, racially based disparities in terms of service provision, employment, housing, welfare and access to land are just some of the issues which need to be addressed, and they in turn provide the context for a persistently high rate of violent crime (chapter 7). As the ANC recognizes, apartheid resulted in a particular kind of imposed impoverishment: 'in every sphere of our society – economic, social, political, moral, cultural, environmental – South Africans are confronted by serious problems. There is not a single sector of South African society, nor a person living in South Africa, untouched by the ravages of apartheid' (ANC 1994: 3).

In addition to the specific legacies which it has inherited from apartheid though, the new South African state faces a daunting task by virtue of its position (one shared with most other decolonized states) within a global economy over which it has relatively little control. In the words of former Tanzanian President, Julius Nyerere, '[t]he truth of the matter is that the South African economy manifests the main characteristics of a typical developing economy. These are the predominance of subsistence over commercial activities; the narrow disarticulated production base with ill-adapted technology; the neglected informal sector; the lopsided development due to the urban and racial biases of public policies generally and development policies in particular; a particularly heavily subsidised import-substituting industrialisation strategy; and weak institutional capabilities' (Nyerere 1996: 25). The government's ability to address these development challenges will ultimately determine the success of its own policies and whether the overall quality of life can be improved.

Despite these considerations and the rather bleak picture painted here, there is some scope for cautious optimism. Bearing in mind the dangers of using averaged statistics as suggested above, within the context of Africa, South Africa is a relatively wealthy country. With a per capita GNP of more than US$3,160 in 1995, South Africa is the only country in sub-Saharan Africa, with the exception of Mauritius, to be classified as an 'upper middle income' country by the World Bank (World Bank 1997). There is surely cause for a degree of optimism, and hope that 'given its resources, South Africa can afford to feed, house, educate and provide health care for all its citizens' (ANC 1994: 14). Whether this potential will actually be realized remains to be seen.

## Chapter 7

# POST-APARTHEID 'DEVELOPMENT': Redistribution with growth?

The birth of the 'New South Africa' propelled the ANC from being a popular opposition movement lacking political experience to a political party which had to assume
the reins of power over one of the world's larger, and certainly one of its most complex, economies and societies. ANC politicians understandably embarked on a steep
learning curve as they tried to cope with an inherited and not always co-
operative, largely white and conservative bureaucracy, an enormous development
backlog and the expectation that the country would play a leading role in the region,
continent and, ultimately, the world, as a champion of democracy. For many in the
outside world the birth of a rare success story in Africa was so significant that undue
optimism and expectations were expressed which the new government cannot possibly live up to. Some observers, for instance, understandably wrote of a 'Golden Age
for Africa', suggesting that 'South Africa's vast amount of natural resources and
developed infrastructure could provide the mighty economic engine to drive sub-
Saharan Africa' (Evans 1991: 28). Although, as we shall see in the following chapter,
South Africa may be the only country on the continent which has the potential to
play anything like such a significant role, its internal problems are the ANC government's immediate priority and the task is not proving to be an easy one. A more sober
reflection on the issues facing the country is provided by Nagle:

> A new political era is beginning in South Africa and optimism is widespread,
> especially among blacks. However, there is little reason for such high expectations ...
> many problems remain to be dealt with. Rapid population growth, low standards of
> living, economic recession, redistribution of land and industrial resources, uncertainty
> and violence characterise contemporary South Africa. Indeed, the hopes of the black
> population are another problem as there is a mountain of projects to be undertaken,
> but without the stability or finance to do so. Apartheid might be dead, but inequalities
> remain and it will take more than political reform to redress them (Nagle 1994: 47).

Without radical transformation of economic fortunes, massive job creation and an
enormously expensive commitment to education and health, as we indicated in the

last chapter, 'huge gaps in access to services and quality of life are likely to persist' (Stock 1995: 394). The state's priority has to be simultaneously to foster high levels of economic growth and bring about greater social justice (Smith 1994). To date however, despite the effective removal of all legal forms of discrimination, economic growth is sluggish and, as we have seen, deep social inequalities persist. The vast levels of investment which were anticipated as a result of the political transition have not materialized. The ending of the ANC's temporary accord with the National Party, escalating crime and corruption and sporadic conflict in KwaZulu-Natal, have all blemished the post-apartheid image of a Rainbow Nation. The post-apartheid 'honeymoon' is clearly over. As one outside observer puts it, 'the "miracle" of 1994, when South Africa buried its divided past and inspired awe in the outside world, is starting to unravel' (*The Economist*, 13 December 1997: 17).

This chapter focuses on how the new South African state has attempted to come to terms with the development challenges of the post-apartheid era. It assesses the strategies pursued to bring about economic growth and to address those inequalities which were outlined in the last chapter. Particular attention is paid to a remarkably rapid transition in the ANC leadership's thinking, from the socialist imperatives of the Freedom Charter through the more mildly redistributionist Reconstruction and Development Programme (RDP) to the neo-liberal macro-economic strategy, known as Growth, Employment and Redistribution (GEAR) and the related rhetoric of an 'African Renaissance'.

## Reconciliation and compromise after 1994

In the early days of political transition after April 1994, the notion of reconciliation reigned supreme (Marais 1998). A remarkable alliance developed between former apartheid government officials and ANC leaders who had, until recently, been classified by the state as 'terrorists'. As one of these former 'terrorists', Govan Mbeki (the current President's father) put it, 'the oppressor and the oppressed came together to chart the road to a democratic South Africa' (Marais 1998: 85). Inspired by the example and character of Nelson Mandela, the transition was remarkably peaceful. The promise of language and other rights to minority ethnic groups such as Afrikaners, the acceptance of a Bill of Rights, the forging of the 'Government of National Unity', the absence of any significant 'racial strife', and the message to the local and international capitalist community that it was 'business as usual', together made for a remarkably trouble-free transition. These conditions also provided substance to the image of South Africa as the Rainbow Nation, in which all the different 'racial' elements combine harmoniously – an image projected most effectively by the popular Anglican Archbishop of Cape Town, Desmond Tutu.

One of the most impressive displays of a fragile, but very noticeable, national reconciliation came in 1995 in typical South African fashion, namely through sport. The country's dramatic victory in the Rugby World Cup was the 'most potent political event since the release of Mandela' (Waldmeir 1997: 270). Mandela's astute identification with the national team, his wearing of the Springbok rugby jersey and his personal involvement with the players endeared him to the largely white, Afrikaner, crowd and to the nation as a whole. The team's rugby victory and the later victory of the national soccer team at the Africa Cup of Nations contest, were important building blocks strengthening racial reconciliation and national unity, two trends which, sadly, are not as evident in the early twenty-first century as they were then.

The key stumbling block to a more enduring sense of national solidarity, according to Fine and Rustomjee, has been that the political transition, however much it may have been marked by symbolic moments of national unity and pride, 'has yet to be complemented by a corresponding economic transition' in which conditions are improved for the majority of citizens (Fine and Rustomjee 1996: 4). In order to understand the economic limitations of the transition, we need to recognize three kinds of constraint under which the new government has laboured. The first, and most immediate, concerned the accommodations with the NP that were necessary if a political agreement was to be reached at all. As Michie and Padayachee note, the ANC's electoral victory in 1994 was conditioned by 'the negotiations conducted under the old apartheid regime, which demanded and got various agreements to be adhered to, post election' (Michie and Padayachee 1997: 9). The compromises which the ANC leadership felt that it had to make included the entrenchment of Reserve Bank independence (partially removing government control over monetary policy, interest rates and exchange rates), the continued influence of the NP over many key ministries and guaranteed posts for former apartheid officials in the civil service (Michie and Padayachee 1997).

The second constraint, which is of a more enduring nature, follows from the fact that South Africa experienced the handover of political control from a white minority to a black majority later than any other colonized country. By the time that the ANC gained power, post-independence African Socialist and Marxist-Leninist experiments in government had already been tried elsewhere on the continent, as we will see in the next chapter, and had generally resulted in economic stagnation and disillusionment. As we suggested at the end of chapter 5, the new government in South Africa found itself already inextricably locked into a post-Cold War, Western-dominated global political economy. Neo-liberal economic policies, co-ordinated under what has been called the Washington Consensus, dominated the international environment by the mid-1990s, and effectively barred the ANC from employing radical initiatives of the kind attempted earlier and elsewhere on the continent. The currently deteriorating employment situation, however, would seem to suggest that it might have been more appropriate to pursue a different growth path – one more in line with an emerging 'post-Washington Consensus'. In principle, this new, and at the moment merely potential 'consensus', advocates a greater degree of state involvement in a national economy (Bond 1998).

The third constraint on the ANC government is also very much a consequence of the globalized economy. This is the need to attract international public and private investment if both economic growth and some degree of redistribution are to be achieved. The ANC's recognition of the manoeuvres which are necessary to secure overseas finance was manifested in its 1993 letter of intent to the International Monetary Fund, written to help access an international loan. This letter, discussed more fully below, stressed the importance of wage and inflation controls, trade and industrial liberalization, as well as the virtues of market forces (Marais 1998). It helped bind the ANC into a neo-liberal economic growth path and it indicated that the socialist rhetoric of the Freedom Charter had been abandoned in favour of global orthodoxy. The ANC's shift from socialism to neo-liberalism will now be analysed more closely by examining each major post-apartheid policy initiative in turn.

## The Reconstruction and Development Programme (RDP), 1994–96

In 1994, the ANC released the RDP as a programme of 'people-centred development', designed to address the injustices of the past and promote sustainable development in the future. Although the programme conceived of wide-ranging social transformation, it was couched in apolitical terms, and seemed more like a statement of moral regeneration than a set of policy positions. However, it provided an agenda to which all South Africa's political parties could subscribe (ANC 1994). Foreign endorsement was secured and not inconsiderable foreign grants, amounting to approximately US$200 million, or one-sixth of the first RDP budget, were received by 1995 to assist with the programme (RDP 1995).

During 1994 and 1995, the RDP in effect became the cornerstone of government policy, with all government departments, unions, civic bodies and many non-governmental organizations broadly identifying with its principles. In many ways it became a mantra for government policy, appearing to stand for the total 'transformation' of South Africa. It was widely used as a yardstick against which the success of the government could be assessed. As Blumenfeld puts it, the RDP 'served initially as a powerful symbol and focus for the post-apartheid reconstruction effort. In the immediate post-election period, it provided a mechanism for reconciling widely differing attitudes towards South Africa's development challenges, and its general support afforded legitimacy to the new government's efforts to come to grips with these challenges' (Blumenfeld 1997: 87).

### The goals of the RDP

Deploying the original 1994 RDP document, the ANC sought 'to mobilise all our people and our country's resources toward the final eradication of apartheid and the building of a democratic, non-racial and non-sexist future' (ANC 1994: 1). The programme sought to attain the material objectives of economic growth and basic needs provision, while at the same time addressing past injustices (Michie and Padayachee 1997). The document was full of the rhetoric of 'people-centred development', 'integrated and sustainable development', 'peace and security', 'nation-building' and 'democratization', expressed in five core programmes:

1. Meeting basic needs, such as public works, housing, land reform, transport, nutrition, health care and social welfare
2. Developing human resources through education, training, arts and culture and sport
3. Building the economy and addressing economic imbalances and poverty by encouraging urban and rural development
4. Democratizing the state and society, which meant reforming the various tiers of government, the public service and civil society
5. Bringing about effective management and financing so that each of the above could be implemented.

Some of the more important development targets which the document identified included the creation of 2.5 million jobs over a ten-year period; the building of one million houses by the year 2000, the connection to the national electricity grid of 2.5 million homes by 2000; the provision of running water and sewage to 1 million households; the redistribution of 30 per cent of agricultural land to emerging black farmers; the development of a new focus on primary health care; the provision of ten years of compulsory free education for all children; the encouragement of massive infrastructural improvements through public works, and the restructuring of state institutions by 1997 to reflect the broader race, class and gender composition of society (ANC 1994; Marais 1996).

## *Achievements and failures of the RDP*

In 1994, a unique RDP ministry was established within the office of the President, under the leadership of Jay Naidoo. Utilizing the 'extraordinary degree of acceptability' which the RDP document had achieved, Naidoo's ministry set about channelling the funds which were made available to it (Blumenfeld 1997: 76). Many of the RDP's projects, such as that in the Katorus area in the East Rand townships, improved housing, health care, education and infrastructure, whilst also providing support to small businesses. Rural water supply projects and support for the building of clinics and houses in townships and rural areas would also count among the RDP ministry's successes (Munslow and Fitzgerald 1997). The much-abused, but nonetheless important, school-feeding scheme was extended to feed 6 million children by the end of 1994, at a cost of US$0.8 million (RDP 1995). A survey undertaken amongst township residents in December 1997 established that one-fifth of those questioned believed that they had benefited from the RDP through improved electricity and water supply, education, health care and housing (Corder 1997). However, the same survey indicated frustration caused by the slow delivery of housing and job-creation projects.

These limitations were partly the result of administrative problems which plagued the RDP ministry right from the start. From its inception, there was a lack of clarity over whether it was a 'supra-ministry' or merely a unit to co-ordinate the policies of other ministries. The low rank of its officials, together with conflict with line ministries and delays in releasing funds and implementing projects, were all identified as

Plate 7.1  RDP health provision: the Dora Nginze Hospital, Port Elizabeth

stumbling blocks to more effective delivery (Blumenfeld, 1997). Of the approximately US$2.5 billion allocated to the ministry during its brief history, it probably only spent US$1 billion. But the ministry's administrative problems were compounded by difficulties which were inherent within the RDP programme itself. Soon after its release, the RDP document was criticized by the press and opposition parties, not on the grounds of its spirit and intent, which, as we have indicated, were widely applauded, but rather because of the vagueness of the strategies proposed and their tremendous cost implications.

The *Sunday Times* described the RDP document as 'pursuing the unachievable' (*Sunday Times*, 13 March 1994). One of its fundamental weaknesses was that the goals it set out were statements of intent rather than pragmatic development

Plate 7.2  RDP-financed infrastructure improvements Rini, Grahamstown

Plate 7.3  Mass state housing provision, East London

strategies. Rapoo argued that the RDP programme was characterized by an 'integrated approach (but) incoherent policies', and it did not seem to be based on any clear economic model of the South African economy (Rapoo 1996: 5). Compared with the post-independence national development strategies pursued in developing countries around the world, and despite having the superficial appearance of being a 'Five-Year Plan', the RDP lacked the ideological and doctrinaire commitment and the prescriptive planning and rigidity found in other countries. Furthermore, the fact that many targets that were set were quite unrealistic and the ministry was understaffed, left the RDP open to ideological struggle between radicals and pragmatists within the ANC. This contributed to the lack of coherent organization in the planning of projects (Blumenfeld 1997). Far from being a totally bottom-up, people-driven process, the RDP programme became more top-down with only very shallow participation and consultation taking place (Marais 1996).

An indication of the ANC leadership's early recognition of the RDP's limitations came in 1995 when the original document was rewritten in more moderate and market-oriented language as a government white paper (RSA 1995a). In this new version, many of the more ambitious development targets which had been set out only the year before were omitted. The implication was that a greater degree of pragmatism was already creeping in. During the following year, 1996, the RDP ministry was actually closed and its functions transferred to other ministries. Correspondingly, its budget was subsumed within the government budget as a whole (Corder 1997). In 1999, even the special RDP parliamentary committee ceased to operate (*Mail and Guardian*, 20 August 1999).

The effective abandonment of the RDP reflects a significant shift towards economic pragmatism and a less overt pursuit of broad socio-economic development objectives. This in turn is the result of the government's increasing free-market, as opposed to interventionist, leanings, and the basic impracticalities of economic self-reliance in a modern, global economy (Munslow and Fitzgerald 1997). Although the

ANC still professes its adherence to the goals of the 1994 RDP document, explicit reference to that document has declined markedly in official communications. The government's move away from the RDP as a policy-guiding mechanism was most notably manifested in the reduced allocation, in real terms, of funds for housing, agriculture, rural development and water and sanitation in the 1999 budget (*Mail and Guardian*, 19 February 1999).

Despite its practical ineffectiveness, however, the RDP has played a critical role in fostering a more democratic culture and establishing ANC legitimacy, both in South Africa and abroad. The almost universal domestic and international acceptance of the programme in principle, if not in its practicalities, did a great deal to ensure stability during the immediate transition period, and to enhance the credibility of the new government. For the ANC's own constituency, the RDP became a symbol of change, winning broad support for the party's objectives and giving substance to its expressed desire to transform society and the economy (Blumenfeld 1997). As Munslow and Fitzgerald conclude, 'there is no doubt that the RDP played a pivotal role in ensuring the successful transition from separate development towards a more sustainable development future' (Munslow and Fitzgerald 1997: 59).

## The Growth, Employment and Redistribution strategy (GEAR): a shift to economic orthodoxy

While the RDP was a broad-based social and economic development strategy, the ANC government prior to 1996 lacked a clearly defined macro-economic strategy. It soon became evident that this was impeding access to international loans and retarding overseas investment. The new government adopted such a macro-economic strategy, known as Growth, Employment and Redistribution (GEAR) in 1996 – and it was one which accorded with internationally 'acceptable' principles of neo-liberal economic management. This acceptance of Western free-market development prescriptions marked a dramatic turnaround in ANC thinking. Until its unbanning in 1990, the closest thing that the ANC had to a manifesto was the socialistic Freedom Charter, written at the height of apartheid in 1955. As we have seen, this promised a major redistribution of capitalist assets through the nationalization of key economic sectors. After the unbanning of the ANC and the release of Nelson Mandela, ANC policy was still ostensibly based on the principle that growth would occur through redistribution and nationalization. The ANC leadership publicly expressed its belief that 'apartheid had left a legacy of such distortions in the operation of the South African economy that extensive state intervention was required for reconstruction' (Wittenberg 1997: 176). The launch of GEAR in 1996, however, highlighted what has been called the ANC's 'short walk to economic orthodoxy' (Marais 1998: 146).

Business interests in South Africa, the major international financial bodies such as the IMF and World Bank and Western governments immediately welcomed GEAR. But for the opponents of neo-liberalism within the tripartite alliance which the ANC had established with the trade unions and the Communist Party, it marked a serious renunciation of the ANC's revolutionary ideals (McKinley, *Mail and Guardian*, 31 October 1997). These opponents shared Marais' view that GEAR was 'unabashedly geared to service the respective prerogatives of national and international capital and the aspirations of the emerging black middle class', rather than to effect a long-overdue redistribution of wealth (Marais 1996: 147).

Since its implementation, the Left in South Africa has continued to be severely critical of GEAR, for understandable reasons, and most notably because of its failure to create jobs and improve conditions for the majority. However, it is doubtful whether at the present historical juncture, pursuing an alternative strategy would necessarily have achieved better. Within a Western-dominated, post-Cold War international environment, which is hostile to socialistic 'experiments', the risk of international isolation, disinvestment and restricted foreign investment would probably have nullified the potential of any interventionist, redistributive approach. The absence of any tried and tested alternatives which could, realistically, have been copied from elsewhere and applied locally underscores this reality. In the following section the increasingly free-market orientation of the ANC is examined in more detail before the key features of GEAR and the response which it has received are described.

### From RDP to GEAR: the move to economic orthodoxy, 1991–1996

Despite its lingering rhetorical commitment to a mass redistribution of wealth, the ANC leadership's ideological shift had clearly already occurred by the time it came into power in 1994. Several significant events in the early 1990s set the seal on emerging policy. According to Matisonn, writing in the influential *Mail and Guardian*, the collapse of state socialism in Eastern Europe, the evident problems of state-run economies (to which the ANC exiles in Lusaka, Zambia were exposed directly), together with the realization that economic isolation was not a viable option and that significant investment was going to be needed if the country was to achieve mass welfare and real economic growth, all encouraged a fundamental policy shift (*Mail and Guardian*, 6 November 1998). Behind-the-scenes negotiations with overseas powers and global financial institutions influenced the ANC's nascent economic policy as soon as it became clear that the movement would be playing a role in government. Indeed, the ANC leadership was inundated with economic plans and analyses of future growth scenarios produced both by the private sector in South Africa and the World Bank (Marais 1996).

The effect of such external 'advice', and of the ANC's own internal deliberations, was first apparent in 1991 when, in an address on the 'role of the state' at the University of Pittsburgh, Mandela stated that 'the private sector must and will play the central and decisive role' (Matisonn in *Mail and Guardian*, 6 November 1998: 16). Two key events in 1993 firmly demarcated the new policy position. In that year the ANC effectively rejected the policy findings of the Macro-Economic Research Group which the party itself had set up to formulate post-election economic policy. The Group had called for Keynesian-style management and intervention. Although this may well have been necessary in the more depressed areas of the country, the entire approach was rejected by the then ANC executive member and later Finance Minister, Trevor Manuel, and declared not to be ANC policy (Wittenberg 1997). Of equal significance was the secret letter of intent, mentioned above, which was sent to the IMF to unlock a US$850 million loan. That letter apparently committed the ANC to accepting that an increase in the government deficit would jeopardize the future of the country. Furthermore, it committed the signatories to reducing debt to 6 per cent of GDP, maintaining a high interest policy, effecting a cut in government spending and ensuring openness to international competition (Matisonn in *Mail and Guardian*, 6 November 1998).

By 1994 the shift from the earlier socialist ideas was becoming more publicly evident. President Mandela announced in that year of transition that 'in our economic policies ... there is not a single reference to things like nationalisation, and this is not accidental. There is not a single slogan that will connect us with any Marxist ideology' (Marais 1998: 146). The release of the politically 'neutral' RDP, which, as we have seen, did not lock the state into specific policies of economic management, gave further substance to the public impression that a fundamental policy shift had taken place. While still maintaining that their initial redistributionist policies were economically and socially desirable, the ANC had evidently recognized that pursuing them would have unleashed a negative local and international reaction (Marais 1998). According to Matisonn 'it seems that the ANC was looking at a larger picture, the need to show its commitment to a regime that would give the investors confidence' (Matisonn in *Mail and Guardian*, 6 November 1998: 17).

Until late 1995 the new government appeared to have been content with its broadly humanitarian RDP policy and with largely implicit support for the free market. However, from 1995, economic problems became evident. The economy was not achieving the desired growth rates, delivery of RDP targets, as we have seen, was frustratingly slow and investors and international financial institutions were demanding greater clarity on national economic policy. The need for a new growth strategy was evident. In 1995 the Office of the Deputy President put together a first National Growth Strategy which, in some ways seemed to be a modification and extension of the RDP. However, this strategy was seen as lacking a macro-economic thrust and was never fully implemented (Matisonn in *Mail and Guardian*, 6 November 1998). So, a new macro-economic strategy, GEAR, was devised. It was based largely on the thinking emanating from what are, broadly speaking, economically orthodox institutions, both in South Africa and overseas. Particularly influential were the South African Reserve Bank, the World Bank, the Development Bank of Southern Africa and the University of Stellenbosch. Together, they helped the ANC to lay the groundwork for an internationally acceptable policy (Bond 1996).

## Implementing GEAR

GEAR was designed to achieve high rates of economic growth, to expand the private sector, to improve output and employment, to achieve fiscal reform and to encourage trade and investment. Simultaneously, it sought to achieve redistribution and an improvement in basic living conditions as a result of generally revitalized economic performance (RSA 1996a). GEAR shared some similarities with the RDP, in that both were bold statements of ideals which did not rigidly lock the government into the pursuit of a defined strategy. GEAR, however, much more explicitly than the RDP, rests on the belief that the expansion of the private sector will drive the economy, whilst the role of the state will largely be a facilitatory one. The redistribution goals of the RDP will therefore be attained by a more circuitous route.

In order to encourage economic growth and the expansion of the private sector, the government undertook to reduce state spending and the budget deficit; reduce corporate taxes and foreign exchange controls; control inflation; promote privatization, and encourage wage restraint: all goals which are prescribed as universal panaceas for 'development' by US-dominated institutions such as the World Bank and IMF (RSA, 1996a; see chapter 8). According to GEAR, the results of such an approach would be an economic growth rate of 6 per cent per annum by the year 2000, which in turn would generate up to 400,000 jobs per annum, boost exports by over 8 per cent per annum and lead to a drastic improvement in social conditions. It was anticipated that in the period from 1996 to 2000, 1.35 million jobs would be created. Critical to the attainment of these goals was the belief that there would be significant increases in private investment and non-gold exports, together with increased state expenditure on social infrastructure (RSA 1996a; Marais 1998).

Unsurprisingly, GEAR was welcomed by the corporate sector in South Africa and by foreign economic powers and financial institutions. However, the left wing and labour movement in South Africa reacted with a dismay which has, at times, threatened the ANC–trade union alliance. Critics of GEAR have pointed to the contradictions entailed in making a commitment to the expansion of social infrastructure whilst at the same time promising a curb on public spending. Similarly, the goal of job creation is apparently undercut by a commitment to the rationalization of the civil service (Marais 1998). Nevertheless, the ANC has determinedly wedded itself to the strategy, possibly precisely in anticipation of such internal opposition. In 1996 the Finance Minister declared that GEAR was 'non-negotiable' (Marais 1998).

## GEAR and South Africa's Economic Fortunes

In hindsight, GEAR has failed to attain many of its goals. Although changes in corporate tax levels (cut to 30 per cent in 1999), foreign exchange controls and interest rates have yielded some degree of increased investment, many of the strategy's long-term targets have not been met. On the positive side, by September 1999, the monthly inflation rate was 1.9 per cent, its lowest level in 31 years, and investment expanded to reach $1.7 billion in 1997, double the 1996 figure (*Business Times*, 14 February 1999; Nelan in *Time Magazine*, 24 May 1999; *Eastern Province Herald*, 20 October 1999). In addition, according to the Minister of Trade and Industry, Alec Erwin, South Africa survived the 1997–98 Asian crisis and maintained growth despite international insecurity (SABC, 22 October 1998). However, the anticipated 6 per cent growth rates in GDP remain illusory. In 1995, the economy grew by only 3.3 per cent, by 1998 it was actually contracting at 0.3 per cent, and a reversal to a

0.5–1 per cent growth rate was anticipated for 1999 (SABC, 22 February 1999; *Sunday Times*, 28 February 1999).

What is far more sobering, however, is the exceptionally poor performance achieved in the job market. Instead of achieving the creation of 200,000 new jobs in the first 18 months of GEAR's existence, 80,000 were in fact lost (SABC, 16 November 1997). South Africa is currently suffering from the phenomenon of 'jobless growth' – a term devised to describe a situation in which the economy is expanding, mainly in capital-intensive sectors, but simultaneously shedding jobs in the less economically viable, labour-intensive sectors. An example of this scenario has been provided with the development of a new aluminium smelter at Richards Bay, which could contribute a noteworthy 1.5 per cent of national GDP, but which will create only some 1,500 new jobs (Fitzgerald *et al.* 1997).

By 1999, it was estimated that the economy had actually shed some 500,000 jobs since 1994 and 350,000 since the inauguration of GEAR in 1996 (*Sunday Times*, 4 April 1999). This was at a time when it was estimated that the number of people entering the job market increased by 2.5 per cent per annum (Michie and Padayachee 1997). In terms of the country's international stature and potential for investment, the situation is not that favourable. According to the 1999 World Competitive Report, out of the 54 countries examined, South Africa secured 47th place (*Business Times*, 18 July 1999). GEAR clearly has not lived up to the government's expectations. Although it has helped to ensure macro-economic stability and thus enhanced the government's international status, it has done little to address internal problems of rocketing unemployment and limited economic growth. Indeed, it would be fair to say that conditions for the average South African have deteriorated significantly since 1996, albeit not just because of GEAR.

## The economy and job creation

### *Broad economic trends and unemployment*

Aside from any effects of government policy, South Africa's recent economic performance has been affected by a range of influences beyond the government's direct control. These include the national and international recessions of the 1970s and 1980s, a major drought in the early 1990s, the decline in world market prices for products such as wool, the lowering of trade barriers, which has devastated labour-intensive production, particularly in the clothing and textile sectors and, of critical importance, a recent fall in demand and the price of gold, which had been the traditional mainstay of the economy (Lundahl 1997b). The worst-affected sector in terms of employment is that of mining, where the job loss in the 10-year period from 1987 to 1997 was estimated to be in the order of 250,000 (SABC, 7 July 1997). Total gold production has nearly halved since 1970 (du Toit 1998).

The consequences of such job loss on selected mining towns (such as Welkom in the Free State, which has lost 100,000 jobs), and on the migrants' former homeland source areas, has been devastating (Department and Provincial and Local Government, 2000). A net result of these trends is a *de jure* unemployment rate of 36 per cent (StatsSA 1999a), although figures of over 50 per cent are quoted in many parts of the country. In the deep rural areas and the former homelands in particular, figures in excess of 80 per cent have been noted (DBSA 1991). In these areas, on average, 53 per cent of people depend on remittances and pensions for survival, with only 3 per cent of households relying on farming as their main source of income (StatsSA 1999b). These problems are further aggravated by a variety of deep-rooted economic problems, including the limited capacity of the industrial sector to achieve international competitiveness following decades of tariff protection and support, the traditional over-reliance on the mining economy, low domestic savings rates and an unevenly distributed culture of non-payment for services in urban areas (Nattrass 1996; *Sunday Times*, 21 February 1999).

In terms of international trade, however, conditions have been improving. By 1998, the country was recording a trade surplus of almost US$300 million per annum, with total exports amounting to some US$24 billion (SABC, 24 January 1999). Following drawn-out negotiations to secure a free trade deal with the European Union, an agreement was finally signed in October 1999 (SABC, 14 October 1999; see chapter 8), an achievement which will no doubt favour exporters, but could well threaten still further less efficient domestic producers (see chapter 8).

By the end of 1998 there were signs of some significant economic growth in certain sectors. In that year the government reported that 20 mega-projects, each costing about US$100 million, were being undertaken (Government Communication and Information Services 1998). Lowering the corporate tax rate and opening up the economy has encouraged an increase in foreign investment. In parallel, the tourism sector is booming, with the number of visitors doubling from 1994 to almost 4 million a year. Based on projections made in 1995, the country is, however, only attracting less than 0.5 per cent of the global tourist trade (RSA 1995b). The building of hotels and the promotion of a range of tourism activities are very evident responses to this new boom industry which relies on the country's incredible natural and cultural beauty and diversity (Nelan in *Time Magazine*, 24 May 1999). Significant potential for expansion exists in this sector, but the uncertainties caused by crime are serious impediments to the sustained growth of the industry. On the negative side the slow downward slide in the value of the rand for several years, and the decrease in domestic investment are worrying trends (Nelan in *Time Magazine*, 24 May 1999).

### *Policies to promote economic development and the 'jobs summit'*

Government industrial and trade policy has focused on increasing South Africa's export orientation, which is reflected in improved trade statistics and the securing of lucrative foreign contracts, for example in vehicle sales. Another core thrust of government policy has been targeted support for the small, medium and micro-enterprise sector, which is estimated as providing 44.8 per cent of private sector employment (RSA 1997). Support has been given to emerging entrepreneurs through various organizations such as Ntsika, and government has encouraged the emergence of local business service centres. Unevenly distributed skills levels and the lack of access of new businesses to financial support and credit have, however, limited the overall success of such strategies. Other spatially targeted employment-generating strategies are detailed in the section on regional development below.

One of the more noteworthy recent economic successes of the government was the result of an ironic continuation of the apartheid state's military concerns. This was the announcement of an arms deal in November 1998. The deal, which will see the country purchase US$5 billion worth of ships, submarines, helicopters and jet fighters from a range of European countries, is linked to significant counter-trade deals, which could lead to investments and purchases worth US$18 billion, generating some 65,000 jobs (*Business Day*, 19 November 1998). Whether all the anticipated spin-offs will materialize is yet to be seen, and the anticipated gains are questioned by some commentators (Mulholland in *Sunday Times*, 25 July 1999).

As this overview of recent economic performance suggests, creating jobs has to be one of the central considerations of government in post-apartheid South Africa. The worsening employment situation and the fact that the free market is not providing for the anticipated expansion in job opportunities have become issues of serious concern. Achieving the necessary economic growth to generate adequate employment is the greatest challenge for the government according to Blumenfeld (1997), whilst Venter (1997) regards the persistence of poverty and unemployment as a 'time bomb'. The failure to generate adequate jobs has led to a situation where in the 1980s at least 30 per cent of South Africa's labour force was engaged in some form of informal activity (Rogerson 1995). At a regional level, unemployment is aggravated by the tendency of new employment opportunities to focus on the metropolitan areas, while the rural areas and small towns suffer from the loss of former regional development support and need to rely more on pro-active local or community development initiatives. GEAR is clearly not providing the mechanisms necessary to boost employment in such areas (*Mail and Guardian*, 20 August 1999).

Given the seriousness of the prevailing employment crisis, it is surprising how long the government took to respond, and then only in a somewhat limited fashion. Following repeated delays, a Jobs Summit was finally called in October 1998. At that meeting government, business, the private sector, trades unions and various representatives of civil society agreed on a variety of programmes to try to expand the employment base of the country. Heralded by President Mandela as the most important event since the first democratic election, the meeting enabled government to secure consensus with the private sector and partially to address tensions with organized labour (*Sunday Independent*, 1 November 1998). In the declaration which emerged from the summit, it was, interestingly, acknowledged that GEAR targets would not be met, and it was stated that government was committed to structural reform in a socially sustainable way (*Eastern Province Herald*, 31 October 1998). It was also announced that government would allocate approximately US$16 million per month to address unemployment, to accelerate housing construction and to fund various public works programmes (*Mail and Guardian*, 30 October 1998). In addition, in the February 1999 budget, it was announced that nearly US$130 million was to be allocated to support existing job creation programmes, such as Working for Water (which clears alien vegetation from catchments), land care and municipal infrastructure programmes (*Mail and Guardian*, 19 February 1999). While such positive action is to be welcomed, there is a sense that significantly more needs to be done, given the scale of the unemployment problem. According to one critic, as long as the government remains wedded to the market and to tight controls, 'the Jobs Summit is a cruel deception ... it raises expectations that will not be met' (Barrel in *Mail and Guardian*, 30 October 1998).

### *Regional development and public works programmes*

Regional development has long been pursued in South Africa, albeit until the 1990s, primarily as a mechanism designed to uphold apartheid through the 'independence' of the former homelands (Rogerson and Pirie 1979; summary box 8, pages 227–8). By the late 1980s some 2,500 firms or 14 per cent of the national total were in receipt of state assistance. In 1991, former levels of state support were terminated in favour of a significantly scaled-down 'aspatial' strategy which provided tax-related incentives to qualifying industrial firms in most parts of the country (Nel 1999). Within any analysis of regional development in South Africa, it is critical to note that the diminution of former levels of support, though perhaps economically rational, has left the residents of the rural peripheries in an even more impoverished situation. In former homelands such as Ciskei, something in the order of 10,000 jobs remain from an original maximum of 26,000 in 1991. The absence of alternative strategies to generate employment in such areas has significantly increased marginalization and suffering in the rural periphery (Nel 1999; Fox and Nel 2000).

Since 1996 the government has pursued four strategies of regional development. The first has focused on the provision of tax relief to firms selected on the basis of what they produce, their degree of labour-intensity and their location. Known as the 'tax holiday scheme' the strategy does not appear to have had a major impact on industrial employment and was largely in abeyance by 1999. A parallel strategy, the small–medium manufacturing grant, makes provision for financial support and tax relief for smaller enterprises. By the end of 1998, 21,000 jobs had been created through this endeavour (Government Communication and Information Services 1999). On a far grander scale have been the two parallel strategies of Spatial Development Initiatives (SDIs) and Industrial Development Zones (IDZs) (see Fig. 7.1). Whilst the former seeks to promote development in a range of sectors in the proximity of key transport corridors, the latter has identified a series of proposed 'state of the art' industrial zones, which, superficially at least, would seem to share many similarities with the internationally recognized concept of Export Processing Zones. It is important to note that beyond the provision of transport infrastructure and basic facilitation, the government has essentially handed the process over to the private sector. With the exception of the 'Maputo corridor' which links Johannesburg to Maputo in neighbouring Mozambique, it would be difficult to argue that significant progress had been made in achieving the desired levels of growth by 1999. By 1998 some US$13 billion of investment had, in principle, been committed by the private sector to SDI projects (Government Communication and Information Service 1998). In terms of IDZs, enabling legislation was still pending in 1999 and it is difficult to predict whether prospective investors will be drawn to such estates, where incentives compare poorly with those available internationally in similar zones. At the proposed 'flagship' IDZ, adjacent to the planned deep-water port of Coega near Port Elizabeth in the Eastern Cape, a worst-case estimate was that in return for an investment of US$250 million only 1,000 jobs would be created; this in an area where the population has increased by over 100,000 in the 1990s and the unemployment rate is estimated to be 50 per cent (Fox and Nel 2000).

Figure 7.1 Spatial Development Initiative and Industrial Development Zones, 1996

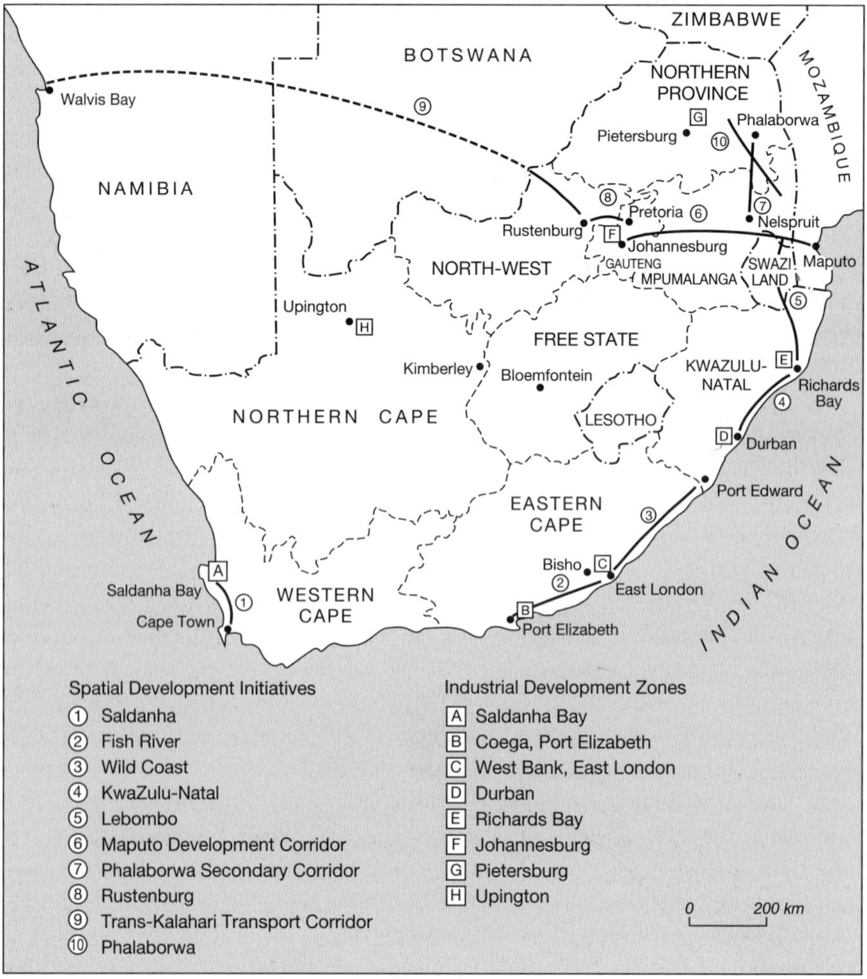
ZIMBABWE
BOTSWANA
MOZAMBIQUE
NORTHERN PROVINCE
Walvis Bay
NAMIBIA
Pietersburg
Phalaborwa
Rustenburg
Pretoria
Johannesburg
Nelspruit
Maputo
GAUTENG
MPUMALANGA
SWAZI LAND
NORTH-WEST
Upington
FREE STATE
Kimberley
Bloemfontein
KWAZULU-NATAL
Richards Bay
LESOTHO
NORTHERN CAPE
Durban
ATLANTIC OCEAN
Port Edward
EASTERN CAPE
INDIAN OCEAN
Bisho
East London
Saldanha Bay
Cape Town
WESTERN CAPE
Port Elizabeth
Spatial Development Initiatives
① Saldanha
② Fish River
③ Wild Coast
④ KwaZulu-Natal
⑤ Lebombo
⑥ Maputo Development Corridor
⑦ Phalaborwa Secondary Corridor
⑧ Rustenburg
⑨ Trans-Kalahari Transport Corridor
⑩ Phalaborwa
Industrial Development Zones
A Saldanha Bay
B Coega, Port Elizabeth
C West Bank, East London
D Durban
E Richards Bay
F Johannesburg
G Pietersburg
H Upington
0 200 km

*Sources*: Rogerson (1997); Nel (2000)

Less dramatic, but of equal significance have been attempts to promote short-term employment and skills training through a variety of public works type programmes, which have variously targeted the provision of rural water supply, housing construction and the aforementioned Working for Water programme. By late 1997, locally employed people had helped to extend water supply to 1.7 million people, using labour-intensive construction techniques (*The Economist*, 13 December 1997). By late 1998, 42,000 people had been employed in various catchments around the country on the Working for Water programme, removing alien vegetation to ensure improved water supplies to storage dams (*Mail and Guardian*, 8 January 1999). During 1999, as result of the Jobs Summit, these strategies were expanded and new target groups were reached through programmes such as labour-intensive road repair programmes. While the significance of such programmes in addressing poverty is indisputable, the lasting impact of such selective skills training and the finite nature of projects such as tree-clearing, do raise concerns over the long-term future of many such programmes once the funding ceases.

## Addressing inequalities

In parallel with the various development and economic strategies in post-apartheid South Africa, important and often impressive progress has been made in confronting a range of social issues. In accordance with the principles of the RDP, and in an effort to improve conditions for all South Africans, some significant advances have been made in areas as diverse as housing, education and health care. Some of the main achievements will be considered here.

As a result of the democratic transition in 1994, basic freedoms have at least been extended to all South African citizens, primarily through a new constitution, agreed in 1996, and an associated Bill of Rights (RSA 1996b). South Africa's constitution, recognizing the universal rights of its citizens and prescribing government activities, is now regarded by many as the most liberal in the world. Principles of democratization and local participation have also been written into local government policy, which assure citizens a right to participate in local level decision-making affecting their lives (RSA 1998a). As a result, extremely high levels of transparency now exist in the public sector, on paper at least, and strenuous efforts are being made to ensure accountability and to eradicate corruption. Efforts to promote community commitment and payment for services within local authority areas, known as the Masakhane campaign, have not, however, lived up to expectations.

As far as improving living conditions and addressing inherited inequalities are concerned, the most impressive gains have been made by the electricity and telecommunications parastatals, which have significantly expanded their coverage, particularly in rural areas. Between 1994 and 1998 the number of houses with electricity connections increased from 31 per cent to 63 per cent, while 1,700 schools and 497 clinics were connected (*Eastern Province Herald*, 10 February 1999). In the same period, the number of telephone lines increased from 150,000 to 386,426 (*Sunday*

*Times Supplement*, 28 February 1999). By 1998, as a result of government programmes, 1.7 million additional households had been supplied with water (Government Communication and Information Service 1998). Despite these impressive gains, as emerging reports are suggesting, in many areas people are too poor to pay the user charges which are now being levied for these services. As a result, some are being disconnected. Training for community management of schemes has also been inadequate and, in the case of water, many villagers have reverted to collecting their supplies manually (*Mail and Guardian*, 26 March 1999).

As we have already noted, one of the major challenges facing the new government in 1994 was an overwhelming shortage of housing. By 1996, the backlog was estimated at 1.5 million houses, with a further 1 million squatter shacks existing in the country (Department of Housing 1996). Despite a slow start, the government's low-income housing programme has achieved some impressive results. By 1999, 3 million people (a quarter of those in need) had been rehoused (SABC 16 January 1999). By that stage the government had built 680,000 houses for low-income residents and had also provided 959,000 housing subsidies (*Mail and Guardian*, 30 April 1999). Although falling short of the RDP's promise of 1 million houses in five years, what has been achieved is

Plate 7.4  Rural public works programme: domestic water supply in Brak Kloof, Eastern Cape

Plate 7.5 New telephones and transmitter in a remote part of the former Ciskei 'homeland'

nonetheless extremely significant. Impressive advances have been made in addressing the housing shortfall and improving the livelihoods of millions of people in a comparatively short space of time. Attendant problems include the small size of units provided and, in certain areas, poor building standards and corruption, which have undercut the general success of the strategy (*Sunday Times*, 14 March 1999).

Improvements in health care were also given a prominent position within the RDP and, under the leadership of Minister Zuma, significant advances were made in the extension of primary health care and the provision of free medical attention to a large proportion of the population. On the negative side, however, allegations over the

Plate 7.6  A squatter settlement in Duncan village outside East London

misuse of funds, reluctance to make available anti-AIDS drugs and disputes with medical aid and drug suppliers have occasioned significant criticism of government health policy (*Mail and Guardian*, 5 February 1999). The purging of the drug-control agency when it disagreed with the government's support for a toxic 'cure' for AIDS called Virodene, was a surprising attack on principles of medical practitioners' independence (*The Economist*, 29 May 1999). In a parallel programme, an ambitious school feeding scheme was introduced in 1994, in an effort to address malnutrition. Despite instances of abuse, some 6 million pupils were already benefiting from the provision by late 1994 (RDP 1995).

Plate 7.7  RDP houses/Mandela homes, Agulhas, Western Cape

It would be fair to argue that education is probably the single most important priority of the government, but it is also one in which some of the most intractable problems have been experienced. The state has at last achieved parity in government spending per pupil and removed racially separate educational provision. But gross disparities remain between different areas, with schools in the predominantly black areas continuing to be characterized by poor facilities and services (Lemon 1999; Nel and Binns 1999). Great economic differences between the areas in which schools are located and the differential levels of investment which parents are able to put into schooling suggest that, at present, many schools in predominantly black areas are unlikely to be significantly upgraded (Lemon 1999). In addition, despite state attempts to even out resource allocation between provinces, provincial autonomy over the use of educational resources may actually widen the gap between the relatively wealthy and the poorer provinces (Lemon and Stevens 1999). On the positive side, 10,000 new classrooms were provided in the period from 1994 to 1999, and 1.5 million more pupils were brought into the education system (*Mail and Guardian*, 21 May 1999).

Attempts are being made to transcend syllabuses which were designed to propagate apartheid's racial divisions. A radical curriculum redesign, known as Curriculum 2005 is being introduced, but is being impeded by lack of funds, very serious shortfalls in the supply of textbooks, minimal retraining of teachers and the sheer rapidity of changes (*Mail and Guardian*, 18 January 1999). Despite the attention which is being paid to education, many problems persist, with a decline in the numbers of tertiary-level students and a fall in the overall pass rate of the final school-leaving certificate from 60 per cent in 1994 to 50 per cent in 1997. Such trends do not bode well for the country's human resource development prospects (du Toit, J. 1998). In 1999 Kadar Asmal, one of the most dynamic senior ANC cabinet members, was appointed as the new Minister of Education. This move suggests that the further reform of the education system is likley to be forthcoming.

Under apartheid many millions of Africans were forcibly removed from their land  and dumped in the former 'homelands' (chapter 5). Some of the primary programmes of the new government include attempts to resettle and/or compensate displaced persons, to promote land redistribution (using state or purchased land) and to promote rural development. Land reform, unlike in other post-independent African countries, does not appear to have been the overwhelming focus of concern in South Africa, possibly because, unlike the rest of Africa, the country's population is predominantly urbanized. Despite this, some noteworthy achievements have been attained. In terms of the Land Reform Pilot Programme, by 1997, 372 projects which would benefit 136,000 people and affect 2.17 million hectares of land had been approved. In parallel, 340 land redistribution projects which would benefit 68,000 households were under way. By 1999 progress had been made, some 22 laws relating to land matters had been promulgated and 53,675 families (354,225 people) had gained access to 6,544,998 hectares of land (*Sunday Times Supplement*, 28 February 1999). Workers housed on farms are now protected from unfair evictions and greater security of tenure has been established for some 6 million people. However, by 1999, only 31 out of the 22,500 land claims filed by people who wished to return to land from which they had been removed, had actually been settled (Government Communication and Information Service 1998). By 1999, less than 1 per cent of land in the country had been redistributed, in spite of the government's initial goal of 30 per cent (*Mail and Guardian*, 27 May 1999). Inexperience and administrative 'red tape' have been blamed for the slow progress in this area.

Given South Africa's appalling legacy of human rights abuses under apartheid and the escalating crime problem, it is inevitable that issues of justice should feature prominently in the current social transformation. The single most important feature of the transition in this regard has been the Truth and Reconciliation Commission (TRC), established under the chairmanship of Archbishop Desmond Tutu to hear submissions from both the victims and the perpetrators of human rights abuses in the apartheid era. The offering of amnesty from prosecution and compensation for victims has been regarded as part of a process of national catharsis, helping all South Africans to come to terms with the past. By 1999, 7,000 people had applied for amnesty (SABC, 27 May 1999). The TRC has exposed many whites for the first time to the atrocities committed by government forces on their behalf, and it has given the relatives of many victims of violence the chance to find out what happened to their loved ones. However, while the ANC's application for blanket amnesty for crimes committed in its military camps was refused on technical grounds, apartheid leaders have largely managed to avoid being held to account (*The Star*, 5 March 1999).

At a broader level, advances in legal protection have been achieved through the creation of a range of legal instruments which include the Constitutional Court, the Office of the Public Protector, the Human Rights Commission and the Gender Equality Commission (*Sunday Times Supplement*, 28 February 1999). Equally important has been the establishment of the Heath Special Investigation Unit which has been empowered to investigate cases of corruption and to recover lost assets and funds. By the end of 1998, the unit had recovered or protected nearly US$150 million of assets and money (*Mail and Guardian*, 12 May 1999).

Plate 7.8  Farmworkers' accommodation, Cathcart, Eastern Cape

## The 'African Renaissance'

The catch phrase of Deputy President – now President – Thabo Mbeki's political philosophy has been the 'African Renaissance'. Though not a defined development plan like the RDP or GEAR, it has become equally important in terms of the political mantras used in contemporary South Africa and has, in fact, subsequently been adopted by many African leaders. Its use, indeed, has become infectious and even US President Clinton has made use of the term. The concept of the African Renaissance focuses on the notion of rebirth and hope for Africa, but is rather vague in terms of focus and content. Vale (1997) argues that this is deliberate as it is only designed to be a rallying point and focus of inspiration and new vision for Africa.

According to Mbeki in his address to the South African Constitutional Assembly in 1996, 'the aims of the African Renaissance are the encouragement of cultural exchange; the "emancipation of African women from patriarchy"; the mobilisation of youth; the broadening, deepening and sustenance of democracy; and the initiation of sustainable economic development – a clear mixture of New Africanism and globalism' (in Akosah-Sarpong 1998: 665). In his presidential inauguration speech in June 1999 Mbeki extended his vision with a reference to the twenty-first century as the 'African Century', one in which quality of life had to be improved for all (SABC, 16 June 1999). However, there is a very real sense that without concrete actions and funds to support such ideals, the 'African Renaissance', like the RDP and GEAR before it, might remain simply an appealing and alluring vision.

Plate 7.9  Thabo Mbeki

## Post-apartheid politics

The political dominance of the ANC has remained largely unchallenged since 1994. In the second, democratic election of 1999, the party secured some 65 per cent of the vote. In June 1999, Thabo Mbeki succeeded Mandela as President of the country. In his inauguration address, he re-committed the government to the pursuit of the policy objectives of both the RDP and GEAR and further emphasized his belief in the importance of the 'African Renaissance' (*The Star*, 30 June 1999). Despite his reputation of being more of a 'manager' than Mandela, initial policy pronouncements do not indicate that significant shifts in government and national policy are to be anticipated (Nelan in *Time Magazine*, 24 May 1999).

The political transition since 1994 has not been entirely without tension. In 1996, the initial Government of National Unity collapsed following the withdrawal of the National Party from their power-sharing arrangement with the ANC. In addition, violent conflicts between and among ANC and IFP supporters have continued in parts of KwaZulu-Natal. By the late 1990s, however, the overall level of conflict appeared to be diminishing. One of the more persistent problems for the ANC government has been the tension in their alliance with COSATU and the SACP. As we have already indicated, the two latter bodies have felt increasingly alienated by the neo-liberal shift on the part of government, the dissolution of the RDP as a ministry, the restricted job-market, and most importantly, the adoption of GEAR in 1996. In fact GEAR forced the ANC leadership into near-confrontation with the SACP and COSATU in 1998. In a stinging criticism of the policy shift, the SACP writer, Dale McKinley, detailed how the 'ANC had sold out the revolution'. He continued to state that the ANC had shown an 'unwillingness and /or inability to recognise that revolutionary struggle cannot be advanced by attempting to reconcile the priorities of people with the priorities of capital' (*Mail and Guardian*, 31 October 1997: 31).

## Race and class in post-apartheid South Africa

In a country whose history has been so inextricably intertwined with the issue of 'racial' difference, it is inevitable that in post-apartheid South Africa racial issues still feature prominently. Although South Africa mercifully did not descend into an abyss of violent, racially defined conflict as many had predicted, 'given its demography and history, race will undoubtedly [continue to] matter in South African society and politics for a long time to come' (Price 1997: 149).

Since 1994 the government has consciously tried to address deeply entrenched poverty which, as we saw in the previous chapter, almost exclusively affects the black majority. It has also pursued strategies designed to ensure greater racial representativeness in the workplace and the economy. Recent years have seen the rapid enlargement of the black middle class in government and the private sector, largely in anticipation of 'affirmative action' legislation. From 1994 to 1997, the number of blacks in senior managerial positions increased from 5 to 7.3 per cent, while the number of professionals rose from 6.2 to 15.3 per cent (du Toit, J. 1998). Nearly 70 per cent of companies were embarking on affirmative action programmes by 1997. Black business control is also increasing, rising from 1 per cent of total market capitalization in 1996 to 5.5 per cent (US$9.2 billion) in 1999 (Nelan in *Time Magazine*, 24 May 1999). A change in the racial composition of the dominant classes is clearly occurring, but still at a rather slow pace. While the vast majority of the destitute, as we have seen, are among the black population, a more racially diversified middle class has developed – one which is united in its maintenance of privilege (Marais 1996). As a result of the progressive 'incorporation' of a professional elite, the former cohesiveness of black urban society, forged particularly in South Africa's segregated urban townships, has been undermined.

In order to bring about changed racial composition of the labour force, the government has piloted through parliament the Employment Equity Act, which, though not prescribing fixed racial quotas, has provoked a significant white backlash. According to parliamentary opposition leader, Tony Leon, the Act will 'bring back apartheid' (SABC, 29 April 1999). Similarly, the *Sunday Times* newspaper (6 December 1998) has referred to growing 'shades of racism' and *The Economist* (29 May 1999) has questioned whether South Africa is 're-racialising'. Inevitably, there is tension in the labour market between Africans on the one hand and whites and Coloureds on the other, who feel passed by on the promotion ladder. The reality of emerging 'reverse discrimination' (Behren in *Time Magazine* 7 June 1999: 30) runs the danger of increasing 'racial' tension. Fortunately, instances of overt hostility are not that common, although there have been some disturbing attacks in what are now racially integrated schools, whilst in the rural areas, racial discrimination and animosity often still persists. The race-related shooting of several soldiers at the Tempe

Plate 7.10 Black taxis: an example of black entrepreneurship

military base in Bloemfontein in 1999 and allegations of racism at other defence force bases hint at the existence of smouldering racism among former South African Defence Force (SADF) and ANC soldiers (E-Tv News, 28 October 1999).

## The post-apartheid government record

There can be no doubt that, even if the ANC government has not attained all the ideals which it set for itself in 1994, some very impressive advances have been made. In his speech to parliament in February 1999, President Mandela referred to the dramatic scale of the ongoing development process (*Sunday Times*, 7 February 1999). On a daily basis since 1994, 1,300 homes have been electrified, 750 phones installed and 1,700 people given access to clean water. Development at this level can have few if any parallels anywhere in the world and is testimony, that in certain sectors at least, the government has held true to some of its promises to bring about fundamental improvements. Achievements in the areas of basic human rights and health care are all equally positive. Just as important are the very real efforts being made by government to be more open and accountable. According to a recent government report, 'we are proud not only to have turned our back on a terrible past, but the new nation has started to take root in all spheres of life' (Government Communication and Information Services 1998: 3). Regrettably, however, and in large part due to factors beyond the government's control, satisfactory levels of economic growth have not been achieved. The degree to which the government has succeeded in attaining its own ideals can be gauged from a comparison between the broad goals set out by the RDP in 1994 and subsequent available data. The target of creating 2.5 million jobs over a ten-year period clearly will not be attained in the current climate of job losses and 'jobless growth' indicated above. The proposal to build 1 million homes by the year 2000, has not been achieved, but with the completion of 680,000 houses by early 1999 the target should be attained within two to three years (*Mail and Guardian*, 30 April 1999). At the current rate of increasing electricity provision, even though the number of consumers has doubled in five years, it is unlikely that the target of electrifying 2.5 million homes by 2000 will be attained until 2001 or 2002. The objective of providing running water to 1 million households, however, may well be met, with over 3 million people being connected by 1998 (Nelan in *Time Magazine*, 24 May 1999). Statistics for sewage connections are not available, but the target here will almost certainly not be met. The redistribution of 30 per cent of agricultural land to emerging black farmers definitely will not occur. By 1999, only 1 per cent of land had actually been transferred (*Mail and Guardian*, 22 May 1999). The planned focus on primary health care has taken place and the goal of providing ten years of compulsory free education has almost been attained. Massive infrastructure provision through public works is being undertaken, albeit very slowly, and the restructuring of state institutions by 1997 to reflect the race, class and gender divisions in South African society is an ongoing process (ANC 1994; Marais 1996).

Whilst the RDP objectives, except for those relating to job creation, have to a large degree, nearly been attained within five years, the same is certainly not the case with GEAR. Although there have been fundamental economic reforms and the economy has been liberalized, investment levels are not as high as was hoped and private sector expansion has been sluggish. The economy has not attained the envisaged 6 per cent growth rate and was hovering at around only 1 per cent in 1999. Related to this, the economy has definitely not generated the envisaged 1.35 million jobs within five years. Instead, as we have seen, jobs have been lost on a massive scale (RSA 1996a).

The net result is a mixed balance sheet. Given the limited foreign investment which has taken place and global economic instability, the government does not seem to have fared too badly since 1994. With the dramatic exceptions of job creation and rapid economic growth, many of government's social and infrastructural objectives are attainable and current programmes are definitely helping to improve basic living conditions. In future, issues of job creation and economic growth will need to be accorded far more serious attention than they have received to date. In the meantime, the vast majority of South Africa's black poor remain relatively impoverished, with many suffering from absolute deprivation when measured by any international standard.

## South Africa's post-apartheid difficulties

At this stage in its history, and in common with many of its neighbours, South Africa faces a range of daunting obstacles to development. Undoubtedly, the key ones revolve around the economy and the consequences of a dismal employment record. The official unemployment estimate of 36 per cent (StatsSA 1999a) and the mass retrenchments which have taken place in the first five years after the 1994 elections have led to increasing impoverishment. The results of such processes range from aggravated nutritional deficiencies to spiralling crime levels. As has already been suggested, important spatial inequalities exist in the country. Addressing poverty and creating jobs is vital if the country is to avoid the downward spiral into the state of near-collapse so pessimistically predicted by Venter (1997). At this juncture the Jobs Summit and general private sector expansion are not making any significant difference and, unless urgent action is taken, increased human suffering, crime and instability are inevitable. The popularity of unions is reflected in the fact that the country is one of the few on earth where unionism is growing and the parallel passing of more humane labour laws makes it expensive for employers to hire labour and difficult to fire them. Such a situation may impede employment creation and encourage increasing capital intensification (*The Economist*, 29 May 1999).

Another challenge which is of enormous proportions is the AIDS crisis. In view of its situation in the worst-affected continent, where the disease has already devastated the populations of numerous countries, it is hard to be optimistic about South Africa's future in this regard. According to recent reports, 22 per cent of pregnant women tested in hospitals are HIV-positive. The highest figure, 33 per cent, was recorded in KwaZulu-Natal (SABC, 3 March 1999). Up to 50 per cent of patients in medical wards in King Edward hospital in Durban are now AIDS victims (SABC, 12 August 1999). So severe is the actual and potential crisis that in 1999 the United Nations Development Programme reduced their estimate for the average South African's life expectancy from 64 to 54 years (SABC, 12 August 1999). Slow government response to the crisis, the cost of treatment and inadequate health care budgets are not helping to address the situation to any significant degree at present. This is in spite of the fact that the disease has the potential to kill one-third of South Africans (*The Economist*, 29 May 1999).

A further intractable problem, as we have noted earlier, is that of crime. Lawlessness has become a feature of everyday society, and every newspaper and news bulletin is filled with details of horrific assaults, robberies and corruption. The World Health Organization claims that South Africa is the 'world's most violent society not at war', while a local newspaper has christened the country the 'world's crime capital' (Adedeji 1996: 217; *Business Times*, 14 February 1999). Johannesburg has become notorious as a dangerous place to live, but the Western Cape has also seen the rise of militant gangsterism, especially in the Coloured townships around Cape Town, and sporadic violence continues in parts of KwaZulu-Natal. A range of aggregated statistics illustrates how serious the situation is in the country as a whole, although it should not be forgotten that criminal acts are spatially concentrated, largely in the urban areas, and especially within the relatively impoverished townships. In 1995, 20,000 people were murdered, at a rate of 65 persons per day. This murder rate is six times higher than that of the USA, a country some seven times larger than South Africa (*The Economist* 13 December 1997; Marais 1998). Worst-case estimates indicate that up to one in three South African women are likely to be raped (*The Economist* 29 May 1999). Between 1990 and 1997, the number of rapes doubled to 120 per 100,000 per annum, the murder rate went up from 48 to 58 per 100,000 and robberies nearly doubled to 290 per 100,000 (du Toit 1998). These statistics sadly reflect only reported cases. In addition, some 1,400 police were killed in the line of duty between 1994 and 1999 and there are an estimated 4 million illegal guns in the country (SABC, 17 May 1999; *Time*, 24 May 1999).

Between 1997 and 1999, 82 per cent of businesses were subjected to some form of crime (*Business Times*, 14 February 1999). The overall economic cost is estimated to be enormous, including US$5 billion per annum of lost investment and tourist revenue. The increased emigration of skilled people as result will entail further costs. Corruption and extortion within the public sector in particular is also acute. According to the Democratic Party, nearly US$4 billion was extorted between 1994 and 1999. Corruption has been uncovered even at the level of provincial cabinets and senior officials. In 1999 over 92,000 cases were being investigated (Behren, *Time Magazine*, 7 June 1999).

Many South Africans, and especially those who are most exposed in the urban townships, live with levels of fear which have been described by psychologists as being equal to that in a wartime situation. As we saw in chapter 5, there are multiple explanations for the widespread resort to violent crime, including the brutalization of non-whites under apartheid, the development of an anti-government and anti-social culture in the years of struggle against apartheid and, particularly during the recent insurrection, the proliferation of guns and gangs. As Morrell points out, in urban areas, and especially during the insurrection, 'the means by which youth masculinity was primarily asserted was by violence – against other gangs, against workers, against symbols of authority and against women' (Morrell 1998: 627). Massive unemployment, huge and very evident inequalities in wealth and the poor success rate of a police force, which was trained until recently more in counter-insurgency than in detective work, have not assisted to bring down the crime rate (Adedeji 1996).

The police generally still lack legitimacy in black communities, are riddled with corrupt members and are severely demoralized. The staggering figure has been revealed in parliament that only one in seven murderers and only one in 50 car high-jackers are convicted (SABC, 17 February 1999). Along with these more immediate causes of criminal activity we should also consider the broader features of social change which have shifted many groups' codes of morality. As Lonsdale points out, 'modern African history has been a disturbing experience ... Colonial rule's contrasts between oppression and opportunity; the growing differentials between malnourished migrants and the more prosperous ... the rising rate of population growth and greater numbers of impatient young; perhaps especially the social corruptions of urbanization – have challenged the moral assumptions of most people on the continent.' If this is the case for the continent in general, in South Africa, with its especially divisive recent history, many poorer people have been prompted 'to reconsider the moral bases of their social relations' (Lonsdale 1998: 294). Although this does not explain attacks on people who share their poverty, for some criminals, a resort to crime appears as legitimate as is the continued prosperity of some of their victims.

Those who can afford the expense rely on sophisticated protection systems to guard against robbery and assault, whilst many individuals and families among the middle class, both white and black, have emigrated in search of a more prosperous and less stressful life. At present there are three times more security guards than there are policemen in South Africa and the security industry is one of the few 'boom' industries (*The Economist*, 29 May 1999). Along with the new legislative protections for labour have come a range of humane laws designed to ensure human rights for convicted criminals. However, as many critics as well as victims of crime point out, the police often seem powerless to act, while prisoners enjoy protection, health care and support to which their victims are seldom entitled. When convictions are secured, many South Africans feel that the sentences handed down are frequently too lenient. This, together with frequent jailbreaks and the regular shortening of sentences, has persuaded some critics that there is no longer an effective judicial deterrent to crime (*Eastern Province Herald*, 10 April 1999). According to Behren, 'in South Africa the state has all but abrogated its role as a custodian of the law. The criminal class rules' (Behren in *Time Magazine*, 7 June 1999: 30).

One of the consequences of crime and instability is an accelerated 'brain drain', with the country having lost enormous numbers of highly qualified professionals, including doctors, nurses, accountants and engineers. Although accurate statistics do not exist, the fact that 20 per cent of South Africans registered as chartered accountants with the

Plate 7.11 Security precautions around white houses – walls and armed response signs: Johannesburg's northern suburbs

South African Board live overseas is an indication of what has happened (*Business Times*, 20 November 1998). An estimated 8 per cent of South Africa's professionals were living in the USA by 1999, whilst the total number of South Africans who had emigrated to that country was estimated at being between 100,000 and 250,000 (*Sunday Times*, 1 August 1999). The loss of such skilled workers further retards the development process and frequently has implications for job losses among less skilled workers.

Other challenges which face the state include the continued 'culture of non-payment' for services in the townships and the 'collapse of the culture of learning' in many schools, both of which reflect the forms of resistance to which township residents resorted during apartheid's final decade. Low teacher morale, frequent non-attendance at school, poor facilities, the absence of effective discipline, deteriorating facilities, poor supplies, gangsterism and intimidation, all place extra pressures on the human resource base of the country and constrain future employment and economic growth (*The Economist*, 29 May 1999).

After five years of democratic rule, we can certainly conclude that South Africa is a strikingly different place, in many respects, from how it had been in the early 1990s. The remarkably peaceful nature of the transition period, political reconciliation and the restoration of economic growth (albeit mediocre) are all very positive achievements. Tremendous gains have been made in the extension of basic needs and services to a large proportion of the population. One of the key reasons for this relative success was the personality and leadership of former President Mandela, whose own suffering, tolerance and pursuit of reconciliation won domestic and international acclaim. 'Mandela [was] the indispensable man in his country's emergence from the darkness of apartheid into the light of democracy. To whites and blacks alike he is the symbol of racial reconciliation, and he has been the referee among fiercely competing claims to power' (Nelan in *Time Magazine*, 24 May 1999: 38).

Ensuring that there was a peaceful transition was absolutely vital, but launching the country on a growth path from which all can benefit has been far more difficult. The mixed performance of the RDP and GEAR, burgeoning unemployment, the increasing threat of AIDS and the lack of resources backing the notion of an African Renaissance, are all causes for concern. Most frustrating of all is the general constraint of trying to achieve national development in an increasingly interconnected and unequal world, in which the new state is as dependent as the old on Western finance and investment. The fixation with neo-liberal economic management, though a practical course of action in terms of securing international acceptance and loans, is unfortunately showing no signs of addressing the deep-rooted poverty of the marginalized majority. The continued holding of economic power by what is slowly becoming a racially integrated upper and middle class, emerging black elitism and the ANC's embracing of neo-liberal policies would seem to be entrenching class boundaries and undermining the ideal of a harmonious 'Rainbow Nation'. As the *Mail and Guardian* puts it, 'ANC economic policy is much more in tune with the demands of globalisation than those of millions of desperately poor people who put their trust in the party' (23 July 1999: 32). Waldmeir concludes: 'when South Africa stepped through the looking glass, it did not emerge in Wonderland. It emerged in the real world, where poverty is the biggest challenge to all democratic governments, and where there are tougher problems to solve than apartheid' (Waldmeir 1997: 283).

# Chapter 8

# BEYOND THE LIMPOPO:
## South Africa in Africa
## and the wider world

As the analysis of the constraints on the ANC in the last chapter demonstrates, we cannot consider South Africa's future without an understanding of its situation within an international political economy. In this chapter, we widen the book's focus to consider South Africa's position and role within the southern African region, the African continent and the international community at the beginning of the twenty-first century. Since the demise of apartheid in South Africa, increasing attention has been paid in particular to the experiences which the country has shared with the rest of the African continent, and to the lessons which should be learnt from Africa as a whole if the post-apartheid government is to achieve its stated development objectives (see, for example, Mamdani 1996). Here we explore some of these experiences and hint at some of these potential lessons.

### Africa: decline and re-birth?

While postcolonial African states remain situated within a global economic system, the terms on which they engage with that system often differ greatly from those that were enjoyed by their colonial predecessors. It is sadly the case that many African countries will be, in relative terms, far worse off at the beginning of the twenty-first century than they were at independence in the 1960s. Both colonial and postcolonial states have targeted resources selectively according to race, ethnicity, region and gender, but in most cases the former were more successful in securing benefits for their narrowly defined constituents. A lamentable record of poor governance manifested in political instability, endemic corruption and civil strife, together with environmental catastrophes such as drought, have had disastrous economic and social effects in many countries since independence.

The disappointing record of many post-independence African governments cannot, however, be conceived outside of its historical context. As has recently been argued by Mamdani (1996) and others, 'the modes of power which operate in contemporary Africa are derived from the colonial state' (Ahluwalia 1999: 327). Greenstein (1998) is at pains to point out that 'the modern African state has been marked by the legacy of European and settler domination. Shaped by the imperatives of rule over the indigenous masses ... specific forms of state policies came to the fore, elements of which persisted long after the collapse of colonialism.' The legacy of European conquest and administration has thus left 'the African colonial state a political species with a singular historical personality' (Greenstein 1998: 16–17, citing Young 1994). This 'personality' was forged through the common African experiences of rapid and disruptive imperial conquest from the late nineteenth century (and as we have seen, earlier in the case of South Africa), ruthless colonial extraction of resources, determined settler efforts to prise peasants off the land and onto the labour market, autocratic and frequently brutal European governance, and the maintenance and exacerbation of ethnic boundaries by colonial states attempting to 'divide and rule'. Greenstein continues: 'None of these factors is entirely unique to Africa, but together they created a specific political trajectory that is still relevant today' (Greenstein 1998: 17). What European imperialism bequeathed to postcolonial African states was above all authoritarianism, selective patronage, administrative inefficiency in respect of those groups outside the favoured minority of citizens, and corruption in the interests of maintaining minority privilege. Mamdani (1996) argues that the granting of independence to most African states from the 1960s enabled this favoured minority of citizens to become deracialized. Rather than fostering a white racial elite, most states now operated largely on behalf of an urban African elite, composed of those who had led the nationalist independence struggle and their clients. However, social relations in the countryside – the former 'reserves', or in South Africa's case, 'homelands' – were much less affected by independence. Under colonial indirect rule, rural social relations were based not on citizenship, or mass participation in political decision-making, but on the authority of chiefs who maintained ossified versions of customary law among their ethnically defined subjects. Segregationist and apartheid South Africa, as we have seen in chapters 4 and 5, was characterized by such urban–rural divisions in the mode of authority, just as much as was the rest of sub-Saharan Africa.

Elsewhere in the continent, Mamdani (1996) argues, independence and the deracialization of the urban elite has not led to democratization because the new African-led governments continue to rely on ethnically defined authorities in the rural areas to mobilize support. This, in turn, has encouraged a form of politics in which 'urban politicians establish patron–client relationships with rural constituencies', resulting in corruption, tribalism and patrimonialism, a term defined by Fauré as the imperative of 'survival for the ruler, his regime and its "hangers-on"' (Ahluwalia 1999: 318; Fauré 1989). This form of politics has been destructive of democratic development initiatives. With despotic chiefly rule still prevalent in the rural areas of most African countries, post-independence development, as we will see below, has generally taken the form of a 'top-down agenda enforced on the peasantry' (Mamdani 1996: 287). This has been the case even under professedly socialist governments, where strong single-party states have mediated urban and rural relations.

This colonial legacy, of course, was inherited by most African states some decades ago and theoretically there has been time to effect major changes at state level.

However, the influence of former colonial powers did not end with the transfer of formal independence, and even states such as Uganda which have recently striven to become more thoroughly democratic, have had to comply with a number of Western demands which limit their autonomy. After two decades of neo-liberal structural adjustment programmes emanating from such Western-dominated institutions as the World Bank and the International Monetary Fund (IMF), for the majority of ordinary Africans, postcolonial 'development', in the shape of a recognizable improvement in their quality of life, has just not occurred (Riddell 1992).

In fact, some writers have described the 1980s in particular as the 'lost decade' as far as development in Africa was concerned (Onimode 1992: 1). In too many sub-Saharan African countries life expectancy is still less than 50 years (UK 76.8 years; USA 76.4 years), infant mortality is over 100 per 1,000 live births (UK 6 per 1,000; USA 5 per 1,000) and the adult literacy rate is still below 50 per cent (UK 99 per cent; USA 99 per cent) (UNDP 1998). At a time of increasing globalization in so many different ways, Africa is peripheralized relative to other regions of the world in respect of many, if not most, recognized development indicators. Perhaps global inequalities are most starkly displayed in the area of telecommunications, where it is a staggering fact that there are more telephones in New York City than in the whole of sub-Saharan Africa (APIC 1996).

Considering the position of Africa in the global economy, although the continent in 1998 actually had the world's fastest-growing economy, Botswana, it also included most of the slowest growers (Collier 1998). Some economists have referred to the progressive 'marginalization' of the continent, with its steadily falling share of world exports (reflecting a deterioration in producer prices) and the declining proportion of direct private investment into developing countries which is destined for Africa. It could be argued that 'the only international economic sphere in which Africa has remained non-marginal is aid' (Collier 1995: 541). In fact, for a number of African countries aid from a variety of donor agencies, together with remittances of migrants, actually represents their main form of participation in the world economy. Sub-Saharan Africa receives roughly five times the amount of net official transfers per head of population than does south Asia, such that in 1991 the region was the destination for about one-third of global net aid transfers (Collier 1995: 543). When official development assistance is considered as a percentage of GNP, African countries are at the top of the world list, with official development assistance to Guinea Bissau and Mozambique, for example, representing 67.5 per cent and 59.8 per cent respectively of their GNP in 1996 (World Bank 1998: 230–1). As Claude Ake asserts, an essentially Western perception of Africa is that it 'is not so much marginal as irrelevant; at any rate relevant only as a nuisance – a nuisance forever complaining about being exploited, forever begging for help, constantly making a mess of its own affairs and looking to others to clean it up' (Ake, quoted in Ede 1993: 29).

Since the 'wind of change' brought independence to much of the continent during the 1960s, Africa has steadily lost the niche in the world economy which it had under colonialism (Agnew and Grant 1997). Since 1980, whilst the value of global trade has increased threefold, very few African nations have achieved a significant increase in their exports. On the contrary, many countries have actually experienced a decline in the value of their exports, as for example in Nigeria, which had exports valued at US\$25,057 million in 1980, but only at US\$15,610 million in 1996. Even the relatively stronger economies of the countries comprising the Southern African Customs Union (Botswana, Lesotho, Namibia, South Africa and Swaziland) also suffered more than a 25 per cent reduction in the value of their exports over the same period (from US\$25,539 million to US\$18,132 million) (World Bank 1998: 228–9).

Considering levels of foreign direct investment (FDI), which are a useful indicator of links with the global economy, African countries are among the lowest in the world. In fact, a large number of African countries receive FDI amounting to less than 0.5 per cent of the value of their GNP which, when consideration is given to the generally low levels of GNP in Africa, means that the real value of FDI is very small. World Bank statistics show that the three West African states of Burkina Faso, Niger and Togo actually received no FDI at all in 1996 (World Bank 1998: 230). This dearth of foreign investment may be due to the fact that many potential investors have remarkably superficial knowledge and stereotypical views about Africa and, furthermore, they often see little incentive to invest in obtaining more detailed information. As Collier observes, 'Africa is currently attracting only those investments which cannot be located elsewhere, such as mineral extraction or production for the (tiny) domestic market. The major internationally footloose investments are simply bypassing Africa as a location' (Collier 1998: 22).

Africa's 'condition' and its progressive marginalization have been further exacerbated by the increased attention from the world's media which was directed in the late 1980s and early 1990s to the momentous changes occurring in Eastern Europe with the collapse of communism. South Africa's Foreign Minister in the last minority government, Pik Botha, commented in 1990 that 'Africa is concerned because the opening of markets in the USSR and Central and Eastern Europe might draw away investors and reduce interest in the economic development of this continent' (Botha 1990: 215). It was revealed that in 1990 aid to Eastern Europe increased significantly while regional allocation for Southern Africa, which had until recently been a proxy Cold War battleground between the US-led Western and the Soviet-led Eastern blocs, fell in real terms by 30 per cent (Owoeye 1994). More specifically, with regard to relationships between the European Union and both South and Southern Africa, it seems certain that the end of the Cold War has led to Western Europe focusing more on the development and strategic importance of Eastern Europe, resulting in a gradual decline in the geopolitical significance of the South, and an associated reduction in developing country bargaining power and access to development funds (Asante 1996).

The poor economic record of many African countries in the 1970s and early 1980s prompted the World Bank and IMF to make a case for the introduction of structural adjustment policies. These organizations were concerned about levels of state involvement in production and regulating economic activity, as well as the existence of overvalued exchange rates with large and prolonged budget deficits. It was further suggested that protectionist trade policies and government monopolies reduced competition, whilst agricultural exports, one of the largest suppliers of foreign exchange, were heavily taxed (World Bank 1994a). Although the nature of structural adjustment programmes (SAPs) across the continent does vary, several basic elements are invariably present – most notably: currency devaluation, the removal/reduction of the state from the workings of the economy, the elimination of subsidies in an attempt to reduce expenditures, wage restraints, and trade liberalization (Potter *et al.* 1999: 168).

Such policies are often seen as being encouraged, not least, in order to bring about better conditions for the repayment of Western loans, and they are frequently imposed in disregard of the consequences of reduced social service expenditure by debtor governments. Much attention has been directed towards evaluating the specific impacts of SAPs and this is still the subject of widespread and ongoing debate. In Ghana, for example, often heralded by the World Bank as one of Africa's success stories in responding to structural adjustment, whilst average annual growth in GDP was 3.8 per cent throughout the 1980s, people and environment undoubtedly suffered, with large numbers of public sector workers being retrenched and rapidly growing timber exports leading to a 25 per cent reduction in the forest area. On the basis of a detailed survey of a number of SAPs, Riddell argues, 'There can be little doubt that the effects of SAPs in Africa have led to worsened conditions … Poverty has increased … The quality of life has declined as prices have risen, as infrastructures have crumbled, as services have deteriorated, and as employment opportunities have been reduced. Almost everyone has suffered, but rural peasants, urban slum dwellers, female-headed households and the children of the poor have felt the negative effects of adjustment most severely, especially when their conditions are exacerbated by drought and conflict' (Riddell 1992: 66). Susan George is particularly critical of the IMF, which she suggests is oblivious to the reality of Africa and its people: 'The Fund lives in a never-never land of perfect competition and perfect trading opportunities, where dwell no monopolies, no transnational corporations with captive markets, no protectionism, no powerful nations getting their own first' (George 1988: 56).

Meanwhile, at the southern tip of the world's poorest continent, in South Africa, the euphoria which greeted the release of Nelson Mandela from prison in 1990 and the subsequent successful transition from apartheid to multi-racial democracy in 1994, contrast starkly with the turmoil and poverty which confront so much of the rest of Africa. In the self-congratulatory mood which accompanied the birth of the 'Rainbow Nation' and attracted the close attention of the world's media, Thabo Mbeki, the then Deputy President and heir apparent to Nelson Mandela, proclaimed to South Africa's Constitutional Assembly on 8 May 1996, and again a year later in a landmark speech delivered to the US Corporate Council on Africa, that an 'African Renaissance' was a real possibility, in which the current period of crisis might be seen as a time of opportunity, 'which the New Africa must seize for its own advantage' (Akosah-Sarpong 1998: 665). As we have seen, as key elements in this renaissance, Mbeki spoke strongly of the need for the emancipation of women and the mobilization of youth. He also expressed hope for sustainable economic development, together with the broadening, deepening and sustenance of democracy, with decision-making 'trickling down' to the level of the actual people affected.

President Mandela pursued the 'African Renaissance' theme a year later in a speech to the parliament in Zimbabwe, and soon after presiding over talks which led to the collapse of the Mobutu regime in Zaire, Mandela said that he was, 'convinced that our region and our continent have set out along the new road of lasting peace, democracy, social and economic development' (*The Independent*, 21 May 1997: 14). Interestingly, many thousands of miles away in Africa's internally troubled and most populous country, a group of Nigerian policy analysts and intellectuals gathered on 27 July 1998 in the capital city, Abuja, to mark the 80th birthday of Nelson Mandela as the harbinger of the 'African Renaissance'. US President Bill Clinton's 11-day visit to six African countries in March 1998 gave further impetus to the idea of an 'African Renaissance'. The tour was significant in that it was not only his longest-ever foreign trip, but was only the second ever by a US president to Africa, following Jimmy Carter's visit 20 years ago. Five of the six countries visited during the tour were all 'better-performing' states – Ghana, Senegal, Uganda, Botswana and South Africa – which no doubt were identified as having a potentially significant role in spearheading such a renaissance. In the case of Rwanda, the sixth country visited in Clinton's African tour, the President merely 'touched down' at the airport and paid his respects after the genocide.

Such events, and the relative strength of South Africa in relation to the rest of Africa, suggest that a key question to ask here is to what extent might the 'new' South Africa be both willing and able to play a crucial, if not leading, role in promoting an 'African Renaissance'? Furthermore, does South Africa have the ability and inclination to foster development and reverse the marginalization of the continent, such that in time Africa will no longer be synonymous with aid, poverty and hunger? Adedeji has no doubts about South Africa's future role: 'Instead of looking furtively at the rest of the world – second-guessing how that world may perceive and judge its actions – South Africa should become a self-confident force, reinvigorating Africa's drive and resourcefulness and playing a leadership role in this most important endeavour: putting Africa's house in order' (Adedeji 1996: 13).

## South Africa: economic powerhouse of Africa

At the time of President Clinton's state visit to South Africa in March 1998, as part of his tour of African countries, the respected South African newspaper *Mail and Guardian* commented, 'President Bill Clinton is spending 11 days on this continent, but he is really making two different trips: one to Africa and one to South Africa. South Africa is of Africa, but different too. It is no financial basket case dependent on foreign aid and its citizens never grumble that their standard of living was better in colonial days. Its economy dwarfs that of any other African country. Some 3000 American companies do business here. And in Nelson Mandela, South Africa has a president whose global stature rivals – perhaps even surpasses – Clinton's' (*Mail and Guardian*, 27 March 1998: 3).

There is no doubting the fact that if we take aggregate statistics (the problems with which we identified in chapter 6, the South African economy 'towers above' that of other African countries. Table 8.1 compares those eight African countries with the largest economies, according to size of Gross National Product. It can be seen that South Africa's GNP in 1997 was US$130.2 billion, which was almost twice as large as the next biggest African economy, Egypt (US$71.2 billion), and more than three times the total size of the economies of the 11 other countries comprising the Southern African Development Community (SADC).

Furthermore, South Africa's Gross National Product represents over 42 per cent of the total GNP for all countries in sub-Saharan Africa. With a per capita GNP of US$3,400 and a world ranking of 45, South Africa is the highest-placed country in Africa and, in terms of the rest of the world, comes just below Venezuela (US$3,450, Rank 44) and Poland (US$3,590, Rank 43). In fact, with a population of 39 million (1997) and a total GNP of US$138.9 billion, Poland has some interesting similarities with South Africa. Considering per capita GNP, South Africa is well above many Eastern European countries, such as Romania (US$1,420), the Russian Federation (US$2,740) and Ukraine (US$1,040).

In addition, unlike so many other African countries, South Africa has a relatively well-developed infrastructure, with good road, rail and air networks. The country has more than one-quarter of Africa's railway track and one-fifth of its paved roads (Binns 1998). South African Airways (SAA) is the largest Africa-based airline, which has already forged commercial alliances with a number of other large international airlines and is likely to do the same with certain African carriers. At a time of rapid developments in global communications, in terms of the level of connections to the Internet, in January 1997 South Africa was ranked 16th in the world, behind Denmark, but ahead of Austria (Ahwireng-Obeng and McGowan 1998: 19). In electricity production South Africa actually generates more than half of all the electricity in the continent and has sufficient spare capacity to last until 2007 (Smets 1996). The country is also especially rich in mineral resources, producing in 1993 some 54 per cent of the world's platinum, 33 per cent of chromium, and 27 per cent of gold (Africa Institute of South Africa 1995).

Table 8.2 presents comparative statistics on eight selected African countries, including South Africa. Examining a range of non-economic variables, South Africa also appears to be much better off than most other African countries, with a life expectancy at birth of 64 years, an infant mortality rate of 50 per 1,000 live births and an 18 per cent adult illiteracy rate, one of the lowest on the continent. The Human Development Index (HDI) provides a particularly useful indicator of living standards. Devised by the United Nations Development Programme in 1990, HDI indicates the quality of life, rather than merely focusing on economic criteria, such as per capita income. Three key variables are used to calculate HDI: life expectancy at birth; educational attainment, as measured by a combination of adult literacy and combined primary, secondary and tertiary enrolment ratios; and standard of living, as measured by Gross Domestic Product per capita. The index does not measure absolute levels of development, but instead a rank order is presented from highest to lowest HDI, from which the relative position of individual countries can be ascertained. The 1998 Human Development Report (UNDP 1998) ranks 174 countries

Table 8.1  South Africa and other significant African economies

| Country | Population (million) 1997 | Land area (000 km²) 1995 | Population density (people per km²) 1997 | GNP (US$ billion) 1997 | GNP per capita (US$) 1997 | GNP per capita rank 1997 |
|---|---|---|---|---|---|---|
| Algeria | 29 | 2,382 | 12 | 43.8 | 1,490 | 67 |
| Congo DR | 47 | 2,267 | 19 | 5.1 | 110 | 131 |
| Côte d'Ivoire | 15 | 318 | 44 | 10.2 | 690 | 87 |
| Egypt | 60 | 995 | 58 | 71.2 | 1,180 | 72 |
| Morocco | 28 | 446 | 59 | 34.4 | 1,250 | 70 |
| Nigeria | 118 | 911 | 122 | 30.7 | 260 | 119 |
| **South Africa** | **38** | **1,221** | **30** | **130.2** | **3,400** | **45** |
| Tunisia | 9 | 155 | 58 | 19.4 | 2,090 | 59 |
| 11 other SADC countries* | 98 | 5,578 | 80 (av.) | 42.1 | 1,216 (av.) | n/a |
| Sub-Saharan Africa | 614 | 23,628 | 25 | 309.1 | 500 | |
| UK | 59 | 242 | 243 | 1,220.2 | 20,710 | 15 |
| USA | 268 | 9,159 | 29 | 7,690.1 | 28,740 | 6 |

Source: World Bank (1998)
Note: *excluding South Africa

Table 8.2   Comparative statistics for selected African countries

| Country | Land area (000 km²)* | Population (million) 1995* | Population density (people per km²) 1997* | Average annual population growth rate (%) 1990–97* | Life expectancy at birth 1995** | Infant mortality (per 1000 live births) 1996** | GNP per capita (US$) 1997* | GNP per capita rank 1997* | Human development index 1995** |
|---|---|---|---|---|---|---|---|---|---|
| **South Africa** | 1,221 | 38 | 30 | 1.7 | 64 | 50 | 3,400 | 45 | 0.717 |
| Botswana | 567 | 1.5 | 3 | n/a | 52 | 40 | 3,260 | n/a | 0.678 |
| Egypt | 995 | 60.0 | 58 | 2.0 | 65 | 57 | 1,180 | 72 | 0.612 |
| Ghana | 228 | 18.0 | 75 | 2.7 | 57 | 70 | 370 | 106 | 0.473 |
| Kenya | 569 | 28.0 | 47 | 2.6 | 54 | 61 | 330 | 109 | 0.463 |
| Mali | 1,220 | 10.0 | 8 | 2.8 | 47 | 134 | 260 | 118 | 0.236 |
| Nigeria | 911 | 118.0 | 122 | 2.9 | 51 | 114 | 260 | 119 | 0.391 |
| Zimbabwe | 387 | 11.0 | 28 | 2.3 | 49 | 49 | 750 | 85 | 0.507 |
| UK | 242 | 59.0 | 243 | 0.3 | 77 | 6 | 20,710 | 15 | 0.932 |
| USA | 9,159 | 268.0 | 29 | 1.0 | 76 | 5 | 28,740 | 6 | 0.943 |

*Sources*: *World Bank (1998), **UNDP (1998)

for which 1995 data are available. Canada is shown as having the highest HDI (0.960), whilst the West African state of Sierra Leone, which has suffered from instability and civil war during much of the 1990s, has the lowest (0.185). In fact, of all the countries listed, the 15 countries with the lowest HDI values are all in Africa. South Africa is ranked 89 out of 174 with an HDI of 0.717. Whereas several North African countries (Algeria, Libya and Tunisia) have higher HDI values than South Africa, there is no other sub-Saharan African country which is ranked higher than South Africa.

But, as we saw in chapter 6, it is too easy to be 'blinded' by such national-level statistics which can conceal considerable social and spatial variation within the boundaries of a single country, and possibly none more so than in the case of South Africa. In South Africa's particularly unequal society, the HDI value for the white population may well be twice as great as that for the black population. In fact, it has been suggested that in 1992, towards the end of the apartheid era, the white minority population had an HDI similar to that of Spain (0.878), whilst the black population, constituting the vast majority of South Africa's population, had an HDI just above that of Congo Brazzaville (0.462) (UNDP 1994).

## Perceptions of South Africa in Africa

Many white South Africans will freely admit in conversation that South Africa is a 'Third World' country. This view was echoed in 1990 by Pik Botha, Foreign Minister in F. W. de Klerk's National Party government, at a conference in Pretoria on the theme 'Southern Africa towards the year 2000'. He commented: 'And may I, in saying that we are still evolving from an agricultural and mining economy to a manu-facturing economy, also say that South Africa ought to be classified as a developing country and not a developed country … and we ought to qualify for international advantages granted to developing countries' (Botha 1990: 216).

Despite the expression of such sentiments, many white South Africans frequently display a strong reluctance to accept the fact that they are 'Africans', a term which in the past has been reserved for both South Africa's and the continent's black African population, rather than simply *all* the inhabitants of Africa. There has in fact been a long-standing dualistic tendency among South African politicians, academics and others, such that in one sense they display a somewhat myopic and parochial view in focusing their attention entirely on South Africa, whilst in another sense, as we have seen in the case of modern discourses of urban planning in chapter 4, for instance, they have tended to look over and beyond the rest of the African continent towards their traditional links with Europe and North America. The academic fraternity are no exception to the rule. As Greenstein comments, 'The involvement of South African academics in the study of any African country other than their own has been insignificant' (Greenstein 1998: 14). For many South Africans (and by no means just the white population), civilization and development go no further than the Limpopo River along the country's northern border with Zimbabwe. Whereas the white popu-lation understandably maintains strong links with its European (principally British and Dutch) heritages, there is a strong historical case for considering South Africa as very much a part of Africa. Lonsdale asserts that, 'In the long view South Africa's exceptional industrial and political history since the mineral discoveries is no more surprising or instructive than any other such singularity in Africa's many pasts ... (such as) ancient Egypt, medieval Ethiopia, the Gold Coast's "model colony" or contemporary Rwanda or Somalia' (Lonsdale 1998: 287). Through, for example, Khoisan language groups, population movements and traditional agro-pastoral communities living in dispersed settlements with male control of household wealth in livestock, there are many long-standing connections between east and southern Africa (Lonsdale 1998). Yet, as Greenstein argues, 'Without doubt, the most important gap in the comparative literature on South Africa is the dearth of material that looks at its history and society in the African setting' (Greenstein 1998: 2). David Simon urges South Africa's political, economic and scholarly elites, 'to shed their traditional inward-looking preferences, which were greatly exacerbated by apartheid isolation' (Simon 1998: 5).

Meanwhile, Adedeji is strongly critical of the past record of South Africa's relationships with the rest of the continent, suggesting that,

> Virtually ever since the establishment of what became the Republic of South Africa, it has stood aside and apart from the rest of the continent. From 1910, when the Union of South Africa was established and became an independent dominion in the British Empire, the leaders of the new nation, invariably European descendants, saw their country as a European outpost. In this regard, South Africa was exactly like the other British dominions of Australia, Canada and New Zealand. It was a country where the colonial settlers were intent on either eradicating or subjugating the natives. Thus the White settlers in South Africa were, by and large, no different from White settlers anywhere else (Adedeji 1996: 5).

The enduring connections with Europe and North America are still reflected in South Africa's trading links (Table 8.3). In 1995, for example, South Africa's main trading partners were Germany, the USA, Japan and the UK, although significant new trade links were developing with Latin America, Eastern Europe and Asia, most notably China and South Korea. Two-way trade with China has increased significantly in recent years, amounting to a total of US$1.3 billion in 1997. Whilst trade with Taiwan was probably still greater at around US$1.5 billion, South Africa's decision in October 1996 to open diplomatic relations with Beijing could well change the balance in favour of China (EIU 1998: 37). But Europe still overwhelmingly dominates South Africa's trading relationships, with the EU accepting 25 per cent of South Africa's exports in 1996 and supplying 42 per cent of its imports (Mills 1998a).

The South African Department of Foreign Affairs (DFA) has suggested that any reticence that might have seemed apparent in relation to South Africa becoming more involved in both Southern African and continent-wide affairs was based on a concern for guarding against over-domination. Such a concern would not be surprising given the apartheid economy's tendency to absorb cheap labour from the entire subcontinent, especially for the gold mines, and the former government's strategic endeavours to reduce all 'frontline states' in the region to economic dependency on

Table 8.3   South Africa's main trading partners, 1997

| Exporting to | % total | Importing from | % total |
| --- | --- | --- | --- |
| UK | 10.1 | Germany | 13.7 |
| USA | 6.0 | USA | 12.3 |
| Japan | 5.5 | UK | 11.5 |
| Germany | 4.4 | Japan | 7.4 |
| Italy | 3.7 | Italy | 4.0 |

*Source*: EIU (1998)

South Africa. However, the current government promises a new approach to the region. As the DFA explains,

> Perhaps initially, because of our past experience and fear of being accused of maintaining a 'big brother' syndrome, we did not see ourselves as playing a leading role in the (Southern African) region, but now we have come to understand that there is an expectation from Africa and the rest of the world that we have a role to play, a role of contributing peace and stability in our continent and to the African economic renaissance. Our perceived reluctance to have a 'hands on' approach to our region and to be pro-active in our continent has to some extent been viewed by our neighbours and friends with some suspicion and a great deal of caution (DFA 1997).

Perceptions of Africa and Africans are slowly changing among South Africans, for example among businesses, which are keen to develop new markets in the region and the continent, and the media, which is notably more dispassionate and extensive than during the apartheid era (Simon 1998). But it is not uncommon to hear concerns expressed about South Africa 'degenerating' to become more like the popular image of the bulk of sub-Saharan Africa.

We will now examine in more detail the past experience and future potential of South Africa's relationships, firstly with the international community in Africa and beyond, and secondly, more specifically, with the southern African region.

## Relationships with the international community

### Organization of African Unity (OAU)

On 3 February 1960, British Prime Minister, Harold Macmillan, addressed a joint sitting of the South African parliament in Cape Town, declaring that 'a wind of change is blowing through this continent', and stressing both Britain's commitment to a policy of rapid decolonization and its displeasure with Prime Minister Verwoerd's

*[handwritten margin notes:]*
*ORGANISATION of AFRICAN UNITY (OAU)*
*established 1969 by 32 h. of states*
*— weak economically*
*— disadvantaged in terms of power rels*
*— African countries v. heavily reliant of aid etc*

enforcement of apartheid policies. A month later, on 21 March 1960, 69 people were killed and a further 180 wounded when police opened fire on the peaceful anti-Pass Law demonstration at Sharpeville. As we saw in chapter 5, the so-called 'Sharpeville massacre' was followed by the apartheid government banning the African National Congress (ANC), the South African Communist Party (SACP) and Pan African Congress (PAC), which in turn led to the initiation of the black 'armed struggle' and the subsequent imprisonment of Nelson Mandela and other key leaders.

It was in the aftermath of these rapidly unfolding events in South Africa that the Organization of African Unity (OAU), was established in Addis Ababa in May 1963 by the heads of state of 32 majority-ruled independent countries (Griffiths 1995). A key objective of the OAU Charter, which was proclaimed as 'the major radical achievement' at the organization's founding conference, was the liberation of Africa from colonialism and apartheid (Mayall 1991: 29). Unfortunately, the record of the OAU has not been as successful as the original signatories to the Charter envisaged, since 'the extent to which the OAU can play a role in the international system, as in other areas, is limited in many ways' (Mathews 1989: 36). From an economic perspective, the fact that Africa is still largely composed of relatively weak and 'underdeveloped' states, representing the greatest concentration of the poorest countries of the world, continues to disadvantage the OAU in terms of power relationships with influential external actors such as the states of Europe and North America, on which African countries remain heavily dependent for both trade and aid. The Lagos Plan of Action (LPA), which was adopted with much enthusiasm at the OAU's economic summit in May 1980, aimed to reduce Africa's dependence on outside aid by promoting self-reliance, self-sustaining development and economic growth on the basis of regional integration. Unfortunately, however, the LPA has thus far had little tangible success.

In the political and military arenas the OAU has also had little positive effect in solving various conflict situations, since external (non-African) intervention has often been involved in conflicts, for example in areas such as southern Africa, the Horn of Africa, Chad and Western Sahara. Mathews suggests that these and other conflicts, 'are all symptoms of a deep-seated malaise; namely, the vulnerability of the continent to imperialist machinations and the utter impotence of the OAU in maintaining peace and security in Africa' (Mathews 1989: 48). *Previously OAU not all SA cos apartheid*

From outside the OAU, the apartheid regime in South Africa made several attempts to develop dialogue with black-led African governments, but these initiatives were generally met with vigorous opposition. In April 1975, the OAU Council of Ministers adopted the Dar es Salaam Declaration on Southern Africa, which included the statement that, 'any talk with the apartheid regime is such nonsense that it should be treated with the contempt it deserves' (Mills and Baynham 1990: 179). However, during the apartheid era South Africa did succeed in establishing diplomatic relations as early as 1967 with Hastings Banda's dictatorial regime in Malawi, which even by the 1990s was still South Africa's only such formal link with a black state. Closer links were achieved from 1969 with Botswana, Lesotho and Swaziland with a revision of the Customs Union Agreement (see below). Meanwhile, during the mid-1970s the South African government was involved in Heads of State summit

*[handwritten note at bottom:] only link — Malawi*

1969
Custom Union : SA, Lesotho, Swaziland + Botswana
— close links = transport Botswana reliant on SA transport

**290**  Beyond the Limpopo: South Africa in Africa and the wider world
90% of B external trade dependant on S.A transport

meetings with Ivory Coast, Liberia, Senegal and Zambia. Trading links with Zimbabwe also remained strong, such that in 1987 about 90 per cent of Zimbabwe's external trade was dependent on South African transport (Mills and Baynham 1990). Connections were also established with other pariah states, most notably Chile, Taiwan, Paraguay and Israel, the latter being particularly important in assisting South Africa to modernize its security forces and acquire nuclear technology. The significance of Nigeria within Africa was also recognized, in that even before the 1994 democratic elections President de Klerk made a historic visit to the country on 9 April 1992, confirming that trade ties with the Nigerian military regime, although limited, did actually exist.

Soon after the 1994 elections South Africa was admitted to the OAU and President Mandela attended his first Heads of State and Government meeting in Tunis in June 1994. Mandela referred to the struggle against apartheid as 'a glorious tale of African solidarity', but it remains to be seen whether such sentiments will be translated into practical actions and a strengthening of the influence of the OAU. South Africa's true commitment to the OAU has sometimes been questioned, with a suggestion that, 'South Africa increasingly appears to be distancing itself from Africa' (Adedeji 1996: 232). Considering the rather chequered history of the OAU over more than three decades of its existence, it seems that the organization has suffered from an overwhelming crisis of confidence, credibility and relevance. Its frequent failure to be in the forefront of African political and economic affairs might suggest that the organization has been made a scapegoat for the inertia and deficiencies of its entire membership. As Venter argues, 'Its ultimate success is, of necessity, contingent upon the commitment and political will of its member states to transform it into a much stronger continental forum' (Venter 1994: 48). Despite these shortcomings, there is no doubt that the OAU is of huge symbolic importance to the African states and their people and is perceived as the embodiment of the 'African ideal', providing a diplomatic forum for airing the views of Africa on international issues which affect Africans.

At the Addis Ababa summit of African Heads of State and Government in 1990 it was recognized that there was a need to sharpen the focus and, 'strengthen the OAU so that it may also become a viable instrument in the service of Africa's economic development and integration' (Legum 1993: 17). At the same summit it was acknowledged that something positive had to be done to improve the financial wellbeing of the organization, and in particular to deal with unpaid membership subscriptions and other arrears of member states. The accession of South Africa as the 54th member could well do much to strengthen the organization and raise its profile. In fact, 'With its greater economic, scientific and technical resources, and its potential military contribution to peace-keeping, a post-April 1994 South Africa will be in a position to make a major contribution to strengthening the role of the OAU in tackling the multifarious challenges facing the continent' (Venter 1994: 54). As Aly adds, 'South Africa's capabilities and level of growth … well qualify it for the role of co-ordinating efforts at co-operation in the continent' (Aly 1997: 28).

In July 1995, at the OAU Heads of State summit, it was resolved that African states should be prepared to make a bigger contribution to conflict prevention and peace-keeping in the continent. South Africa was one of the eight African countries with which the US discussed this initiative, and careful consideration is being given to the proposal for establishing an African crisis response force (van Heerden 1996). Whilst it is sometimes said that South Africa has too many internal problems to be concerned with issues elsewhere, the ANC government has shown a strong commitment to strengthening the image and effectiveness of the OAU and has a potentially significant role to play in the organization.

## United Nations (UN)

At the beginning of the twenty-first century both post-apartheid South Africa and the post-Cold War United Nations are still trying to find their feet. South Africa actually played quite a significant role in the early days of the UN and was one of the original 26 signatories of the UN Declaration issued on 1 January 1942. Subsequently, in 1945 at the UN Conference on International Organizations in San Francisco, the South African Prime Minister, General Jan Smuts, had personally prepared the first draft of the Preamble to the UN Charter, which was then submitted to the Conference for approval and was largely accepted.

As we have already seen, Sharpeville in 1960 was an important watershed in terms of South Africa's relationships with the international community, including the UN. South Africa was frequently the subject of attention and debate at the General Assembly and other committees, and in 1962 the General Assembly urged member states to break off diplomatic relations with Pretoria and to close their harbours and airports to South African ships and planes. In the same year, Resolution 1761 (XVII) of the General Assembly established a special committee to monitor the country's racial policies (Solomon 1996). In 1963 the Security Council resolved to ban the sale of arms to South Africa and eventually, with Resolution 418 in 1977, the Security Council imposed a mandatory arms embargo.

Although South Africa had been one of the eight founding member states of the UN Economic Commission for Africa (ECA) in April 1958, its membership of the Commission was suspended in 1963 until such time as apartheid policies were abandoned. The General Assembly's Resolution 3151 in December 1973 urged all UN agencies and inter-governmental organizations to deny membership to the apartheid regime and instead encouraged representatives of the liberation movement to participate in their meetings. South Africa was eventually expelled from the UN General Assembly on 12 November 1974, and the ANC and PAC were given observer status and received financial assistance. Whilst the South African government claimed that the UN had no right to get involved in a country's internal affairs, the UN justified its actions because of the gross violation of human rights and South Africa's regional destabilization policy which was seen as an international threat to peace and security. As we saw in chapter 5, sanctions began to bite in South Africa from the mid-1980s, such that by 1989 some 90 per cent of the country's exports were subject to some form of sanction and 100 countries applied restrictions on South African trade (Solomon 1996).

Following the release of Nelson Mandela and the introduction of democratic rule, by the end of 1994 South Africa had once again taken up its seat at the UN and was re-admitted to the various UN specialized agencies. In 1997 a seat was also taken up at the UN Commission for Human Rights in Geneva, an agency in which South Africa is well qualified to play a significant future role. Since its re-admittance to the UN, South Africa has played an active role in a number of other important initiatives, such as bringing an indefinite extension to the Nuclear Non-Proliferation Treaty in 1995, in reforming the UN Conference on Trade and Development (UNCTAD) and in help-ing to create an African Nuclear Free Zone in 1996 (Mills 1998a). In fact, the UN has indicated that it sees considerable further scope for South Africa's participation in the organization, suggesting that it has the potential to 'become a catalyst for the rapid development of not only the southern African region but the rest of the continent' (King, quoted in Solomon 1996: 71). In 1990, the South African government had been extremely isolated, with only 30 overseas states represented, but since then diplomatic relations have been established or upgraded with 120 countries, such that by mid-1998 it had established relations with 150 countries, barring just 15, which included North Korea, Iraq, Somalia, Haiti, Liberia and Sierra Leone (Mills 1998a, 1998b).

However, post-apartheid South Africa has had its disagreements with the UN, and particularly the USA, concerning its leadership of the Non-Aligned Movement, and specifically its relationships with some of the world's pariah states – those at the opposite end of the political spectrum from the apartheid state's former tacit allies.

The new South African government has strenuously sought to develop contacts in Africa and the South in general, diversifying links which had traditionally focused on the North. Mandela in particular pursued what has been called 'tightrope diplo-macy', seeking to forge new contacts with governments which assisted the ANC in exile, such as Libya and Cuba, without at the same time seriously alienating Western governments (*Sunday Times*, 14 February 1999). Mandela's visits to Libya, both before and after the Commonwealth conference in Edinburgh in 1997, caused quite a stir and prompted speculation that he was trying to mediate in putting an end to sanctions against Libya. These sanctions were imposed because of the reluctance of the Libyan leader, Colonel Muammar Gaddafi, to hand over the men suspected of arranging the bombing of Pan Am flight 103 over Lockerbie (Scotland) in December 1988 in which 270 people from 21 countries died. On his return trip to Libya after the Edinburgh conference, Mandela was advising his so-called 'brother leader' to moderate his language and support UN efforts to resolve conflicts. Mandela stressed that his visits to Libya were primarily to thank Gaddafi for the strong support he had given to the ANC during the struggle against apartheid, though in fact Mandela played a key role in brokering the trial of the Lockerbie suspects (West Africa 1997).

Another long-standing and controversial diplomatic issue has concerned relation-ships between the US and Cuba. At the UN General Assembly South Africa was in a clear majority in opposing the US-imposed embargo on Cuba, and also refused to condemn Cuba for shooting down two civilian planes piloted by Cuban Americans in March 1996. South Africa's views on this matter were reflected in the Deputy

Foreign Minister, Aziz Pahad's, visit to Cuba in August 1996 and the 'twinning' of Gauteng with Havana (Lemon forthcoming). South Africa's relationships with Cuba are met with very different reactions within South Africa where, 'Black public opinion sees Comrade Fidel as a hero; the powerful White minority see him as a tin-pot dictator' (Solomon 1996: 75). The USA has expressed its misgivings over such contacts, but Mandela has responded forcefully by criticizing the USA's self-assumed role as 'global policeman'. During the visit of President Clinton to South Africa in March 1998, Mandela made it clear that he would not be bullied into supporting a campaign to isolate Cuba, Libya and Iran. Mandela said, 'The US as leader of the world should set an example to all of us to help eliminate tensions throughout the world. And one of the best ways of doing so is to call upon its enemies to say, let's sit down and talk in peace' (West Africa 1998: 388). Despite this incident, however, and probably because of the important regional position which the US sees South Africa as occupying, a bi-national commission established at Vice-Presidential level has helped to ensure a cordial and mutually rewarding working relationship between officials of the two countries.

South Africa's open post-apartheid foreign policy has been facilitated by President Mandela's own 'international superstar' status. Perhaps any other national leader would have had difficulty getting away with such an approach. In particular, Mandela has shown concern over human rights and democratization, seizing the moral high ground by claiming that 'Human rights will be the light that guides our foreign policy' (Mandela 1993: 88). This approach was clearly exemplified in 1998 when a South African delegation was sent to Northern Ireland to assist the peace process, and also in earlier attempts to mediate between Morocco and the Polisario Front concerning the future of Western Sahara. In 1996 Mandela pledged solidarity with the Saharawi Democratic Republic's (SDR) struggle for self-determination. Mandela was also highly critical of the military regime of General Sani Abacha in Nigeria and spoke out strongly at the 1995 Commonwealth Heads of Government Meeting in Auckland. In 1994, Mandela personally met with General Abacha, but when no progress was made he sent first Desmond Tutu and then Deputy President Thabo Mbeki. Deputy Foreign Affairs Minister Aziz Pahad was then despatched to keep up the pressure (van Heerden 1996).

## The European Union and the Lomé Convention

During the apartheid era, the European Economic Community (EEC) pursued a particularly strong line against the South African regime, introducing sanctions, promoting the suspension of full-democratic ties and supporting the frontline states. The EEC gave considerable amounts of aid to the Southern African region to both strengthen Southern African Development Co-ordinating Conference (SADCC) countries against the destabilization strategies being employed by the apartheid regime (see below) and also to reduce the region's dependence on South Africa. All SADCC members were signatories to the Lomé Convention, a neo-liberal arrangement which affords a group of 71 African, Caribbean and Pacific (ACP) countries preferential trade and aid links with the EU. The next such agreement, if there is one, will be Lomé V, which is due to start in 2000.

In November 1994 South Africa applied for membership of the Lomé Convention, but the EU refused to grant full membership because certain South African products were seen as being too competitive in EU markets and because the World Bank regards South Africa as a semi-developed country. However, limited membership was granted in 1997, and ACP leaders resolved to boost intra-ACP co-operation, partly through organizations like the Southern African Development Community (SADC), the new name for the SADCC (see below). In the meantime, the EU proposed a Free Trade Agreement (FTA) with South Africa, but such an agreement would inevitably apply to the BLSN (Botswana, Lesotho, Swaziland and Namibia) countries due to the terms of the Southern African Customs Union (SACU) treaty (see below). Under such an arrangement, these countries might lose a share in the South African market for their exports and be left with less advantageous access to EU markets than they have enjoyed hitherto as full Lomé members. So, in January 1998 the BLSN states asked the EU to undertake a full sector-by-sector impact study of the consequences for them of such an FTA whilst negotiations are still under way.

However, in a farewell present to President Mandela before his retirement, EU leaders at last agreed on 24 March 1999 to sign the trade and co-operation agreement with Pretoria. The deal, which will come into force in 2000, is unprecedented between the EU and a distant trading partner in terms of market access, particularly the 100 per cent access in manufacturing goods. The EU, South Africa's largest trading partner, will gain access to 86 per cent of South Africa's manufacturing market (*Irish Times*, 26 March 1999: 4). Such an important trade agreement further demonstrates that South Africa is deeply embedded in global networks that are Western-dominated, capitalist and neo-liberal.

It is likely that considerable emphasis will be given in a new Lomé agreement to private sector development at all levels, which will be incorporated into specific country or regional-level programmes. Another priority will be strengthening regional integration to achieve economies of scale by broadening markets and encouraging investment and technology flows to and within the region. It seems that 'The European Commission is at a turning point in its approach to development co-operation and in particular the attitude towards the role of the private sector. The macro 'enabling' environment and competitiveness are key issues. The Commission will try to help the private sector in ACP countries to develop its skills and its relations with foreign investors. Governments will be supported to improve public administration, its judiciary concerning private enterprise and its economic management so that they are able to attract investors both from inside and outside their territories' (Lowe 1999: 47). Considering the nature of future EU links with South and Southern Africa, van Heerden argues that, 'It is our view that EU trade must be made much more sensitive to development needs', rather than the imperatives of private sector profitability *per se* (van Heerden 1996: 4).

## *The Commonwealth*

South Africa's links with the United Kingdom are both strong and long-standing. In 1998 there were some 1.1 million UK passport holders in South Africa (Mills 1998a). Despite these links, member countries' antagonism to apartheid, especially after Sharpeville, and South African Prime Minister Hendrik Verwoerd's proclamation in 1961 that his country should become a republic, led to the forfeiting of its membership of the British Commonwealth, thus further isolating it from the international community.

South Africa was re-admitted to the Commonwealth after the democratic elections in 1994. The Commonwealth, which is often vaunted as the largest organization of its kind in the world outside the United Nations, had a total of 54 members by 1997. The significance of South Africa's presence in the community has received considerable publicity. One journalist has commented, 'The boost which the arrival of true democracy in South Africa has given to the Commonwealth in helping to transform the organisation's prospects and make it more relevant is still a palpable factor in creating the sense of well-being at [Commonwealth Heads of Government] summits, and Mandela is its symbol' (West Africa 1997: 1754). In November 1999 the Commonwealth Heads of Governments meeting was held in Durban, with South Africa's newly elected President, Thabo Mbeki, acting as host.

In May 1998, in an effort to boost UK–South Africa trading links, the British Overseas Trade Board launched an important three-year trade and investment campaign entitled 'Britain and South Africa – partners in opportunity'. The campaign focuses on key sectors such as health care, automotive components, leisure and tourism, railways and consumer goods, and aims to make British companies aware of the opportunities in South Africa. The initiative also involves the establishment of both a 'gateway website' to provide easy access to a wide range of websites on trade and investment between Britain and South Africa, and also a Britain–South Africa partnership programme for small and medium-sized enterprises (Laing 1999).

## South Africa and the southern African region

Given South Africa's relative economic strength and developed infrastructure, there is little doubt that it could play a very significant role in the future development of the southern African region. Nevertheless, there are also fears that South African capital could play a rather less constructive role now that it has easier access to regional resources and markets.

Since the introduction of apartheid in 1948, South Africa's relationships with neighbouring countries and the southern African region as a whole have taken various forms – economic, political and military. Some of the negative aspects of these relationships have received much attention. As Durning argues, 'The logic of South Africa's 'total strategy' first developed in the seventies, was to safeguard apartheid by fomenting chaos and economic dependency in its adversaries' camps' (Durning 1990: 29; see chapter 5). South Africa's relative strength, and control over port and transport networks encouraged the apartheid state in 1979, under Prime Minister P. W. Botha, to guarantee the economic dependency of a group of satellite states: Angola, Botswana, Lesotho, Malawi, Mozambique, Namibia, Swaziland, Zaire, Zambia, and Zimbabwe (Adedeji 1996: 10). This 'constellation of states', together with the many acts of aggression undertaken in the apartheid era, might be viewed as the 'bantustanization of the South African periphery', in which the South African government made a concerted attempt to extend the internal policy of bantustans or so-called 'homelands' into the neighbouring region, in order to ensure continued South African hegemony (Adedeji 1996; Rich 1994a). However, the constellation strategy was effectively countered by the frontline states which formed the Southern African Development Co-ordination Conference (SADCC) in 1980 (see below).

It was in Angola and Mozambique that South Africa's involvement and destabilization policies were most apparent. The two countries achieved independence from Portugal in 1975 and adopted socialist development models. In response, Durning argues that South Africa acted like, 'a cornered animal, lashing out against them ferociously' (Durning 1990: 29). During the 1980s, South Africa supplied weapons to the guerrillas of the Mozambique National Resistance (RENAMO), whilst in Angola the apartheid government supported Jonas Savimbi and his UNITA (Union nacional para a independencia total de Angola). This pressure led to Mozambique signing the Nkomati Accord with South Africa in 1984, in which the Mozambique government promised to stop supporting incursions into South Africa by anti-apartheid guerrillas, and to expel them from the country, in return for South Africa agreeing to end its support for RENAMO.

There were also military incursions into countries such as Botswana, Lesotho and Zambia and attempts to destabilize the Zimbabwe government, with training given to the Zimbabwe People's Revolutionary Army and other anti-government forces. According to the United Nations Economic Commission for Africa (UNECA), between 1980 and 1990, the 'destabilization' of the frontline states – Angola, Botswana, Mozambique, Tanzania, Zambia and Zimbabwe – cost those countries 1.5 million lives and US$60 billion (Durning 1990). The destabilization policies also further damaged South Africa's relations with the Western powers, leading in the USA, for example, to Congress's passing of the 'Comprehensive Anti-Apartheid Act', despite President Reagan's opposition, in 1986. As Simon has shown, many of South Africa's neighbours are keen to receive some sort of 'peace dividend' in the form of investment and technical assistance as reparation for both years of destabilization and also their strong support for the ANC during the struggle against apartheid (Simon 1998).

Whilst the RDP had very little to say about South Africa's future role in Africa and further afield, it did actually emphasize the importance of South Africa's position and future development in the context of the wider southern African region:

> In the long run, sustainable reconstruction and development in South Africa requires sustainable reconstruction and development in southern Africa as a whole. Otherwise

the region will face continued high unemployment and underemployment, leading to labour migration and brain drain to the more industrialised areas. The democratic government must negotiate with neighbouring countries to forge an equitable and mutually beneficial programme of increasing co-operation, co-ordination and integration appropriate to the conditions of the region (ANC 1994: 116–17).

More recently, an important emphasis in the GEAR strategy (see chapter 7) is the development of trade and particularly export promotion within the Southern African region (Ministry of Finance 1996). Other commentators have also emphasized the importance of greater collaboration between the countries of southern Africa. As van Heerden, for example, argues, 'The peoples of southern Africa are fully aware of the mutual advantage that can be derived through close interaction and co-operation. As South Africans we know this to be a fact that our development cannot be achieved in isolation from the rest of the region. Our common vision is for a region characterised by the highest possible degree of economic co-operation, mutual assistance and joint planning of regional development initiatives' (van Heerden 1996: 3).

The statistics presented in Table 8.4 indicate the overwhelmingly dominant position which South Africa occupied among the 12 SADC countries in 1997. South Africa had the highest GNP per capita (US$3,400) and a total national GNP of US$130.2 billion dollars, *S. A needs neighbour for export market*

However, the importance of South Africa's trading activities in the rest of Africa should not be underestimated, since in 1993 the African continent as a whole actually represented South Africa's largest export market, accounting for 31.7 per cent of the country's exports (Davies 1997). The value of South Africa's exports to Africa, excluding countries in the Southern African Customs Union (SACU), increased fivefold between 1987 and 1994, from US$456 million in 1987 to US$2,431 million in 1994 (Kotelo 1995: 11). By early 1995 the South African government had already established 22 trade missions in African countries and it was recognized that sub-Saharan Africa represents a potentially enormous market for South African businesses.

With regard to the Southern African region, South Africa's trade with its neighbours represents a relatively small percentage of its total trade with the world, although this trade has been growing rapidly over the past few years. Following the 1994 democratic elections, trade from South Africa to the rest of the Southern African region expanded from US$1.8 billion in 1994 to US$2.8 billion in 1995, with 30 per cent being in manufactured goods. However, imports from the region into South Africa remained fairly static, indicating a progressively more unbalanced trade *unbalanced* scenario (van Heerden 1996: 4). In fact, regional imports from South Africa exceed exports to South Africa by five to one. The RDP document recognized this imbalance of trade, stating that, 'a democratic government must develop policies in consultation with our neighbours to ensure more balanced trade' (ANC 1994: 117). As we will see below though, many South African exporters will continue to undermine such a balance.

Table 8.4   Twelve SADC countries, 1997

| Country | Population (million) 1997 | Land area (000 km²) | Population density (people per km²) 1997 | GNP (US$ billion) 1997 | GNP per capita (US$) 1997 | *GDP average annual growth (% 1991–97) |
|---|---|---|---|---|---|---|
| Angola | 11.0 | 1,247 | 9 | 3.8 | 340 | 3.4 |
| Botswana | 1.5 | 567 | 3 | 4.9 | 3,260 | 7.5 |
| Lesotho | 2.0 | 30 | 65 | 1.4 | 670 | 6.4 |
| Malawi | 10.0 | 94 | 104 | 2.3 | 220 | 6.4 |
| Mauritius | 1.0 | 2 | 553 | 4.3 | 3,800 | 6.7 |
| Mozambique | 19.0 | 784 | 22 | 1.7 | 90 | 7.6 |
| Namibia | 2.0 | 823 | 2 | 3.6 | 2,220 | 5.1 |
| South Africa | 38.0 | 1,221 | 30 | 130.2 | 3,400 | 3.3 |
| Swaziland | 0.9 | 17 | 55 | 1.3 | 1,440 | 3.7 |
| Tanzania | 31.0 | 884 | 34 | 6.6 | 210 | 5.4 |
| Zambia | 9.0 | 743 | 12 | 3.6 | 380 | 6.8 |
| Zimbabwe | 11.0 | 387 | 28 | 8.6 | 750 | 2.7 |
| Total | 136.4 | 6,799 | 76 (av.) | 172.3 | 1,398 (av.) | 5.4 (av.) |

Sources: World Bank (1998), *African Development Bank (1998)

Botswana, Lesotho, Swaziland territories of SA
until (1966) + (1969) → Namibia - administration SA
independence 1990
South Africa and the southern African region    **299**

## The Southern African Customs Union (SACU)

The Southern African Customs Union (SACU) is the oldest of the regional 'group-ings' in Southern Africa, and indeed reputedly the world's oldest still-surviving customs union. Its origins can be traced back to the nineteenth century, when in 1889 the Customs Union Convention was signed by the Cape Colony and the Orange Free State. In 1893, the British High Commission territories of Bechuanaland and Basu-toland (now Botswana and Lesotho) joined the Convention, with Swaziland joining ten years later in 1903. With the formation of the Union of South Africa in 1910, a new customs union agreement was negotiated between South Africa and the three High Commission territories. This negotiated agreement between Britain and South Africa was primarily aimed at potentially incorporating the High Commission Terri-tories into South Africa and it worked overwhelmingly in favour of the South African government, allocating that country 98.7 per cent of the joint revenue while Basu-toland received 0.88 per cent, Bechuanaland 0.27 per cent and Swaziland 0.15 per cent (Cattaneo 1990; McCarthy 1992).

The 1910 agreement lasted for almost 60 years until 1969, by which time the High Commission states had gained their independence – Botswana and Lesotho in 1966 and Swaziland in 1968. Understandably, there was some concern among these states about the fact that the revenue-sharing formula was still weighted heavily in South Africa's favour. A new Agreement was signed in December 1969 and, unlike the 1910 Agreement, it emphasized the promotion of 'economic development', with all member states receiving 'equitable benefits' from trade among themselves and with other countries. Namibia, which had been administered as part of South Africa since 1915, formally became a member after gaining its independence in 1990. A Common Revenue Pool (CRP) was established to pool customs, excise, import surcharges and sales duties for the five member countries and this is managed by the South African Reserve Bank, which divides the pool according to annual imports, production and consumption of dutiable goods (Sidaway and Gibb 1998: 174). However, there is a two-year time lag in payments from the CRP, in effect giving South Africa an inter-est-free loan at a time of high inflation and interest rates.

SACU has survived for so long both because it provides member countries with revenue and it also relieves them of the burden of having to collect some of their own customs duties. SACU receipts actually make a substantial contribution to individual state revenue, such that in 1992 receipts by the BLSN states represented between 11 and 20 per cent of their GDP and between 22 and 47 per cent of government revenue (Sidaway and Gibb 1998: 174). However, 1993/94 figures indicate that South Africa still received a 62 per cent share of total SACU revenue (Leistner 1997: 116). As Sidaway and Gibb observe, 'SACU is a product of its imperial times ... (and) has therefore embodied a complex, but nevertheless profound and enduring, dependency relationship between South Africa and its near neighbours' (Sidaway and Gibb 1998: 172). In 1994, South Africa exported US$3.9 billion of goods to SACU and had a US$2.8 billion trade surplus with the Union (Burton 1995). In fact, in the early 1990s South Africa exported to the rest of SACU more than five times the value of goods and services that it imported from its SACU partners.

Trade data indicate that 40 per cent of total South African manufactured exports went to BLSN in the early 1990s, and BLSN complain that South Africa's industrial protection policy results in them having to pay prices which are well above world market levels for South African goods. It is claimed, for example, that South African prices for goods such as motor vehicles and household appliances are nearly twice as much as average world prices (Leistner 1997). It has therefore been suggested that in the post-apartheid era the biggest benefit to South African capital of the government's continuing membership of SACU is 'the existence of a captive market for its internationally uncompetitive exports' (Sidaway and Gibb 1998: 177). The BLSN countries may be independent, sovereign states, 'but from the point of view of regional trade they are totally dependent on South Africa' (Ahwireng-Obeng and McGowan 1998: 10).

Negotiations for a reconstituted SACU began in November 1994 and it seemed that the main focus of attention would be on renegotiating the somewhat sensitive revenue-sharing formula. South Africa is concerned about its declining share of the revenue pool, whilst the BLSN states argue that they are receiving far too little and are actually subsidizing South Africa, since the growth in their Gross Domestic Product has been greater than South Africa's since the mid-1980s. There is a good deal of suspicion about South Africa's motives within SACU, such that Botswana and Namibia threatened to leave the Union in 1997, whilst Zimbabwe, Mozambique, Malawi and Zambia showed interest in joining. SACU leaders have persistently complained about the delays and selfish attitude of South Africa in the protracted renegotiations, suggesting that the Union 'is an institution left behind by departed "colonial masters" and unfairly privileges South Africa above the other members' (Ahwireng-Obeng and McGowan 1998: 183).

## The Southern African Development Community (SADC)

Another important regional organization mentioned above, the Southern African Development Co-ordination Conference (SADCC), originated among the so-called 'frontline states' of Angola, Botswana, Mozambique and Tanzania, who in the 1970s were collectively concerned about the Rhodesian and South West African conflicts. This attempt at regional co-operation was formalized with the Lusaka Declaration of April 1980, when nine southern African states came together in SADCC: Angola, Botswana, Lesotho, Malawi, Mozambique, Swaziland, Tanzania, Zambia and Zimbabwe. Namibia joined the group after gaining independence in 1990.

The principal aims of SADCC were to strengthen links, promote regional co-operation and integration and reduce economic dependence on South Africa. The Lusaka Declaration, however, stressed that SADCC was not a 'common market' and, 'this rejection of common-market principles was consistently adhered to' (Gibb 1995: 223). It has been suggested that SADCC was 'one of Africa's few successful attempts at regional co-operation' (Morna 1990: 49). As a tangible result of SADCC's endeavours, air links and micro-wave telecommunications between member countries have been improved, and there has been some considerable success in obtaining aid from international bodies such as the World Bank, European Community and, most notably, the Scandinavian countries. But, on the negative side, relatively little progress has been made in reducing external dependence and in particular the region's dependence on South Africa.

In the early 1990s, with the end of the Cold War and the emergence of democratic rule in South Africa, the emphasis among SADCC member states gradually switched from specific projects to the co-ordination of sectoral plans and programmes. The Treaty of Windhoek in October 1993 formalized the replacement of SADCC with SADC – the Southern African Development Community – and the ten countries were joined by a further two, South Africa in 1994 and Mauritius in 1995, making a community of 12 with a total population of over 136 million, a combined land area of 6 million square kilometres and a total GNP of US$172.3 billion (Table 8.4 on page 298). In 1996, President Mandela was elected chair of SADC for three years. The following year, SADC somewhat surprisingly admitted a further two countries, the Democratic Republic of Congo (formerly Zaire) and Seychelles, thus enlarging the community to 14 member states (see Fig. 8.1). South Africa strongly supported admitting Congo, not least because of the considerable business interest of the South African state-owned utility Eskom in Congo's hydro-electric power resources. In fact even before it was admitted to SADC, hydro-electric power from southern Congo was being used in SADC states.

Between 1991 and 1997 economic growth in SADC as a whole (excluding Congo and Seychelles) averaged 5.4 per cent per annum (Table 8.4 on page 298), although there was some considerable variation in average annual growth rates from the 'highfliers', such as Mozambique and Botswana (7.6 per cent and 7.5 per cent respectively), to Zimbabwe, South Africa and Swaziland, with annual economic growth rates of 2.7 per cent, 3.3 per cent and 3.7 per cent respectively.

Despite optimism about SADC's considerable potential both in the region and within Africa, the organization has come in for a good deal of criticism and scepticism from different quarters. Matshaba, for example, has suggested that, 'Misconceptions are still rife among the people of southern Africa, even by senior government officials about the intention and purpose of the Southern African Development Community (SADC). On the one hand, says the community's former executive secretary, Simba Makoni, ordinary people think the grouping is meant for permanent secretaries, ministers and presidents. And on the other hand, officials perceive it as an organisation to seek donor aid' (Matshaba in *Daily News* (Gaborone), 3 November 1994: 2). Sidaway also reports that when SADC invited member states to suggest an appropriate official logo for the community, a satirical article in Zimbabwe's weekly *Financial Gazette* commented (in alluding to the community's dependence on Western aid for financing projects) that SADC could be best symbolized by, 'a begging bowl. [For] What has SADC achieved apart from mugging gullible Scandinavian countries? SADC members are today more dependent upon South Africa than they were in the early 1980s' (*Financial Gazette* 1994, quoted in Sidaway 1998: 556). It seems to many that 'The core business of the SADC lies in the putting together of bids for development programmes and in soliciting finance for these. In practice, all SADC-mediated development programmes function on the basis of external aid' (Sidaway 1998: 559). In 1995 alone, various donors pledged SADC over US$4,000 million of aid and development funding, almost three-quarters of which (US$2,991 million) was earmarked for 208 identified transport and communications projects (Sidaway 1998: 561; Table 8.4 on page 298).

SADC's recognition by the World Trade Organization will depend upon the implementation of a free trade area, since, 'there is a recognition that SADC's bargaining power in international forums will be improved if the organisation can prove that it is serious about regional economic integration' (Bertelsmann 1998: 5). However, in fairness, SADC has achieved some notable successes (particularly in the fields of transport and telecommunications), and despite the very different historical and political records of the member states, the Community has successfully held together, building its deliberations on achieving consensus.

South Africa's agenda for the future development of SADC undoubtedly recognizes the opportunity for South African capital to exploit resources and markets in the region, a development which is seen as a potential threat by the other member states. There has also been some concern about South Africa's disregard for procedure, for example in the protocol surrounding the first SADC conference held in

Figure 8.1  The 14 Southern African Development Community (SADC) Countries

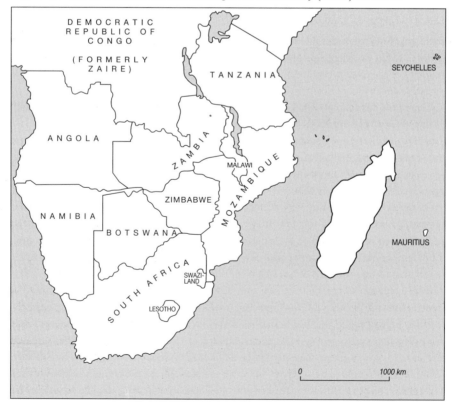

South Africa in 1996. There is a strong element of suspicion directed at South Africa by the other SADC members because of South Africa's massive economy and strong world profile, particularly since 1994, and not least because members are all too aware that SADCC was originally established to counter the apartheid regime. As one South African diplomat has commented, 'There is a sense in which SADC is joining South Africa, not that South Africa is joining SADC' (quoted in Sidaway 1998: 568).

The overwhelming dominance of South Africa within SADC could well lead to pressures for some form of import restriction from neighbouring countries if trade with South Africa is all in one direction. Furthermore, with much concern about unemployment levels in South Africa, it is likely that opportunities for migrants from neighbouring countries might decline, and refugees from countries such as Mozambique (numbering some 350,000 in 1992) could be deported in an attempt to preserve jobs for black South Africans (Rich 1994b: 43). A further concern is that, with a relatively more developed South Africa, this could lead to a damaging 'brain drain' among educated professionals from neighbouring countries, as such people are attracted to higher salaries in South Africa. Worries about South Africa's potentially domineering role within SADC were exacerbated in 1998, when the South African army was sent in, with disastrous results, to secure 'order' during an attempted coup in Lesotho.        S.A army attempted to secure order in Lesotho

## PTA/COMESA

The Preferential Trade Area for Eastern and Southern African States (PTA), now known as the Common Market for Eastern and Southern African States (COMESA) is another and very much larger regional grouping than either SACU or SADC. The organization was established following initiatives by the OAU and the UN Economic Commission for Africa in an attempt to strengthen regional integration and create an African common market by the year 2000. Although the PTA treaty was signed by ten states in 1981 and was envisaged as the first step to a free trade area and ultimately a common market, the organization only started functioning in 1993. By 1997 COMESA had 23 members – the SADC states: Lesotho, Namibia, Swaziland, Zimbabwe, Zambia, Malawi, Angola, Mozambique, Tanzania, Mauritius (but not South Africa), together with Burundi, Comoros, Djibouti, Eritrea, Ethiopia, Kenya, Madagascar, Rwanda, Seychelles, Somalia, Sudan, Uganda and Zaire. COMESA covers a vast area, with a wide diversity of economic and political characteristics among its member states.

Although under the apartheid regime both the African National Congress (ANC) and Pan African Congress (PAC) were granted membership of COMESA, somewhat surprisingly, the new democratic South African government decided not to join, indicating its concern about the organization's level of efficiency and viability and its increasingly overlapping programmes with SADC (Adedeji 1996). This represented a major blow to the status and potential of COMESA and has helped to give more emphasis to SADC among those countries which are members of both bodies. As Leistner comments, 'Some observers even believe it may have signalled COMESA's death knell' (Leistner 1997: 120). Interestingly, the African Development Bank (1993) has argued that the smaller and tighter structure of SADC is more likely to lead to regional integration than the much larger and diverse COMESA. At the 1994 Heads of State Summit, SADC agreed that its members should consider withdrawing from COMESA and by April 1997 Mozambique and Lesotho had actually announced their intention to do so. There is some confusion about the reasoning behind South Africa's reluctance to become a member of organizations such as COMESA and, indeed, the African Development Bank. With regard to the latter, Adedeji asks, 'How can such a mighty economy, in relative terms, be reluctant to become a member of Africa's only economic development institution?' (Adedeji 1996: 232). Adedeji believes such moves reflect a policy of South Africa distancing itself from Africa in favour of developing more profitable links with Asia and its enthusiasm to work with India and Australia in establishing an Indian Ocean Rim regional co-operation arrangement.

## South African capital and the wider region

Despite all of the hopes which have been expressed, some writers, such as Dietrichsen, from the South African Department of Foreign Affairs, have expressed their concern about South Africa's ability to foster development in the wider region: 'Although South Africa's foreign debt is low by world standards, the country's own development needs are such that South Africa could not possibly become a substantial donor of development assistance' (Dietrichsen, quoted in Mills 1994). Foreign Minister Alfred Nzo, in his address to the UN General Assembly in June 1994, also reminded the international community of his country's domestic priorities: 'Uppermost in our minds ... are the responsibilities which our new Government of National Unity has towards the people of South Africa. Our primary goal is to strive to create a better life for all our people ... [as a result] ... South Africa will have extremely limited resources for anything which falls outside the Reconstruction and Development Programme' (Nzo 1994: 12–13).

Far from being a source of positive economic development in the region, some analysts fear that post-apartheid South Africa's capitalist enterprises will simply exploit the region's resources more effectively, whilst generating little in the way of local development. Terms like 'hegemony' and 'neo-imperialism' are frequently used in relation to both the South African government's and its capital's position and policies vis-à-vis other African countries, and there is concern that if South Africa takes a greater role this could reinforce the power relations which developed during the apartheid era. Just as Nigeria has often been accused of being the 'big brother' in West Africa and is much criticized within the West African region, so South Africa could become the 'big brother' of the whole continent. In economic terms, there are also close parallels between Nigeria and South Africa, with the former dominating the Economic Community of West African States (ECOWAS) whilst, as we have seen, South Africa in turn dominates both SACU and SADC.

There is no doubt that since 1994 South African businesses have recognized the potential for penetration into other parts of Africa. As Johann Wingard, vice-president of the South African Chamber for Agricultural Development in Africa commented, 'Our economic border is no longer the Limpopo River – there are unbelievable opportunities out there' (quoted by Pickard-Cambridge in *Business Day*, 22 April 1998: 13). The aggressive expansion of South African businesses is exemplified in the operations of three very different companies chosen from a long list. For example, the Anglo-American Corporation now has mining operations in Angola, Botswana, Ghana, Guinea, Ivory Coast, Mali, Namibia, Senegal, Tanzania, Zambia and Zimbabwe. Meanwhile, SA Breweries is involved in beer and soft drink production in Botswana, Lesotho, Mozambique, Swaziland, Tanzania and Zambia. Finally, the hotel chain, Protea, has establishments in Botswana, Egypt, Kenya, Malawi, Mauritius, Swaziland, Uganda and Zimbabwe (Ahwireng-Obeng and McGowan 1998: 26–8).

This largely non-reciprocal South African business expansion into other parts of Africa reflects the considerable relative expertise of South African businesses, together with more investor-friendly environments generated by structural adjustment programmes. Iheduru draws an analogy between this business expansion, encouraged by the GEAR strategy, and the 'constellation of states' policy during the apartheid era, referring to, 'a constellation of Southern African economies led by private capital' (Iheduru 1996: 22). In the future it is possible that 'trade diversion will occur, in which South African manufactured exports and services will gradually replace SADC's and other African states' traditional sources of supply from the North because of their lower transportation costs and African-friendly characteristics' (Ahwireng-Obeng and McGowan 1998: 23). Just as neo-liberalism is constraining opportunities for redistribution within South Africa (as we have seen in chapter 7), so it is simultaneously allowing South African capital to exploit new opportunities beyond the national borders. The strong international profile of South African business has been reflected, for example, in the relocation in April 1997 of the African Business Round Table (ABRT – founded in 1990 by the African Development Bank) from Abidjan, Ivory Coast to Johannesburg. The relocation of ABRT was due to the realization that 'major investors and international organisations were located in South Africa' (*Business Day*, 25 April 1997: 3).

While South African enterprises seek out profitable opportunities elsewhere in Africa, South Africa itself continues to be a 'magnet' within the region, and even more so with more penetrable international borders and the ending of the siege mentality typified by the apartheid era. South Africa could well become the major destination for migrant skilled labour from the rest of Africa, which could further impoverish those states while contributing to South Africa's neo-imperialist status in the continent. In 1991, there were 165,825 workers from outside the borders of South Africa (excluding those from the so-called 'independent' homelands) who were employed in gold and coal mines owned by members of the Chamber of Mines, representing 42 per cent of the total mine labour force. These numbers were very much greater in the 1970s and are continuing to fall, with the closure of mines and the deteriorating domestic unemployment situation (Chamber of Mines 1991: 60).

In addition, the number of official refugees in South Africa was around 10,000 in 1995, but illegal migration from neighbouring states rose from 45,000 in 1988 to nearly 100,000 in 1993. Estimates of the total number of such migrants in the country vary widely, from 2 million to 9 million, but the South African police apparently believe the latter is the more accurate figure (Mills 1995: 61). Traditionally, such migrants have been mainly Mozambicans, but recently some of the streets of South Africa's cities have been occupied by street hawkers from as far away as Senegal, Nigeria and Kenya. There is also a serious 'brain drain' migration of doctors, engineers, teachers and other intellectuals leaving their countries of origin for South Africa and its immediate neighbours which have more robust economies (Botswana, Swaziland and Namibia). Some observers advocate strict measures to stem the flows of migrants to South Africa, for example forced repatriation, intensified patrolling of frontiers or even the use of electric fences, reminiscent of the apartheid era (Solomon 1995: 75). As the South African economy and job opportunities strengthen relative to those of neighbouring countries, the 'magnet effect' is likely to further increase, leading to greater regional inequality and a variety of associated pressures.

So what are South Africa's medium- and long-term visions of its relationships with the rest of Africa? Will post-apartheid South Africa be serious about turning its attention towards regional development in Africa and the southern African region, or will it choose instead to foster its long-standing relations with the First World countries? (Aly 1997). Whilst it is impossible to pre-judge future political decisions, there is no doubt that much scepticism abounds. Ahwireng-Obeng and McGowan are particularly critical of the post-apartheid record thus far:

> Many thought that a free and democratic South Africa would fundamentally change this relationship with the rest of the Southern African region. So far this has not happened … Since liberation *what has South Africa actually done* to promote more balanced and equal development in the region? Our answer is: very little. This is deeply ironic given the widespread African belief that the end of apartheid would promote an African renaissance … It has been said, not in jest, that President Bill Clinton's foreign policies seek to make the world safe for American capitalism. President Mandela and particularly Vice-President Mbeki appear to be trying to do the same for South African capital (Ahwireng-Obeng and McGowan 1998: 190).

What seems to have happened, despite the political changes in South Africa, is that the power of a global neo-liberal discourse has allowed a basic symbiosis between the South African state and South African capital to continue through the abandonment of apartheid and into the post-apartheid era.

### South Africa and Africa – what future within the global political economy?

As we have seen from the evidence presented here, the power which the 'new' post-apartheid South African state wields depends upon the scale of our analysis. By virtue of its industrialized economic base, it is in a very powerful position within the southern African region and, indeed, in the African continent as a whole. Although it has been able to negotiate relatively good trade terms within some global arenas, it carries far less weight within a world economic and political system dominated by the Western metropoles. Nevertheless, with an economy that towers above all other sub-Saharan African economies, an infrastructure and resource base which in many ways are comparable to many European countries, and a new-found political status until recently enhanced by one of the most charismatic and universally popular leaders the world has ever produced, the South African state entered the twenty-first century in a very strong position relative to other African states. Yet, as we saw in chapter 6, within the country's borders there is some appalling poverty, gross social and spatial inequalities and a soaring crime rate, which together could prove to be the undoing of the 'Rainbow Nation'. These difficulties could themselves have disastrous repercussions on the state's performance and potential on the international scene.

It remains to be seen whether, having effectively abandoned the redistributionist RDP in favour of GEAR – a strategy clearly more favoured by the West – the new black leaders will be as successful in harnessing international capital to their political project, as their Afrikaner predecessors were during the 1960s. Furthermore, if the new leaders are successful, can such capital then be operationalized to reduce poverty and inequality as envisaged in the RDP? As we will argue in the following chapter, this seems unlikely at present. In considering the future role and position of South Africa in the African continent, we can gain certain insights from the past experiences of other African states. States following both left-wing and neo-liberal development trajectories have been equally beset with problems and had their internal stability challenged, we would argue, not simply because of colonial legacies and competition for scarce resources within their borders, but also because of the postcolonial global networks of finance and trade in which they are inescapably situated. Post-independent Ghana under Nkrumah, and Côte d'Ivoire under Houphouët-Boigny in the 1970s and 1980s, followed very different development models, yet both became deeply indebted and had to resort to assistance from the World Bank and IMF.

Under the charismatic leadership of Kwame Nkrumah, Ghana gained independence in 1957 and set the pace for other countries in sub-Saharan Africa to follow. Nkrumah preached the virtues of Pan-Africanism and saw the outside world as dominated by imperialist machinations intent upon neo-colonial exploitation. He stressed the importance of a socialist development strategy, in which concern for social redistribution and employment generation outweighed consideration of growth and efficiency. Nkrumah's socialism, however, entailed the same characteristics of 'modernity' in state planning that characterized the West's post-independence development agenda. The state was assigned the role of mobilizer, allocator and manager of resources and opportunities, and the bureaucracy inevitably grew to manage these state-sector activities. The main thrust of Nkrumah's domestic policy emphasized capital-intensive industrialization to reduce external dependence, rather than improving the conditions of the poor rural farmers, although comprehensive free education and health care systems were introduced.

Along with many Western 'developmentalists', Nkrumah believed that agriculture was the 'handmaiden' to industry, and that revenues from the agricultural sector should be used to fuel urban and industrial investment. Paradoxically, such ideas were reminiscent in particular of Rostow's anti-communist 'stages of economic growth' model. With foreign reserves and a reliable revenue base ensured by monopolizing the marketing of cocoa through the Cocoa Marketing Board, the state could tax cocoa farmers to provide extensive economic infrastructure. An ambitious programme of large state farms was instituted in an attempt to modernize agriculture and replace subsistence-oriented peasant farming.

The ambitious development programme was unsustainable, as initial assets were quickly exhausted and Ghana was plunged into economic crisis. Corruption and mismanagement were rife and the highly politicized bureaucracy was unable to formulate and implement policies efficiently and objectively. In 1966, Nkrumah's regime was overthrown by the military. There were many, perhaps inevitable, contradictions in Nkrumah's development programme. For example, whilst he was openly critical of Western capitalism, he also tolerated foreign companies operating within Ghana and even supported some foreign businessmen financially while deliberately denying Ghanaians the opportunity to undertake large-scale private economic activities. Reflecting on the Nkrumah years, it has been suggested that in one sense the independent state was like the colonial state, indirectly working with foreign capital to inhibit the rise of the local bourgeoisie (Dzorgbo 1998). Perhaps the greatest irony though, was that in seeking to escape neo-colonial domination, a 'development' regime founded on Western notions of modernity and progress was nevertheless adopted. This, of course, raises the difficult question of what other kinds of strategy, pursued independently from any form of Western influence, are available to postcolonial states.

Unlike Nkrumah's Ghana, Côte d'Ivoire under Félix Houphouët-Boigny encouraged the development of a local business class, in a strategy which demonstrated extreme openness to the outside world, rapid economic growth of 7 per cent per annum in the early 1970s, and a liberal policy towards capital movement. In the rural sector there was initially a heavy dependence on peasant farming, rather than creating large mechanized farms, whilst state firms channelled resources to private local businesses through development banks. But Côte d'Ivoire, like Ghana, experienced severe financial difficulties. From the mid-1970s a number of large agricultural development operations were launched and managed by a rapidly expanding public enterprise sector. Much heavy equipment was purchased and there was considerable investment in infrastructure. The collapse of coffee and cocoa prices from the second half of 1978, however, exposed all the components of a serious financial crisis. A marked deterioration in the terms of trade led to an escalating balance of payments crisis, such that from 1980 the government had to ask the World Bank and IMF for assistance, in return for agreeing to adopt a severe programme for financial recovery. The situation deteriorated further in the early 1980s, and in May 1987 the country's insolvency was declared. Whilst a number of major industrial projects were abandoned due to the financial crisis, some notable prestige developments were pursued, such as the massively expensive building of an inland capital city in the President's home town of

Yamoussoukro. An important element of Côte d'Ivoire's financial difficulties came from the public enterprise sector, which was plagued by patrimonialism (Fauré 1989).

From a review of political and economic events in African countries such as Ghana and Côte d'Ivoire come a number of insights into the dilemmas which now face the leaders of the 'new' South Africa. Perhaps just three specific points might be raised here. First, the influence of fluctuating world commodity prices, over which domestic producers have little control, is of considerable relevance in the light of the falling gold price. During the late 1990s, the gold price declined from over US$360 per troy ounce in 1996 to US$255 in September 1999. The World Gold Council estimated that the drop in prices in the six months between March and September 1999 had cost the highly indebted poor countries US$150–200 million in lost export earnings. Understandably, there is much concern in South Africa about possible future trends in the price of gold and the likely impact on economy and society – a concern fostered by a relatively uncompetitive manufacturing sector and a continued partial reliance on primary product exports.

Secondly, as we argued in the introduction to this chapter, patrimonialism, patronage, prestige expenditure and corruption seem to be both widespread and enduring in postcolonial Africa. All too frequently there has been a failure among governing elites to transcend the state forms bequeathed by colonialism and develop 'consensual' politics with clearly specified development objectives which could be widely understood and accepted. 'Lopsided' state structures established by colonial regimes often seem to have been inherited by nationalist black elites with little reformation, and with little capital to effect any such reformation. Those structures continue in many cases, including South Africa's, to favour urban elites (formerly white colonists, now politically significant black citizens) over most rural producers, who are often regarded as less as full citizens (Mamdani 1996). Finally, there are many cases of governments embarking on development plans which, given their limited access to capital, have been over-ambitious in terms of their cost, scale and timetable for their successful implementation. The context for each of these observations, however, is one of a global political economy created through centuries of colonialism, in which power is still weighted heavily towards the West. This is a context to which we will return in the final chapter.

# Chapter 9

# SOUTH AFRICA'S CURRENT TRANSITION IN TEMPORAL AND SPATIAL CONTEXT

In this chapter we aim to pull together some of the many strands of analysis which connect the historical and development sections of this book. The chapter was written partly in order to allow for a reprise of some of the key points made in earlier chapters, but it is also geared towards the construction of a new set of arguments. We hope to demonstrate on the one hand that a deep historical grounding can generate useful caution in our approaches to development, and on the other, that an awareness of contemporary development initiatives can allow us to interpret the past in new and fruitful ways. Specifically, we intend to conceptualize South Africa's current, post-apartheid transition within the context of two prior transitions in the region's economic, cultural, social and political practices. Each of these transitions is conceived as the outcome of shifts not only within southern African societies, but within a broader capitalist network of power. This network in turn is thought of as being dominated by particular, dynamic groups located both in the global peripheries like South Africa and in successive metropoles such as Britain and the USA.

## The argument

It seems obvious that opportunities opened by the abandonment of apartheid are limited by legacies inherited from the past. But, through the analysis in this chapter, we want to go beyond this observation and argue that certain parallels exist between colonial reform associated with the abolition of slavery in the early nineteenth century (chapter 3), the onset of industrial modernity in the late nineteenth and early twentieth centuries (chapter 4), and the current post-apartheid transition in South Africa (chapters 5, 7 and 8). Each of these transitions, we assert, has been occasioned by a shift in the basis of capitalist production initiated not so much in South Africa itself as in successive global metropoles. Each transition has given rise to adjustments of the social boundaries of race, class and, to a certain extent, gender within South Africa, and each such adjustment has been guided by both material and discursive connections between dominant groups distributed across a global capitalist terrain. In the light of these understandings, we agree to some extent with David Harvey when he states more generally that recent and rapid social and political changes, 'when set against the basic rules of [global] capitalistic accumulation, appear more as shifts in surface appearance rather than as signs of the emergence of some entirely new ... society' (Harvey 1990: vii).

We are by no means the first observers to set South Africa within such a global framework of analysis. Historians like Stanley Trapido (1971), Martin Legassick (1974), Anthony Atmore and Shula Marks (1974) have long attempted to draw attention to the global circuits of capitalist practice within which South Africa has been situated. They have highlighted in particular the ways in which the British metropole and the colonial southern African periphery were linked through flows of capital. More recent accounts by Marks (1987), Trapido (1990), Legassick (1993), Andrew Bank (1995), Saul Dubow (1995) and Timothy Keegan (1996) have supplemented these materialist analyses by focusing on the cultural discourses which accompanied capital as it traversed its global circuits. Geographers such as Jenny Robinson (1992, 1996, 1999), Susan Parnell and Alan Mabin (Parnell and Mabin 1995; Parnell 1996) have enhanced the historians' efforts by situating South African urban planning within a global discourse of modernity. The sociologist Kathryn Manzo (1992) has focused almost exclusively on the effects of global power relations on South Africa, but here we want to emphasize, rather more than is the case in her analysis, the ways in which the South African periphery and global metropoles have been mutually constitutive over an extended period (see Stoler 1995; Stoler and Cooper 1997; Lester 1998a).

What we wish to assert here is that in each case of major transformation in modern South Africa's history, local elites have been able to utilize their connections with capitalist elites elsewhere, and particularly in the global metropoles, to re-establish their hegemony. The precise composition of these elites may change during the transitional period. Some who were previously excluded from dominance may attain it, whilst others who were previously privileged may lose that privilege. But, taken as a whole, we suggest, there are powerful economic and cultural mechanisms which enable privilege to be contained and perpetuated, in inter-generational fashion, within certain social boundaries. Through their global connections, we suggest, dominant groups, in South Africa and elsewhere, are effectively able to 'manage' the periodic shifts in capitalist practice which would otherwise jeopardize their own position.

We will now take each of the major transitions in modern South African society in turn, analysing their key characteristics, before concluding by pointing out the parallels which we identify between them.

## The transition to liberal capitalism and the abolition of slavery in the early nineteenth century

As we saw in chapter 3, during the seventeenth century, colonialism in the Cape Colony was overseen by Europe's most successful capitalist enterprise, and the one with the furthest global reach – the Dutch East India Company. Through slavery and the domination of the Khoisan, the colony, like others, was founded upon the production and exchange of commodities produced by forced labour. In order to sustain such a system, colonists and the Company government had to recognize and enforce legal distinctions between status groups. The most obvious and clearly delineated boundary was that between the free and the enslaved, but more ambiguous distinctions were also maintained between burgher citizens and non-burghers including the mixed race 'Bastaards', and between masters and indentured (often captive) servants.

The impetus to transform the Cape's legal and social boundaries came during the 1820s, some 14 years after the colony had been taken under British governance and some decades after Britain had displaced mercantilist Holland as the most powerful metropole within the global economy. In order to understand British officials' desire to transform Cape colonial society at this time, we need to turn our attention to developments not in the Cape itself, but in Britain. Despite their many differences, the increasingly prosperous industrial and commercial middle classes of Britain had been reacting since the late eighteenth century to their continued exclusion from aristocratic political power by developing an alternative 'bourgeois' identity. It was based on a universalist conception of human nature: a recognition of the capacity of each individual, given freedom from confining regulations, to progress spiritually and materially, thus contributing to the greater good of society. Implicitly, such a notion challenged the hereditary structures of exclusion that had concentrated privilege among the aristocratic class in Britain. The bourgeois idea had its explicitly economic counterpart in the free trade philosophy of Adam Smith. This too, called for the abolition of restrictive measures that protected particular interests at the expense of the many.

By the 1820s, their shared principles of reform had allowed bourgeois evangelicals and free trade adherents in both the British metropole and its colonies to form a global humanitarian reformist movement. In Britain, the beleaguered aristocracy, caught between the twin threats of domestic working-class insurrection and the Napoleonic armies, had also come to accept the need for an alliance with the bourgeoisie. From the 1830s, the gates of political privilege were partially opened to admit the 'respectable' middle classes through the extension of the franchise. The fact of middle-class political as well as economic hegemony guaranteed that bourgeois identity, including its evangelical component, would provide the foundation of a new ruling-class discourse in the British metropole. Within this discourse, labour had to be free to seek out employers rather than forced to work for any one individual. By the same token, however, the British working classes would be disciplined into rendering their labour soberly, respectfully and without overt resistance. While direct violence had to be employed at times, the bourgeoisie's alliance with the state and its cultural power enabled it to manage the process of early industrialization in Britain on behalf of the propertied classes as a whole, shielding them from threatened proletarian revolution.

The establishment of a non-racial liberal tradition in the nineteenth-century British Cape Colony cannot be conceived in isolation from these profound metropolitan developments. That 'tradition' began with the abolition of the slave trade in 1807, and then of slavery itself throughout the British empire, in 1834. These accomplishments were the result of bourgeois evangelical pressure brought to bear in both the colonies and the metropole in the interests of universal humanity and free trade. Riding on the back of the abolitionist movement, humanitarians from lower-middle-class British backgrounds in the Cape, such as John Philip, were able to overcome vehement settler resistance and secure the legal liberation of the indentured Khoisan and Bastaards as well as the slaves, through Ordinance 50 of 1828.

By the late 1830s then, a reformist impulse connecting the Cape with the most dynamic centre of the world economy (Britain) had led to the abolition of legislated racial boundaries such as those of slavery, burgher recognition and indenture. However, the long-term implications of this transformation depended on the prescriptions which bourgeois reformers held out for the newly freed labourers. While early-nineteenth-century reformism in Britain had been directed at the eradication of a hereditary monopoly on power, as we have suggested, it was not intended to create a social order free of distinctions. The metropolitan bourgeoisie desired that propertyless British subjects be freed from arbitrary legal tyranny, so that they might aspire individually to social ascendancy and material reward. But bourgeois reformers also wanted to ensure that free labour remained generally compliant within an unequal, yet productive order.

Prescriptions for the freed labour force of the Cape paralleled those for the poor in Britain. After emancipation in 1834, the redistribution of land, for instance, was out of the question, since one of the first lessons of freedom for the slaves was that an inherited distribution of property must be respected. The post-emancipation Masters and Servants Ordinance of 1841 followed precedents set in Britain. There, Master and Servant Laws, imposing criminal punishments for labourers who broke their contracts, were relaxed only in the late 1860s. The Cape Ordinance made no distinctions on the grounds of colour, but it too sanctioned harsh treatment for servants breaking their contracts with, or even being 'disobedient' to, their 'masters'. Such measures ultimately facilitated the assimilation of the established elite – the landed, Dutch-speaking, slave-owning Cape 'gentry' – into a new colonial ruling class dominated by British merchants and officials.

Liberal capitalism, emanating from the British global metropole, did mount a formidable challenge to the existing order of privilege in the Cape during the early nineteenth century. But the challenge was met by the interconnected colonial and metropolitan bourgeoisies. Their notion of a universal humanity, free of inherent and inherited status distinctions, was readily combined with an insistence on governmental respect for the market and for property. Together, these concepts amounted to a discourse by which transition was conservatively managed. The Cape's landless slaves and Khoisan found themselves emancipated from the arbitrary will of the master, but subjected to the regulating forces of the labour market and of the colonial state instead. Meanwhile, a new and consolidated colonial elite had emerged – one which incorporated most of the Dutch-speaking former 'Cape gentry' as well as the new British colonial bourgeoisie.

## The transition to industrial capitalism and segregation in the late nineteenth and early twentieth centuries

If a metropolitan-centred discourse had guided the first of colonial South Africa's major transitions, it was a reformulated global discourse that allowed for the conservative restructuring of the country's social boundaries when they were later threatened by industrialization. By the 1860s, the bourgeoisie in Britain itself had been successfully accommodated within the political elite, and the better-off and more 'respectable' members of the working class were also gradually being incorporated through successive franchise reforms. But this was all the extent of incorporation that could be tolerated. Until such time as the labouring classes as a whole could be considered 'respectable', further reform would render the British social order once again susceptible to socialist disruption. The universalist bourgeois notions which had helped undermine aristocratic hegemony in the early nineteenth century now had to be modified in order that the dominance of the propertied classes as a whole could be consolidated.

Universalist liberalism would be adjusted in Britain during the late nineteenth century with the assistance of ideas which had first been generated at the margins of the empire by planters, settlers and officials in direct conflict with indigenous peoples and former slaves. These ideas, conveyed to the metropole through various media, concerned the supposedly innate and scientifically observable differences between groups of human beings, and notably between racial groups. By the late nineteenth century, social Darwinism, fuelled by settler representations of African, Maori, Indian and Aborigine 'savagery', provided the metropolitan bourgeoisie with an explanatory key not only to the recalcitrant behaviour of these 'non-Europeans' in the colonies, but also to that of the 'disreputable' working classes and other potentially subversive groups in Britain itself (Stoler, 1995). Racial science allowed for the discourse of reform and potential inclusion, which had been dominant in the early nineteenth century, to be converted into one based on the economic and social exclusion of certain biologically defined groups. We will return to the utility of this discourse for South African elites below, but first we must again consider the factors that brought renewed British metropolitan intervention in the region.

By the 1880s, newly industrializing rivals within Europe, and particularly Germany, were threatening to challenge British commercial paramountcy overseas. As a result, a policy of free trade was giving way to imperial protectionism and the European scramble for Africa. In southern Africa, the discovery of vast quantities of minerals initiated a fundamental transition from agrarian to industrial capitalism, fuelled by the influx of European, and especially British capital, as we saw towards the end of chapter 3. Once British suzerainty had been bloodily established over the region, especially through the South African War, the implications of modern industrial capitalism in a colonial context were addressed by the state acting in liaison with major capitalists.

In chapter 4 we argued that, on the one hand, the flow of industrial capital from the metropole had prompted the commercialization of African agriculture. This had allowed an elite of African farmers to accumulate property, acquire Western consumer products and missionary education, and aspire towards citizenship. On the other hand, industrial restructuring had brought a more urgent imperative for a mass, urban-centred African workforce. Modern South Africa's 'native policy' had two major thrusts, each of which was intended to deal with specific destabilizing effects of these changes in African orientation. The first was to exclude more effectively the black middle class. As it leapt over the hurdles of Christianity, property and education, placed in the way of citizenship under the Cape's non-racial franchise, this group in particular threatened the erosion of racial privilege. A significant enfranchised black citizenry was now a real prospect, raising the question, 'How much civilisation was appropriate' for South Africa's blacks? (Stoler and Cooper 1997: 7). Comaroff points out that this was 'a question that was answered in many ways over the long run [and in different colonial territories]. But invariably with an eye to reproducing distinction, discrimination, and dualism' (Comaroff 1998: 334). In South Africa, the prospect of black encroachment on white privilege necessitated a new exclusionary regime, one tied implicitly to the scientific ideas of biological difference which had become accepted in the British metropole, and fulfilled through the staged exclusion of black voters.

The second thrust of South African 'native policy' was the more effective control, supply and regulation of the African labour force. The recently conquered workforce in South Africa had not had generations to appropriate or capitulate to the time discipline and individualism of a commercialized economy. Industrial capital here had to come to terms with 'the still pulsating remains of powerful African kingdoms' (Marks 1987: 395), rather than with the 'tenderized' proletariat of Britain. Through state and capital collusion, as we saw in chapter 4, the practice of segregation was formulated as a response. This practice, however, was not developed in isolation. In its rural aspects, it was informed by indirect rule in other British African colonies and in its urban aspects it was influenced by networks of Western modernist planning. Its imposition meant that African aspirations were channelled away from cultural and political inclusion and along the alternative path of 'retribalization'.

In the British metropole, working men were finally gaining the vote in the late nineteenth and early twentieth centuries, with women achieving this goal in 1918. In South Africa, white men, like their British counterparts, were already considered rational Western citizens and white women would be from the 1930s. But Africans of both sexes would remain collectively defined tribal 'subjects' rather than 'citizens'. This differentiation allowed an established racial order to be modified only marginally, even through a potentially destabilizing period of secondary industrialization which had briefly opened up new opportunities to black elites.

As we saw at the end of chapter 4, a further bout of industrial and urban growth during the Second World War was widely perceived among whites to have thrown their privilege into jeopardy once more. The segregationist 'solution' of the early part of the century was no longer seen as being sufficient to safeguard white, and particularly poor white, supremacy. Again, the threat was countered by strong central state intervention, this time taking the form of apartheid. With this fortified segregationist system, as we suggested in chapter 5, the modern South African state sought not only to shore up white privilege as a whole, but also to make it work in particular on behalf of an ethnically defined Afrikaner constituency.

## Late capitalism and the abolition of apartheid in the late twentieth and early twenty-first centuries

Racially exclusive structures of segregation and apartheid continued, broadly, to sustain an established, if marginally modified, distribution of privilege in South Africa as long as the production of that privilege remained dependent on a cheap, mass labour force; as long as organized resistance could be contained through both repression and the consent of 'tribal' leaders; and as long as 'bifurcated' systems of citizenship (white citizens and black subjects) were regarded as 'normal' within other colonies. However, each of these conditions was undermined at various stages after the Second World War, as a global capitalist order was again reconstructed. The final abandonment of apartheid can be understood as the last resort of dominant South African groups and the state in their attempts to manage the adjustment of social boundaries that this latest capitalist restructuring has necessitated.

In the wake of the Second World War, as we argued in chapter 5, the USA was able to capitalize on the exhaustion of Europe in order to press for an international regulatory regime promising capital freer movement across the globe. Accompanying that regime was a revitalized notion of universal humanity, constructed by the Western powers in a wartime search for colonial allies, and expressed through a rhetoric of human rights and non-racialism. In order to accommodate the colonial nationalist and worker-based movements deploying this notion for their own ends, and with the relative freedom of Western capital to move across the boundaries of sovereign states in mind, independence was progressively granted to Europe's colonies from the late 1940s. Post-war universalist discourse, however, like its early-nineteenth-century incarnation, was accompanied by metropolitan prescriptions. Rather than the bourgeois exhortation to individual discipline and docility, these prescriptions centred on the necessity for Third World 'development' based around free market economics. As Stoler and Cooper put it, the colonial 'notion of the civilising mission gave way after World War II to the notion of development, embodying in a subtler way the hierarchy that civilising entailed' (Stoler and Cooper 1997: 35). The developmentalist paths prescribed by Western donors and 'experts' were intended particularly to direct independent states in Africa and elsewhere away from the lure of capitalism's newly powerful alternative in the shape of Soviet communism.

The apparent success of Marshall Aid assistance to Europe after the Second World War promoted a major shift in US foreign policy, with the 1949 Truman address inaugurating the so-called 'development age' and laying the basis for evolutionary models of growth based on Western experience and ideals in contrast to the Soviet model of revolutionary change. Rostow's classic 'stages of economic growth' model (published under the sub-heading a *Non-Communist Manifesto*) proved to be a particularly influential manifestation of modernization theory, in which the crucial 'take-off' phase was associated with increasing investment in industrialization such as occurred during the British industrial revolution (Rostow 1960). Despite varying experiences, most states in the Third World, including Africa, as we saw in chapter 8, were initially keen to pursue a development path that seemed to offer the reward of Western lifestyle and 'development' levels. Within the model of development which they adopted, poverty was seen as failure to attain a 'normal' process of growth such as that exemplified by the West, rather than as the 'product of a bitter history' of exploitation (Stoler and Cooper 1997: 35, citing Ferguson 1990).

We argued in chapter 8 that the elites which led many African states after winning their nationalist independence struggles were frequently able to employ Western development ideas in such a way as to generate and maintain privilege for certain select groups. In other words, using the developmentalist language of neo-liberal economics, they maintained the same kind of boundaries between 'insider' and 'outsider' groups as had existed in colonial days, with only the complexion of the ruling classes being different. As Mamdani argues, many postcolonial African states inherited their divisive mode of authority from the colonial system. Upon independence, they were deracialized in the sense that privileged 'citizens' were now largely black rather than white, but they did not necessarily become democratized, because ethnically based politics, in which chiefs had authority over subjects, remained dominant in the rural areas (Mamdani 1996). Clapham thus concludes that the state in post-independence Africa was generally held together, 'not by some shared sense of legitimacy or moral values, but rather by the self-interest of its leading beneficiaries, and as a partial association favouring some groups much more than others' (Clapham 1991: 99). Even in Ghana, for instance, where Nkrumah extolled the virtues of an African socialism which eschewed capitalism *and* communism, 'both of which were seen as essentially European political-economic systems', the early post-independent 'development trajectory' closely followed the modernization model (Rapley 1996: 31). Priority was given to urban elites and industrial growth, while the rural-agricultural sector, in which the bulk of the population lived and worked, was regarded merely as a source of revenue to fuel urban investment.

In South Africa, which had already apparently attained Rostow's mythical 'take-off', capitalist development discourse, constructed around the models of Rostow, Friedmann and Perroux, was actively propagated during the 1960s and 1970s in attempts to ensure enhanced levels of growth, and implicitly, the dominance of the white elite (Fair 1982). As was stated in chapter 5, the apartheid state's adaptation to global decolonization in the form of 'Separate Development', involving self-governing bantustans, allowed for a sustained economic boom in the 1960s, during which multinational investment brought the country's economic infrastructure far more firmly within the orbit of Western capital.

The South African state's role as a strategic ally of the capitalist West enabled it to sustain crucial global political alliances during the Cold War, despite condemnation of its overtly racist policies. However, its dependency on globalizing capital nevertheless rendered economic restructuring necessary when recession struck during the

1970s. After successive failures to co-opt a new black middle class in the wake of the Soweto Revolt of 1976, the opportunity for the state and the privileged classes to reformulate from a position of relative strength in South Africa first arose with the end of the Cold War, as we suggested towards the end of chapter 5. On the one hand, the removal of the Soviet 'threat' heralded the loss of the USA's and UK's support for South Africa against sanctions, and thus brought about a more urgent imperative for reform. On the other hand, it also raised new possibilities for negotiating with an ANC which, although it had won internal legitimacy among the majority of blacks during the 1980s insurrection, had been weakened both materially and ideologically by the collapse of Eastern bloc communism.

By the late 1980s, the NP's new-found willingness to accept the universalist, non-racial political rules prescribed by Western states in the post-Second World War period, was to a great extent the consequence of economic crisis, the failure to quell resistance and the lack of any new, more subtly racist initiatives, as we saw in chapter 5. But it was also a product of the belief that white material, and perhaps even political privilege could now best be secured through negotiation. In the event, the NP leadership proved unable to manage the negotiation process in the way that it had hoped. Attempts to weaken ANC support through covert violence failed once ANC negotiators carried out their threat to return to mass action, and greater democratic concessions than those originally envisaged had to be made.

Nevertheless, as the NP leadership realized it would be, the new South African state has been conditioned by the prescriptions of a newly assertive, post-Cold War West. In common with other post-independence African states, it has been able to carve out relatively little room to pursue independent action – an argument which we elaborated in chapter 7. In recognition of dominant global economic discourses, the ANC's initial promises of radical transformation and nationalization, contained in the Freedom Charter and in pre-election manifestos, were watered down by the time of the election in 1994 to take the form of the RDP. This set out development targets and strategies which did not threaten the established economic order and which won domestic and foreign business acceptance. The country's internal negotiation process, the influence exerted by metropolitan governments and international organizations, and an inherited neo-liberal economic order all played vital roles in moderating the ANC's pre-election radicalism. As far as the ANC's socialist members and allies were concerned, the subsequent closure of the RDP office in March 1996, and the dispersal of its programmes into various ministries could be seen as the end of a 'people-driven', 'poverty-focused' development strategy.

By the time of the publication of GEAR as a revised programme which faced up to the country's harsh economic realities in 1996, the ANC leadership's re-alignment along the lines of global orthodoxy had clearly taken place. GEAR has been subversively referred to by some critics as standing for 'Greed Entirely Avoids Redistribution'. The strategy was, as we have seen in chapter 7, based on predictions generated by four economic models devised by the Reserve Bank, the World Bank, the Development Bank of Southern Africa and the Bureau for Economic Research at the University of Stellenbosch. Its adoption, according to Pillay, indicates that 'our politicians ... have succumbed to the view that the new global order and its concomitant [neo-liberal] economic regime cannot be challenged or even controlled' (Pillay 1996: 34). GEAR is an economic strategy which, like those of the late-twentieth-century global metropoles, rests on the assumption that markets are reliable, well-functioning institutions. As Manzo puts it, in South Africa as elsewhere, the 'market' is now 'the privileged subject around which all struggles for economic justice must orient themselves' (Manzo 1992: 253). With its key emphases on export performance, foreign investment, competition and control of wage increases and interest rates, there is much in GEAR that resembles the Thatcherite policies of Britain in the 1980s – policies which led to a well-documented increase in social inequality within the UK (Joseph Rowntree Foundation 1995). In fact, at the press conference announcing GEAR, Thabo Mbeki actually quipped, 'Just call me a Thatcherite' (quoted in Bond 1996: 30).

The introduction of GEAR represents a defeat for the ANC left wing. There is much concern that the stark inequalities which were the main target for attention in the RDP proposals (chapter 6), could become even greater with the change in policy direction. Bond sees parallels with South Africa's darkest past, asserting that, 'Even though the failure of the inherited apartheid economy to deliver the goods is not in dispute, GEAR promises that more of virtually the same (neo-liberal) policies will somehow generate durable growth' (Bond 1996: 23). As we suggested in chapter 7, Thabo Mbeki's 'African Renaissance', while its rhetoric contains a ring of African socialism, is unlikely to produce any change in these *de facto* economic policies. It is certainly not going to generate the jobs which so many Africans in particular desperately need now and in the foreseeable future. As a result, 'The real possibility exists that a major new axis of inequality will solidify between a minority of unionized, skilled, and semi-skilled African workers and a majority of [the] unskilled, under-employed, unemployed, and unemployable' (Bundy 1992: 35).

Despite criticism of its strategies, the mediocre growth levels achieved and the (at best) slow alleviation of poverty which is occurring, it is difficult to conceptualize that the South African state could have pursued an alternative course of action, given the demise of state socialism internationally and the neo-liberal discursive prescriptions of the global economic powers. Even with a radical and widely welcomed revision of its political constitution, South Africa's current 'transformation' has been more of a transition to a new social and economic order which is 'acceptable' to key metropolitan and local constituencies, than a radical break with past socio-economic structures. It is these key local and global constituencies which make it so difficult for the new South African state to deploy the universalist notion of 'development' in a way which acts against the 'exclusions and inequities that have historically [and elsewhere] been associated' with the term (Stoler and Cooper 1997: 35).

The ending of racial discrimination and introduction of policies of affirmative action have certainly facilitated the effective incorporation of the black bourgeoisie as part of the dominant classes, and the alienation of many whites who previously enjoyed state patronage in South Africa. For the first time, the South African government has also made an explicit commitment to equalizing conditions for its male and female citizens. The absorption of blacks into the country's financial, commercial and administrative elite has, however, been limited (in many cases to those with prior access to private education), and the revision of gender boundaries is a mammoth task which no government, let alone one with so many other agendas, can achieve in isolation. The result is a society characterized ever more conspicuously by gendered class as opposed to racial distinctions.

Alterations in the boundaries of privilege and the achievement of non-racialism have yet to make a significant material impact on most of the enfranchised but disempowered majority, especially perhaps, on those black women who have not gained entry to the middle classes. It is unlikely that the ANC will be able to assist its core constituency of black voters in the same way as proved possible for the NP in the twentieth century (see chapter 4). As Bundy points out, 'The historic "solution" to the "Poor White Problem" owed much to the specific economic conditions of the 1930s and 1940s; but it was due, too, to conscious and far-reaching programmes of affirmative action, job creation, and social welfare. Latter-day enthusiasm for market forces and a "lean" state will have to be challenged if parallel policies and programmes are to have any purchase in the decades ahead' (Bundy 1992: 37). Western-led notions of modernity encouraged massive government intervention on behalf of targeted constituents in the 1930s and 1940s, not just in South Africa, but in the global metropoles too. However, as Evans notes, 'South African democracy has been launched in an international climate hostile to "big government" programmes and in a global economy that prompts states to remain competitive by reducing expenditures on social welfare programmes and lowering wages' (Evans 1997: 305).

One likely scenario for South Africa's future is that the recent abolition of legal racial boundaries will ensure economic efficiency and productivity for a flexibly defined set of privileged social groups within a late capitalist global order. Such groups would include white, and an increasing number of black, entrepreneurs, who, as we saw in the last chapter, now have new fields of profitable endeavour opening up to them within the southern African region as a whole. Their activities will help perpetuate South African economic dominance, some would even say neo-colonialism, within the southern African region in the post-apartheid era (see chapter 8). In some respects, the current neo-liberal global economic climate gives them greater freedom of action than the South African government itself enjoys, given its dependence on cultivating favourable responses from institutions such as the IMF, World Bank and EU. Certainly GEAR and the idea of 'African Renaissance' do not seem to offer any obstacles to South African capital's continued penetration of regional and continental economies.

An alternative and, perhaps less likely prospect more closely resembles the experience of some post-independence states to the north, such as Namibia, Zimbabwe and Kenya. There, governing elites have inherited former state structures without retaining the close articulation with metropolitan capital enjoyed by their colonial predecessors. In this case, state patronage may underwrite the prosperity of a new, largely black bourgeoisie without that bourgeoisie being integrated productively within the global economy. In either case – limited economic growth and market-led redistribution, or general economic stagnation and selective state patronage – one has a sense of *déjà vu* concerning the prospects for the mass of South Africa's black poor.

Like the incorporation of the Cape Colony within a British imperial network in the nineteenth century, South Africa's post-apartheid immersion within a globalizing, late-capitalist political economy, has been accomplished through an insistence on governmental respect for the market and for property. The Cape's landless slaves and Khoisans found themselves freed from the arbitrary will of the master during the early nineteenth century, but subjected to the regulating forces of the labour market and of the colonial state. South Africa's contemporary poor are liberated from the humiliating and often fatal practices of apartheid to find themselves living in a 'Rainbow Nation', in which the notion of a universal humanity of diverse culture is respected, but where significant state-led redistribution is out of the question. The abandonment of apartheid has given rise to a South African state which is explicitly committed to non-racialism, thus 'normalizing' South Africa's position within the global terrain. But 'normalization' implies acquiescence to free-market doctrines that impede large-scale, effective redistribution (Manzo 1992). Shifts in the discourse and practice of capitalism, pioneered and still centred in the Western metropoles, continue to provide agendas for change, even if they do not go so far as to determine the precise nature of change, in 'peripheries' such as South Africa.

Given all of this, it is clear that post-apartheid 'development' is not simply a matter of overcoming the legacies of apartheid. When it left apartheid behind, South Africa did not leave behind the structures and processes which generate inequality. It did not enter some 'neutral' post-modern, post-industrial and post-apartheid arena in which it could transcend inherited inequalities and construct a totally revised political discourse and economic structure. Rather, it remains inescapably embedded within a globalized, Western-dominated, capitalist system, which continues actively to produce inequalities. Thinking about South Africa's current transition in broad historical and spatial terms, then, as this book has sought to do, is far from merely an indulgence.

# GLOSSARY

**ABRT** African Business Round Table
**ACP** Africa, Caribbean, Pacific
**ANC** African National Congress
**BLSN** Botswana, Lesotho, Swaziland, Namibia
**Boer** literally, farmer. Term used until 1960s to describe Afrikaners
*Broederbond* brotherhood
*Bywoner* landless Afrikaner squatters/tenants
**COMESA** Common Market of Eastern and Southern Africa
**COSATU** Congress of South African Trade Unions
**CRP** Common Revenue Pool
**DRC** Dutch Reformed Church
**FDI** Foreign Direct Investment
**FTA** Free Trade Area
**GEAR** Growth, Employment and Redistribution
**GDP** Gross Domestic Product
**GGP** Gross Geographic Product
**GNP** Gross National Product
*hlonipa* respect for elders
**ICU** Industrial and Commercial Workers' Union
**IFP** Inkatha Freedom Party
**IDZ** Industrial Development Zone
**IMF** International Monetary Fund
*Landdrost* district official
**LP** Labour Party
**LPA** Lagos Plan of Action
*Mfecane* period of violence and centralization among African polities in early nineteenth century

**NP** National Party
**OAU** Organization of African Unity
**OFS** Orange Free State
**PAC** Pan-Africanist Congress
**PTA** Preferential Trade Area
**RDP** Reconstruction and Development Programme
**RENAMO** Mozambique National Resistance Movement
**SABC** South African Broadcasting Commission
**SACP** South African Communist Party
**SACU** Southern African Customs Union
**SADC** Southern African Development Community
**SADCC** Southern African Development Co-ordinating Conference
**SAP** South African Party
**SDI** Spatial Development Initiative
**UITLANDERS** foreigners
**UNCTAD** United Nations Conference on Trade and Development
**UNECA** United Nations Economic Commission for Africa
**UNITA** Unio Nacional para a Independencia Total de Angola
**UP** United Party
*Verligte* 'enlightened' Afrikaner nationalist
*Verkrampte* conservative, reactionary Afrikaner nationalist
*Volk* people or nation
*Voortrekker* participant on the Great Trek, or migration of Afrikaner farmers from the Cape Colony in the 1830s and 1840s

# BIBLIOGRAPHY

**Adam, H. and Giliomee, H.** (1979) *Ethnic Power Mobilised: Can South Africa Change?*, Yale University Press, New Haven.

**Adam, H. and Moodley, K.** (1986) *South Africa Without Apartheid: Dismantling Racial Domination*, University of California Press, Berkeley.

**Adedeji, A.** (ed.) (1996) *South Africa and Africa: Within or Apart?*, Zed Books, London.

**Adedeji, A. and Otite, O.** (eds) (1997) *Nigeria: Renewal from the Roots?*, Zed Books, London.

**Africa Institute of South Africa** (1995) *Africa at a Glance*, 1995/6, AISA, Pretoria.

**African Development Bank** (1993) *Economic Integration in Southern Africa*, vol. 1, African Development Bank, Abidjan.

**African Development Bank** (1998) *African Development Report 1998*, Oxford University Press, Oxford.

**Agnew, J. and Grant, R.** (1997) 'Falling out of the world economy? Theorizing Africa in world trade', in Lee, R. and Wills, J. (eds) *Geographies of Economies*, Arnold, London.

**Ahluwalia, P.** (1999) 'Citizenship, subjectivity and the crisis of modernity', *Social Identities*, **5**, 3, 313–29.

**Ahwireng-Obeng, F. and McGowan, P. J.** (1998) 'Partner or hegemon? South Africa in Africa', *Journal of Contemporary African Studies* **16**, 1, 5–38 and **16**, 2, 165–95.

**Akosah-Sarpong, K.** (1998) 'A continent in transition', *West Africa*, **4195**, 7–27 Sept., 665–6

**Alden, C.** (1998) *Apartheid's Last Stand: The Rise and Fall of the South African Security State*, Macmillan, London and St. Martin's Press, New York.

**Ally, R.** (1994) *Gold and Empire: The Bank of England and South Africa's Gold Producers, 1886–1926*, Witwatersrand University Press, Johannesburg.

**Altbeker, A. and Steinberg, J.** (1998) 'Race, reason and representation in National Party discourse, 1990–92', in Howarth, D. R. and Norval, A. (eds) *South Africa in Transition: New Theoretical Perspectives*, Macmillan, London and St. Martin's Press, New York.

**Aly, A. H. M.** (1997) 'Post-apartheid South Africa: The implications for regional co-operation in Africa', *Africa Insight*, **27**, 1, 24–31.

**ANC (African National Congress)** (1994) *The Reconstruction and Development Programme*, Umanyano Publications, Johannesburg.

Anderson, B. (1991) *Imagined Communities: Reflections on the Origin and Spread of Nationalism*, Verso, London and New York.

APIC (Africa Policy Information Center) (1996) *Africa on the Internet: Starting Points for Policy Information*, APIC, Washington DC.
http://www.igc.apc.org/apic/bp/inetall.html

Asante, S. K. B. (1996) 'The European Union – Africa (ACP) Lomé convention', *Africa Insight*, 26, 4, 381–91.

Ashforth, A. (1990) *The Politics of Official Discourse in Twentieth-Century South Africa*, Clarendon Press: Oxford.

Atkins, K. E. (1993) *The Moon is Dead! Give us our Money! The Cultural Origins of an African Work Ethic, Natal, South Africa, 1843–1900*, Heinemann, Portsmouth, NH and James Currey, London.

Atmore, A. and Marks, S. (1974) 'The imperial factor in South Africa in the nineteenth century: towards a reassessment', *Journal of Imperial and Commonwealth History*, 3, 105–39.

Bank, A. (1995) *Liberals and their Enemies: Racial Ideology at the Cape of Good Hope, 1820–1850*, unpublished PhD, Cambridge University.

Bank, A. (1996) 'Of "native skulls" and "noble Caucasians": phrenology in colonial South Africa', *Journal of Southern African Studies*, 22, 387–404.

Banton, M. (1987) *Racial Theories*, Cambridge University Press, Cambridge.

Barber, J. and Barratt, J. (1990) *South Africa's Foreign Policy: The Search for Status and Security, 1945–1988*, Cambridge University Press, Cambridge.

Barratt Brown, B. (1995) *Africa's Choices: After Thirty Years of the World Bank*, Penguin, Harmondsworth.

Beall, J. (1990) 'Women under indentured labour in colonial Natal, 1860–1911', in Walker, C. (ed.) *Women and Gender in Southern Africa to 1945*, David Philip, Cape Town and James Currey, London.

Beavon, K. S. O. and Rogerson, C. M. (1982) 'The informal sector of the apartheid city: the pavement people of Johannesburg', in Smith, D. (ed.) *Living Under Apartheid: Aspects of Urbanization and Social Change in South Africa*, George Allen and Unwin, London.

Beinart, W. (1982) *The Political Economy of Pondoland, 1860–1930*, Cambridge University Press, Cambridge.

Beinart, W. (1987a) 'Conflict in Qumbu: rural consciousness, ethnicity and violence in the colonial Transkei', in Beinart, W. and Bundy, C., *Hidden Struggles in Rural South Africa: Politics and Popular Movements in the Transkei and Eastern Cape, 1890–1930*, Ravan Press, Johannesburg.

Beinart, W. (1987b) '*Amafelandawonye* (the die-hards): popular protest and women's movements in Herschel District in the 1920s', in Beinart, W. and Bundy, C., *Hidden Struggles in Rural South Africa: Politics and Popular Movements in the Transkei and Eastern Cape, 1890–1930*, Ravan Press, Johannesburg.

Beinart, W. (1992) 'Political and collective violence in southern African historiography', *Journal of Southern African Studies*, 18, 3, 455–86.

Beinart, W. (1994) *Twentieth-Century South Africa*, Oxford University Press, Oxford and New York.

**Beinart, W. and Bundy, C.** (1987a)
'Introduction: "away in the locations"', in
Beinart, W. and Bundy, C., *Hidden
Struggles in Rural South Africa*: *Politics and
Popular Movements in the Transkei and
Eastern Cape, 1890–1930*, Ravan Press,
Johannesburg.

**Beinart, W. and Bundy, C.** (1987b)
*Hidden Struggles in Rural South Africa*:
*Politics and Popular Movements in the
Transkei and Eastern Cape, 1890–1930*,
Ravan Press, Johannesburg.

**Beinart, W. and Coates, P.** (1995)
*Environment and History: The Taming of
Nature in the USA and South Africa*,
Routledge, London and New York.

**Beinart, W. and Delius, P.** (1986)
'Introduction', in Beinart, W., Delius, P.
and Trapido, S. (eds) *Putting a Plough to
the Ground*: *Accumulation and Dispossession
in Rural South Africa, 1850–1930*, Ravan
Press, Johannesburg.

**Belich, J.** (1986) *The Victorian
Interpretation of Racial Conflict: the Maori,
the British, and the New Zealand Wars*,
McGill-Queen's University Press,
Montreal and London.

**Bell, M.** (1993) '"The pestilence that
walketh in darkness". Imperial health,
gender and images of South Africa *c.*
1880–1910', *Transactions of the Institute of
British Geographers*, **18**, 327–41.

**Bell, R. T.** (1973) *Industrial
Decentralisation in South Africa*, Oxford
University Press, Cape Town.

**Berger, I.** (1987) 'Solidarity fragmented:
garment workers of the Transvaal,
1930–1960', in Marks, S. and Trapido, S.
(eds) *The Politics of Race, Class and
Nationalism in Twentieth Century South
Africa*, Longman, London and New York.

**Bergh, J. S. and Visagie, J. C.** (1985) *The
Eastern Cape Frontier Zone 1660–1980: A
Cartographic Guide for Historical Research*,
Butterworths, Durban.

**Bertelsmann, T.** (1998) South Africa's
foreign policy towards the Southern African
region, unpublished paper presented at
African Studies Association of the UK
Conference, School of Oriental and African
Studies, London, 14–16 September.

**Bickford-Smith, V.** (1994) 'Meanings of
freedom: social position and identity
among ex-slaves and their descendants in
Cape Town, 1875–1910', in Worden, N.
and Crais, C. (eds) *Breaking the Chains:
Slavery and its Legacy in the Nineteenth-
Century Cape Colony*, Witwatersrand
University Press, Johannesburg.

**Bickford-Smith, V.** (1995) *Ethnic Pride
and Racial Prejudice in Victorian Cape
Town*, Witwatersrand University Press,
Johannesburg.

**Binns, T.** (1990) 'Is desertification a
myth?', *Geography*, **75**, 2, 106–13.

**Binns, T.** (1998) 'Geography and
development in the "new" South Africa',
*Geography*, **83**, 1, 3–14.

**Binns, T. and Nel, E. L.** (1999) 'Beyond
the development impasse: The role of local
economic development and community
self-reliance in rural South Africa',
*Journal of Modern African Studies*, **37**, 3,
389–408.

**Bloch, G.** (1991) 'The development of
South Africa's manufacturing industry', in
Konczacki, Z. A., Parpart, J. L. and Shaw,
T. M. (eds) *Studies in the Economic History
of South Africa, Vol 2: South Africa,
Lesotho and Swaziland*, Frank Cass,
London.

Bloomberg, C. (ed. Dubow, S.) (1990) *Christian-Nationalism and the Rise of the Afrikaner Broederbond in South Africa, 1918–48*, Macmillan, London.

Blumenfeld, J. (1997) 'From icon to scapegoat: the experience of South Africa's RDP', *Development Policy Review*, **15**, 65–91.

Boeseken, A. J. *et al.* (1922) *Geskiedenis – Atlas Vir Suid-Africa*, Nasou Beperk.

Bond, P. (1996) 'GEARing up or down?', *SA Labour Bulletin*, **20**, 4, 23–30.

Bond, P. (1998) 'Moving toward – or beyond? – a "post-Washington consensus" on development', *Indicator SA*, **15**, 4, 36–41.

Bonner, P. (1979) 'The 1920 black mineworkers' strike: preliminary account', in Bozzoli, B. (ed.) *Labour, Townships and Protest: Studies in the Social History of the Witwatersrand*, Ravan Press, Johannesburg.

Bonner, P. (1982) 'The Transvaal Native Congress, 1917–1920: the radicalisation of the black petty bourgeoisie on the Rand', in Marks, S. and Rathbone, R. (eds) *Industrialisation and Social Change in South Africa: African Class Formation, Culture and Consciousness, 1870–1930*, Longman, London and New York.

Bonner, P. (1983) *Kings, Commoners and Concessionaries: The Evolution and Dissolution of the Nineteenth-Century Swazi State*, Ravan Press, Johannesburg.

Bonner, P. (1990) '"Desirable or undesirable Basotho women?" Liquor, prostitution and the migration of Basotho women to the Rand, 1920–1945', in Walker, C. (ed.) *Women and Gender in Southern Africa to 1945*, David Philip, Cape Town and James Currey, London.

Bonner, P. (1993) 'The Russians on the Reef: urbanisation, gang warfare and ethnic mobilisation', in Bonner, P., Delius, P. and Posel, D. (eds) *Apartheid's Genesis, 1935–1962*, Ravan Press and Witwatersrand University Press, Johannesburg.

Bonner, P. (1995) 'African urbanisation on the Rand between the 1930s and 1960s: its social character and political consequences', *Journal of Southern African Studies*, **21**, 115–29.

Bonner, P. and Lodge, T. (1989) 'Introduction', in Bonner, P., Hofmeyr, I., James, D. and Lodge T. (eds) *Holding Their Ground: Class, Locality and Culture in Nineteenth and Twentieth Century South Africa*, Witwatersrand University Press and Ravan Press, Johannesburg.

Bonner, P., Delius, P. and Posel, D. (1993) 'The shaping of apartheid: contradiction, continuity and popular struggle', in Bonner, P., Delius, P. and Posel, D. (eds) *Apartheid's Genesis 1935–1962*, Ravan Press and Witwatersrand University Press, Johannesburg.

Bonnin, D., Deacon, R., Morrell, R. and Robinson, J. (1998) 'Identity and the changing politics of gender in South Africa', in Howarth, D. R. and Norval, A. (eds) *South Africa in Transition: New Theoretical Perspectives*, Macmillan, London and St. Martin's Press, New York.

Boonzaier, E., Malherbe, C., Smith, A. and Berens, P. (1996) *The Cape Herders: A History of the Khoikhoi of Southern Africa*, David Philip and Ohio University Press, Cape Town, Johannesburg and Athens.

Botha, R. F. (1990) 'South Africa and Africa', *Africa Insight*, **20**, 4, 215–18.

Bozzoli, B. (1983) 'Marxism, feminism and South African studies', *Journal of Southern African Studies*, **9**, 139–71, reprinted in Beinart, W. and Dubow, S. (eds) (1995) *Segregation and Apartheid in Twentieth Century South Africa*, Routledge, London and New York.

Bozzoli, B., with Nkotsoe, M. (1991) *Women of Phokeng: Consciousness, Life Strategy and Migrancy in South Africa, 1900–1983*, Ravan Press, Johannesburg.

Bradford, H. (1987) *A Taste of Freedom: The ICU in Rural South Africa, 1924–1930*, Yale University Press, New Haven and London.

Bradford, H. (1990) 'Highways, byways and cul-de-sac: the transition to agrarian capitalism in revisionist South African history', *Radical History Review*, **46/7**, 59–88.

Bradford, H. (1996) 'Women, gender and colonialism: rethinking the history of the British Cape Colony and its frontier zones, c. 1806–70', *Journal of African History*, **37**, 3, 351–70.

Brain, J. (1989) 'Natal's Indians: from co-operation through competition, to conflict', in Duminy, A. and Guest, B. (eds) *Natal and Zululand From Earliest Times to 1910: A New History*, University of Natal Press and Shuter and Shooter, Pietermaritzburg.

Bredekamp, H. and Ross, R. (1995) 'Introduction: the naturalization of Christianity in South Africa', in Bredekamp, H. and Ross, R. (eds) *Missions and Christianity in South African History*, Witwatersrand University Press, Johannesburg.

Brink, E. (1990) 'Man-made women: gender, class and the ideology of the *volksmoeder*', in Walker, C. (ed.) *Women and Gender in Southern Africa to 1945*, David Philip, Cape Town and James Currey, London.

Brookes, E. H. and Webb, C. de B. (1987) *A History of Natal*, 2nd edn. University of Natal Press, Pietermaritzburg.

Brooks, A. and Brickhill, J. (1980) *Whirlwind Before the Storm: The Origins and Development of the Uprising in Soweto and the Rest of South Africa From June to December 1976*, International Defence and Aid Fund, London.

Brooks, S. J. and Harrison, P. J. (1998) 'A slice of modernity: planning for the country and the city in Britain and Natal, 1900–1950', *South African Geographical Journal*, **80**, 93–100.

Bundy, C. (1977) 'The Transkei peasantry, c. 1890–1914: "passing through a period of stress"', in Palmer, R. and Parsons, N. (eds) *The Roots of Rural Poverty in Central and Southern Africa*, Heinemann, London, Ibadan, Nairobi and Lusaka.

Bundy, C. (1986) 'Vagabond Hollanders and runaway Englishmen: white poverty in the Cape before poor whiteism', in Beinart, W., Delius, P. and Trapido, S. (eds) *Putting a Plough to the Ground: Accumulation and Dispossession in Rural South Africa, 1850–1930*, Ravan Press, Johannesburg.

Bundy, C. (1987a) 'Mr Rhodes and the poisoned goods: popular opposition to the Glen Grey council system, 1894–1906', in Beinart, W. and Bundy, C., *Hidden Struggles in Rural South Africa: Politics and Popular Movements in the Transkei and Eastern Cape, 1890–1930*, Ravan Press, Johannesburg.

**Bundy, C.** (1987b) '"We don't want your rain, we won't dip": popular opposition, collaboration and social control in the anti-dipping movement, 1908–1916', in Beinart, W. and Bundy, C., *Hidden Struggles in Rural South Africa: Politics and Popular Movements in the Transkei and Eastern Cape, 1890–1930*, Ravan Press, Johannesburg.

**Bundy, C.** (1987c) 'Land and liberation: popular rural protest and the national liberation movements in South Africa, 1920–1960', in Marks, S. and Trapido, S. (eds) *The Politics of Race, Class and Nationalism in Twentieth Century South Africa*, Longman, London and New York.

**Bundy, C.** (1988) *The Rise and Fall of the South African Peasantry*, 2nd edn., David Philip, Cape Town and Johannesburg and James Currey, London.

**Bundy, C.** (1992) 'Development and inequality in historical perspective', in Schrire, R. (ed.) *Wealth or Poverty? Critical Choices for South Africa*, Oxford University Press, Cape Town.

**Bunting, B.** (1986) *The Rise of the South African Reich*, International Defence and Aid Fund for Southern Africa, London.

**Burk, K.** (ed.) (1982) *War and the State: The Transformation of British Government, 1914–19*, George Allen and Unwin, London.

**Burman, S.** (1990) 'Fighting a two-pronged attack: the changing legal status of women in Cape-ruled Basutoland, 1872–1884', in Walker, C. (ed.) *Women and Gender in Southern Africa to 1945*, David Philip, Cape Town and James Currey, London.

**Burton, J.** (1995) 'Short paper on the Renegotiation of SACU and the Free Trade Area between RSA and the EU', unpublished paper, Overseas Development Agency, Pretoria.

**Cain, P. J. and Hopkins, A. G.** (1993) *British Imperialism: Innovation and Expansion, 1688–1914*, Longman, London and New York.

**Cattaneo, N.** (1990) 'Piece of paper or paper of peace: The Southern African Customs Union Agreement', *International Affairs Bulletin*, 11, 543–64.

**CSS (Central Statistical Service)** (1999) *CSS Bulletin of Statistics*, CSS, Pretoria.

**Cell, J.** (1982) *The Highest Stage of White Supremacy: The Origins of Segregation in South Africa and the American South*, Cambridge University Press, Cambridge.

**Chamber of Mines** (1991) *Annual Report 1991*, Chamber of Mines, Johannesburg.

**Charney, C.** (1984) 'Class conflict and the National Party split', *Journal of Southern African Studies*, 10, 269–82.

**Chaskalson, M.** (1991) 'The road to Sharpeville', in Clingman, S. (ed.) *Regions and Repertoires: Topics in South African Politics and Culture*, Ravan Press, Johannesburg.

**Chisholm, L.** (1990) 'Gender and deviance in South African industrial schools and reformatories for girls, 1911–1934', in Walker, C. (ed) *Women and Gender in Southern Africa to 1945*, David Philip, Cape Town and James Currey, London.

**Christopher, A. J.** (1982) *South Africa*, Longman, London.

Christopher, A. J. (1994) *The Atlas of Apartheid*, Routledge, London and New York and Witwatersrand University Press, Johannesburg.

Christopher, A. J. (1999) 'The South African census 1996: first post-apartheid census', *Geography*, **84**, 3, 270–5.

Clapham, C. (1991) 'The African state', in D. Rimmer (ed.) *Africa 30 Years On*, Royal African Society and James Currey, London.

Clark, N. (1993) 'The limits of industrialisation under apartheid', in Bonner, P., Delius, P. and Posel, D. (eds) *Apartheid's Genesis, 1935–1962*, Ravan Press and Witwatersrand University Press, Johannesburg.

Clark, N. L. (1997) 'The economy', in Byrnes, R. M. (ed.) *South Africa: A Country Study*, Library of Congress, Washington.

Cloete, F. (1991) 'Greying and free settlement', in Swilling, M., Humphries, R. and Shubane, K. (eds) *The Apartheid City in Transition*, Oxford University Press, Cape Town.

Cobbing, J. (1988) 'The Mfecane as alibi: thoughts on Dithakong and Mbolompo', *Journal of African History*, **29**, 487–519.

Cobley, A. (1990) *Class and Consciousness: The Black Petty Bourgeoisie in South Africa, 1924–1950*, Greenwood Press, New York.

Cock, J. (1980) *Maids and Madams: A Study in the Politics of Exploitation*, Ravan Press, Johannesburg.

Cock, J. (1990) 'Domestic service and education for domesticity: the incorporation of Xhosa women into colonial society', in Walker, C. (ed) *Women and Gender in Southern Africa to 1945*, David Philip, Cape Town and James Currey, London.

Coetzee, J. M. (1988) *White Writing: On the Culture of Letters in South Africa*, Yale University Press, New Haven and London.

Colley, L. (1992) *Britons: Forging the Nation, 1707–1837*, Pimlico, London.

Collier, P. (1995) 'The marginalisation of Africa', *International Labour Review*, **134**, 4/5, 541–57.

Collier, P. (1998) *Living down the Past: How Europe can Help Africa Grow*, Studies in Trade and Development No.2, Institute of Economic Affairs, London.

Comaroff, Jean (1985) *Body of Power, Spirit of Resistance: The Culture and History of a South African People*, University of Chicago Press, Chicago and London.

Comaroff, Jean and Comaroff, John (1991) *Of Revelation and Revolution: Christianity, Colonialism and Consciousness in South Africa*, vol. 1, University of Chicago Press, Chicago.

Comaroff, John (1997) 'Images of empire, contests of conscience: models of colonial domination in South Africa', in Cooper, F. and Stoler, A. L. (eds) *Tensions of Empire: Colonial Cultures in a Bourgeois World*, University of California Press, Berkeley, Los Angeles and London.

Comaroff, John (1998) 'Reflections on the colonial state, in South Africa and elsewhere: factions, fragments, facts and fictions', *Social Identities* **4**, 321–61.

Comaroff, John and Comaroff, Jean (1997) *Of Revelation and Revolution: The Dialectics of Modernity on a South African*

*Frontier*, vol. 2, University of Chicago Press, Chicago.

Constantine, S., Kirby, M. W. and Rose, M. B. (eds) (1995) *The First World War in British History*, Edward Arnold, London.

Cook, G. P. (1991) 'Cape Town', in Lemon, A. (ed.) *Homes Apart: South Africa's Segregated Cities*, Paul Chapman, London; Indiana University Press, Bloomington and David Philip, Cape Town.

Cooper, F. (1981) 'Peasants, capitalists, and historians: a review article', *Journal of Southern African Studies*, 7, 285–314.

Cooper, F. (1994) 'Conflict and connection: rethinking colonial African history', *American Historical Review*, 99, 5, 1516–45.

Cooper, F. (1996) *Decolonization and African Society: The Labor Question in French and British Africa*, Cambridge: Cambridge University Press.

Cope, N. (1993) *To Bind the Nation: Solomon kaDinuzulu and Zulu Nationalism, 1913–33*, University of Natal Press, Pietermaritzburg.

Coplan, D. (1982) 'The emergence of an African working-class culture', in Marks, S. and Rathbone, R. (eds) *Industrialisation and Social Change in South Africa: African Class Formation, Culture and Consciousness, 1870–1930*, Longman, London and New York.

Corder, C. K. (1997) 'The RDP: success or failure?', *Social Indicators Research*, 41, 183–203.

Cornwell, G. (1996) 'George Webb Hardy's *The Black Peril* and the social meaning of "black peril" in early twentieth-century South Africa', *Journal of Southern African Studies*, 22, 441–53.

Crais, C. (1990) 'Slavery and freedom along a frontier: the eastern Cape, South Africa, 1770–1838', *Slavery and Abolition*, 11, 191–215.

Crais, C. (1992) *White Supremacy and Black Resistance in Pre-Industrial South Africa: The Making of the Colonial Order in the Eastern Cape, 1770–1865*, Cambridge University Press, Cambridge.

Crais, C. (1994) 'Slavery and emancipation in the eastern Cape', in Worden, N. and Crais, C. (eds) *Breaking the Chains: Slavery and its Legacy in the Nineteenth-Century Cape Colony*, Witwatersrand University Press, Johannesburg.

Crais, C. (1998) 'Of men, magic, and the law: popular justice and the political imagination in South Africa', *Journal of Social History*, 31, 49–71.

Crankshaw, O. (1996) 'Changes in the racial division of labour during the apartheid era' *Journal of Southern African Studies*, 22, 633–56.

Crothers, C. (1997) 'The level of poverty in South Africa: Consideration of an experiential measure', *Development Southern Africa*, 14, 4, 505–12.

Crush, J. (1987) *The Struggle for Swazi Labour, 1890–1920*, McGill-Queen's University Press, Kingston and Montreal.

Crush, J. (1994) 'Scripting the compound: power and space in the South African mining industry', *Environment and Planning D: Society and Space*, 12, 301–24.

Crush, J. (ed.) (1995) *Power of Development*, Routledge, London.

**Curtis, F.** (1991) 'Foreign disinvestment and investment – South Africa: 1960–1986', in Konczacki, Z. A., Parpart, J. L. and Shaw, T. M. (eds) *Studies in the Economic History of South Africa, vol. 2: South Africa, Lesotho and Swaziland*, Frank Cass, London.

**Dagut, S.** (1997) 'Paternalism and social distance: British settlers' racial attitudes, 1850s–1890s', *South African Historical Journal*, **37**, 3–20.

**Dagut, S.** (forthcoming) 'Gender, colonial "women's history" and the construction of social distance: middle class British women in later nineteenth century South Africa', *Journal of Southern African Studies*.

**Davenport, T. R. H.** (1966) *The Afrikaner Bond: The History of a South African Political Party, 1880–1911*, Oxford University Press, Cape Town, London and New York.

**Davenport, T. R. H.** (1982) 'The consolidation of a new society: the Cape Colony', in Wilson, M. and Thompson, L. (eds) *A History of South Africa to 1870*, Croom Helm, London and Canberra.

**Davenport, T. R. H.** (1991) *South Africa: A Modern History*, Macmillan, London.

**Davies, R.** (1979) *Capital, State and White Labour in South Africa, 1900–1960: an Historical Materialist Analysis of Class Formation and Class Relations*, Harvester Press, Brighton.

**Davies, R.** (1996) 'South Africa's economic relations with Africa: current patterns and future perspectives', in Adedeji, A. (ed) *South Africa and Africa: Within or Apart?*, Zed Books, London

**Davies, R.** (1997) *South Africa in the SADC: Trade and Investment: Impact on Migration*, http://anc.org.za:80/govdocs/green_papers/migration/davies.html

**Davies, R. J.** (1981) 'The spatial formation of the South African city', *GeoJournal*, Supplementary Issue, **2**, 59–72.

**Davies, R. J.** (1991) 'Durban', in Lemon, A. (ed.) *Homes Apart: South Africa's Segregated Cities*, Paul Chapman, London; Indiana University Press, Bloomington and David Philip, Cape Town.

**Davin, A.** (1978) 'Imperialism and motherhood', *History Workshop*, **5**, 9–65.

**DBSA (Development Bank of Southern Africa)** (1991) *Economic and Social Memorandum, Region D: 1990*, DBSA, Halfway House, Johannesburg.

**DBSA (Development Bank of Southern Africa)** (1994): *South Africa's Nine Provinces: A Human Development Profile*, DBSA, Halfway House, Johannesburg.

**de Kock, L.** (1996) *Civilising Barbarians: Missionary Narrative and African Textual Response in Nineteenth-Century South Africa*, Witwatersrand University Press and Lovedale University Press, Johannesburg.

**Delius, P.** (1980) 'Migrant labour and the Pedi, 1840–80', in Marks, S. and Atmore, A. (eds) *Economy and Society in Pre-Industrial South Africa*, Longman, London.

**Delius, P.** (1983) *The Land Belongs to Us: The Pedi Polity, the Boers and the British in the Nineteenth-Century Transvaal*, Heinemann, London, Ibadan and Nairobi.

**Delius, P.** (1986) 'Abel Erasmus: power and profit in the eastern Transvaal', in

Beinart, W., Delius, P. and Trapido, S. (eds) *Putting a Plough to the Ground*: *Accumulation and Dispossession in Rural South Africa, 1850–1930*, Ravan Press, Johannesburg.

Delius, P. (1989) 'The Ndzundza Ndebele: indenture and the making of ethnic identity', in Bonner, P., Hofmeyr, I., James, D. and Lodge T. (eds) *Holding Their Ground*: *Class, Locality and Culture in Nineteenth and Twentieth Century South Africa*, Witwatersrand University Press and Ravan Press, Johannesburg.

Delius, P. (1993) 'Migrant organisation, the Communist Party, the ANC and the Sekhukhuneland revolt, 1940–1958', in Bonner, P., Delius, P. and Posel, D. (eds) *Apartheid's Genesis, 1935–1962*, Ravan Press and Witwatersrand University Press, Johannesburg.

Delius, P. (1996) *A Lion Amongst the Cattle*: *Reconstruction and Resistance in the Northern Transvaal*, Heinemann, Portsmouth NH; Ravan Press, Johannesburg and James Currey, Oxford.

Delius, P. and Trapido, S. (1982) 'Inboekselings and Oorlams: the creation and transformation of a servile class', *Journal of Southern African Studies*, 8, 214–42.

Dennon, D. and Nyeko, B. (1984) *Southern Africa since 1800*, Longman, London.

Department of Housing (1996) *Annual Report*, Department of Housing, Pretoria.

Department and Provincial and Local Government (2000) *Local Economic development: Manual Series*, Department and Provincial and Local Government, Pretoria.

Dewar, D. and Watson, V. (1982) Urbanization, unemployment and petty commodity production and trading: comparative cases in Cape Town, in Smith, D. (ed.) *Living Under Apartheid*: *Aspects of Urbanization and Social Change in South Africa*, George Allen and Unwin, London.

DFA (Department of Foreign Affairs) (1997) *South Africa's new place in the world*, background document for the parliamentary briefing week, 11 February 1997, Pretoria.

Dlamini, S. N. (1998) 'The construction, meaning and negotiation of ethnic identities in KwaZulu Natal', *Social Identities*, 4, 473–97.

Dooling, W. (1994) '"The good opinion of others": law, slavery and community in the Cape Colony, c.1760–1830', in Worden, N. and Crais, C. (eds) *Breaking the Chains*: *Slavery and its Legacy in the Nineteenth-Century Cape Colony*, Witwatersrand University Press, Johannesburg.

Driver, F. (1992) 'Geography's empire: histories of geographical knowledge', *Environment and Planning D: Society and Space*, 10, 23–40.

Driver, F. (1993) *Power and Pauperism*: *The Workhouse System 1834–1884*, Cambridge University Press, Cambridge.

du Toit, A. (1998) 'The fruits of modernity: law, power and paternalism in the rural Western Cape', in Howarth, D.R. and Norval, A. (eds) *South Africa in Transition*: *New Theoretical Perspectives*, Macmillan, London and St. Martin's Press, New York.

**du Toit, A. B.** (1983) 'No chosen people: the myth of the Calvinist origins of Afrikaner nationalism and racial ideology', *American Historical Review*, **88**, 920–52.

**du Toit, A. B. and Giliomee, H.** (eds) (1983) *Afrikaner Political Thought: Analysis and Documents, volume 1, 1780–1850*, David Philip, Cape Town.

**du Toit, A. E.** (1954) *The Cape Frontier: A Study of Native Policy with Special Reference to the Years 1847–66*, Government Printer, Pretoria.

**du Toit, J.** (1998) *The Structure of the South African Economy*, Southern Books, Halfway House, Johannesburg.

**Dubow, S.** (1986) '"Holding a just balance between White and Black": the Native Affairs Department in South Africa, *c*. 1920–1933', *Journal of Southern African Studies*, **12**, 217–319.

**Dubow, S.** (1987) 'Race, civilisation and culture: the elaboration of segregationist discourse in the inter-war years', in Marks, S. and Trapido, S. (eds) *The Politics of Race, Class and Nationalism in Twentieth Century South Africa*, Longman, London and New York.

**Dubow, S.** (1989) *Racial Segregation and the Origins of Apartheid in South Africa, 1919–36*, Macmillan, London.

**Dubow, S.** (1995) *Scientific Racism in Modern South Africa*, Cambridge University Press, Cambridge.

**Dubow, S.** (1996) 'Introduction', part one in Sachs, W., *Black Hamlet*, Johns Hopkins University Press, Baltimore and London.

**Dubow, S.** (1997) 'Colonial nationalism, the Milner kindergarten and the rise of "South Africanism", 1902–10', *History Workshop Journal*, **43**, 53–85.

**Dubow, S.** (1999) '"The Cape" in the mid-nineteenth century literary and scientific imagination', paper presented to the Societies of Southern Africa in the Nineteenth and Twentieth Century seminar, Institute of Commonwealth Studies, University of London, 9 March.

**Durning, A. B.** (1990) *Apartheid's Environmental Toll*, Paper 95, The Worldwatch Institute, Washington.

**Dzorgbo, D. B. S.** (1998) *Ghana in search of development*, Uppsala University Press, Uppsala.

**Eales, K.** (1989) 'Patriarchs, passes and privilege: Johannesburg's African middle classes and the question of night passes for African women', in Bonner, P., Hofmeyr, I., James, D. and Lodge T. (eds) *Holding Their Ground: Class, Locality and Culture in Nineteenth and Twentieth Century South Africa*, Witwatersrand University Press and Ravan Press, Johannesburg.

**Ede, D.** (1993) 'The OAU: another circus show?', *African Concord*, **8**, 7.

**EIU (Economist Intelligence Unit)** (1996) *South Africa: Country Profile, 1995–96*, EIU, London.

**EIU (Economist Intelligence Unit)** (1998) *South Africa; Country Profile, 1998–99*, EIU, London.

**EIU (Economist Intelligence Unit)** (1999) *South Africa: Country Profile, 1999–2000*, EIU, London.

**Elbourne, E.** (1991) '*To Colonize the Mind*': *Evangelical Missionaries in Britain and the eastern Cape, 1790–1837*', unpublished D. Phil, Oxford University.

**Elbourne, E.** (1995) 'Early Khoisan uses of mission Christianity', in Bredekamp, H.

and Ross, R. (eds) *Missions and Christianity in South African History*, Witwatersrand University Press, Johannesburg.

Eldredge, E. (1993) *A South African Kingdom: The Pursuit of Security in Nineteenth-Century Lesotho*, Cambridge University Press, Cambridge.

Eldredge, E. (1994a) 'Slave raiding across the Cape frontier', in Eldredge, E. and Morton, F. (eds) *Slavery in South Africa: Captive Labour on the Dutch Frontier*, Westview, Boulder and University of Natal Press, Pietermaritzburg.

Eldredge, E. (1994b) 'Delagoa Bay and the hinterland in the early nineteenth century: politics, trade, slaves, and slave raiding', in Eldredge, E. and Morton, F. (eds.) *Slavery in South Africa: Captive Labour on the Dutch Frontier*, Westview, Boulder and University of Natal Press, Pietermaritzburg.

Eldredge, E. (1995) 'Sources of conflict in southern Africa *c.* 1800–1830: the "Mfecane" reconsidered', in Hamilton, C. (ed.) *The Mfecane Aftermath: Reconstructive Debates in Southern African History*, Witwatersrand University Press, Johannesburg and University of Natal Press, Pietermaritzburg.

Ellis, S. (1998) 'The historical significance of South Africa's Third Force', *Journal of Southern African Studies* 24, 261–300.

Elphick, R. (1977) *Kraal and Castle: Khoikhoi and the Founding of White South Africa*, Yale University Press, New Haven and London.

Elphick, R. (1985) *Khoikhoi and the Founding of White South Africa*, Ravan, Johannesburg.

Elphick, R. (1995) 'Writing religion into history: the case of South African Christianity', in Bredekamp, H. and Ross, R. (eds) *Missions and Christianity in South African History*, Witwatersrand University Press, Johannesburg.

Elphick, R. (1997) 'The benevolent empire and the social gospel: missionaries and South African Christians in the age of segregation', in Elphick, R. and Davenport, R. (eds) *Christianity in South Africa: A Political, Social and Cultural History*, James Currey, Oxford and David Philip, Cape Town.

Elphick, R. and Malherbe, V. C. (1988) 'The Khoisan to 1828', in Elphick, R. and Giliomee, H. (eds) *The Shaping of South African Society, 1652–1840*, Wesleyan University Press, Middletown.

Elphick, R. and Shell, R. (1988) 'Intergroup relations: Khoikhoi, settlers, slaves and free blacks, 1652–1795', in Giliomee, H. and Elphick, R. (eds) *The Shaping of South African Society, 1652–1840*, Wesleyan University Press, Middletown.

Escobar, A. (1995) *Encountering Development: The Making and Unmaking of the Third World*, Princeton University Press, Princeton.

Etherington, N. (1978) *Preachers, Peasants and Politics in South East Africa, 1835–1880: African Christian Communities in Natal, Pondoland and Zululand*, Royal Historical Society, London.

Etherington, N. (1979) 'Labour supply and the genesis of South African Confederation in the 1870s', *Journal of African History*, **20**, 235–53.

Etherington, N. (1989a) 'Christianity and African society in nineteenth-century Natal', in Duminy, A. and Guest, B. (eds) *Natal and Zululand From Earliest Times to 1910: A New History*, University of Natal Press and Shuter and Shooter, Pietermaritzburg.

Etherington, N. (1989b) 'The "Shepstone system" in the Colony of Natal and beyond the borders', in Duminy, A. and Guest, B. (eds) *Natal and Zululand From Earliest Times to 1910: A New History*, University of Natal Press and Shuter and Shooter, Pietermaritzburg.

Etherington, N. (1995) 'Old wine in new bottles: the persistence of narrative structures in the historiography of the Mfecane and the Great Trek', in Hamilton, C. (ed.) *The Mfecane Aftermath: Reconstructive Debates in Southern African History*, Witwatersrand University Press, Johannesburg and University of Natal Press, Pietermaritzburg.

Evans, E. J. (1996) *The Forging of the Modern State: Early Industrial Britain, 1783–1870*, Longman, London and New York.

Evans, I. (1997) *Bureaucracy and Race: Native Administration in South Africa*, University of California Press, Berkeley, Los Angeles and London.

Evans, R. (1991) 'A golden age for Africa?', *Geographical Magazine*, 10, 28–30.

Fair, T. J. D. (1982) *South Africa: Spatial Frameworks for Development*, Juta, Kenwyn.

Fauré, Y. A. (1989) 'Côte d'Ivoire: analysing the crisis', in Cruise O'Brien, D. B., Dunn, J. and Rathbone, R. (eds) *Contemporary West African states*, Cambridge University Press, Cambridge.

Ferguson, J. (1990) *The Anti-Politics Machine: 'Development', Depoliticization, and Bureaucratic Power in Lesotho*, Cambridge University Press, Cambridge.

Financial Gazette (1994) 'Media circus with muckraker', 3 February, Zimbabwe.

Fine, B. and Rustomjee, Z. (1996) *The Political Economy of South Africa: From Minerals-Energy Complex to Industrialisation*, Hurst, London.

Fitzgerald, P., McLennan, A., and Munslow, B. (1997) *Managing Sustainable Development in South Africa*, 2nd edn. Oxford University Press, Cape Town.

Fleisch, B. D. (1995) 'Social scientists as policy makers: E. G. Malherbe and the National Bureau for Educational and Social research, 1929–1943', *Journal of Southern African Studies*, 21, 349–72.

Foucault, M. (1977) *Discipline and Punish: The Birth of the Prison*, Penguin, London.

Fox, R. C. and Nel, E. L. (2000) 'De-industrialisation in South Africa's rural periphery: the interplay of institutional and economic trends in the Eastern Cape Province', unpublished paper.

Fox, R., Nel, E. and Reintges, C. (1991) 'East London', in Lemon, A. (ed.) *Homes Apart: South Africa's Segregated Cities*, Paul Chapman, London; Indiana University Press, Bloomington and David Philip, Cape Town.

Fredrickson, G. (1981) *White Supremacy: A Comparative Study in American and South African History*, Oxford University Press, Oxford and New York.

Fredrickson, G. (1995) *Black Liberation: A Comparative History of Black Ideologies in the United States and South Africa*, Oxford University Press, Oxford and New York.

Freund, W. (1988) 'The Cape under the transitional governments, 1795–1814', in Elphick, R. and Giliomee, H. (eds) *The Shaping of South African Society, 1652–1840*, Wesleyan University Press, Middletown.

Freund, W. (1991) 'South African gold mining in transformation', in Gelb, S. (ed.) *South Africa's Economic Crisis*, David Philip, Cape Town and Zed Books London.

Friedman, S. (ed.) (1993) *The Long Journey: South Africa's Quest for a Negotiated Settlement*, Ravan Press, Johannesburg.

Furlong, P. J. (1991) *Between Crown and Swastika: The Impact of the Radical Right on the Afrikaner Nationalist Movement in the Fascist Era*, Wesleyan University Press, Hanover and London.

Gaitskell, D. (1982) '"Wailing for purity": prayer unions, African mothers and adolescent daughters, 1912–1940', in Marks, S. and Rathbone, R. (eds) *Industrialisation and Social Change in South Africa: African Class Formation, Culture and Consciousness, 1870–1930*, Longman, London and New York.

Gaitskell, D. (1990) 'Devout domesticity? A century of African women's Christianity in South Africa', in Walker, C. (ed.) *Women and Gender in Southern Africa to 1945*, David Philip, Cape Town and James Currey, London.

Galbraith, J. S. (1963) *Reluctant Empire: British Policy on the South African Frontier, 1834–1854*, University of California Press, Berkeley.

Garten, J. (1996) 'The big emerging markets', *Columbia Journal of World Business*, **Summer**, 6–31.

Gelb, S. (ed.) (1991) *South Africa's Economic Crisis*, David Philip, Cape Town and Zed Books, London.

George, S. (1988) *A Fate Worse Than Debt: A Radical New Analysis of the Third World Debt Crisis*, Penguin, London.

Gibb, R. A. (1995) 'The relevance of the European approach to regional economic integration in post-apartheid Southern Africa', in Lemon, A. (ed.), *The Geography of Change in South Africa*, John Wiley, Chichester.

Gibb, R. A. (1997) 'Regional integration in post-apartheid Southern Africa: the case of re-negotiating the Southern African Customs Union', *Journal of Southern African Studies*, **23**, 1, 67–86.

Giliomee, H. (1988) 'The eastern frontier, 1770–1812', in Elphick, R. and Giliomee, H. (eds) *The Shaping of South African Society*, 1652–1840, Wesleyan University Press, Middletown.

Giliomee, H. (1989) 'The beginnings of Afrikaner ethnic consciousness, 1850–1915', in Vail, L. (ed.) *The Creation of Tribalism in Southern Africa*, James Currey, London and University of California Press, Berkeley and Los Angeles.

Giliomee, H. and Schlemmer, L. (1989) *From Apartheid to Nation Building*, Oxford University Press, Oxford.

Gilman, S. (1985) *Difference and Pathology: Stereotypes of Sexuality, Race and Madness*, Cornell University Press, Ithaca.

Glaser, D. (1998) 'Changing discourses of democracy and socialism in South Africa', in Howarth, D. R. and Norval, A. (eds) *South Africa in Transition: New Theoretical Perspectives*, Macmillan, London and St. Martin's Press, New York.

Goldberg, D. (ed.) (1990) *Anatomy of Racism*, University of Minnesota Press, Minneapolis.

Goldin, I. (1987) *Making Race: The Politics and Economics of Coloured Identity in South Africa*, Longman, London and New York.

Goldin, I. (1989) 'Coloured identity and Coloured politics in the Western Cape region of South Africa', in Vail, L. (ed.) *The Creation of Tribalism in Southern Africa*, James Currey, London and University of California Press, Berkeley and Los Angeles.

Government Communication and Information Services (1998) *The Building has Begun: Government's Report to the Nation '98*, Government Communication and Information Services, Pretoria.

Government Communication and Information Services (1999) *Realising our Hopes*, Government Communication and Information Services, Pretoria.

Graaff, J. (1990) 'Towards an understanding of bantustan politics', in Nattrass, N. and Ardington, E. (eds) *The Political Economy of South Africa*, Oxford University Press, Oxford.

Greenberg, S. B. (1987) *Legitimating the Illegitimate: State, Markets and Resistance in South Africa*, University of California Press, Berkeley, Los Angeles and London.

Greenstein, R. (1998) 'Identity, race, history: South Africa and the Pan-African context', in Greenstein, R. (ed.), *Comparative Perspectives on South Africa*, Macmillan, London

Griffiths, I. L. (1995) *The African Inheritance*, Routledge, London.

Grundlingh, A. and Sapire, H. (1989) 'From feverish festival to repetitive ritual? The changing fortunes of Great Trek mythology in an industrialising South Africa, 1938–1988', *South African Historical Journal*, 21, 19–37.

Grundy, K. (1991) *South Africa: Domestic Crisis and Global Challenge*, Westview, Boulder.

Guelke, L. (1988) 'Freehold farmers and frontier settlers, 1657–1780', in Elphick, R. and Giliomee, H. (eds) *The Shaping of South African Society, 1652–1840*, Wesleyan University Press, Middletown.

Guelke, L. (1989) 'The origin of white supremacy in South Africa: an interpretation', *Social Dynamics*, 15, 40–5.

Guelke, L. (1991) 'The beginnings of modern South African society: the meeting of two worlds', in Konczacki, Z.A., Parpart, J. L. and Shaw, T. M. (eds) *Studies in the Economic History of South Africa, vol. 2: South Africa, Lesotho and Swaziland*, Frank Cass, London.

Guelke, L. and Shell, R. (1983) 'An early colonial landed gentry: land and wealth in the Cape Colony, 1682–1731', *Journal of Historical Geography*, 9, 265–86.

Guelke, L. and Shell, R. (1992) 'Landscape of conquest: frontier water alienation and Khoikhoi strategies of survival, 1652–1780', *Journal of Southern African Studies*, 18, 803–24.

Guy, J. (1979) *The Destruction of the Zulu Kingdom*, Ravan Press, Johannesburg.

Guy, J. (1982) 'The destruction and reconstruction of Zulu society', in Marks, S. and Rathbone, R. (eds) *Industrialisation and Social Change in South Africa: African Class Formation, Culture and Consciousness, 1870–1930*, Longman, London and New York.

Guy, J. (1990) 'Gender oppression in southern Africa's precapitalist societies', in Walker, C. (ed.) *Women and Gender in Southern Africa to 1945*, David Philip, Cape Town and James Currey, London.

Habermas, J. (1976) *Legitimation Crisis*, Heinemann, London.

Hall, C. (1992) *White, Male and Middle Class: Explorations in Feminism and History*, Verso, Cambridge.

Hall, C. (1996a) 'Imperial man: Edward Eyre in Australasia and the West Indies, 1833–66', in Schwarz, B. (ed.) *The Expansion of England: Race, Ethnicity and Cultural History*, Routledge, London.

Hall, C. (1996b) 'Histories, empires and the post-colonial moment', in Chambers, I. and Curti, L. (eds) *The Post-Colonial Question: Common Skies, Divided Horizons*, Routledge, London and New York.

Hall, C. (1999) 'William Knibb and the constitution of the new black subject', in M. Daunton, M. and Halpern, R. (eds) *Empire and Others: British Encounters with Indigenous Peoples, 1600–1850*, UCL Press, London.

Hall, M. (1987) *The Changing Past: Farmers, Kings and Traders in Southern Africa, 200–1860*, James Currey, London and David Philip, Cape Town.

Hall, P. (1996) *Cities of Tomorrow* (updated edition), Blackwell, Oxford.

Hall, S. (1995) 'Archaelogical indicators of stress in the western Transvaal region between the seventeenth and nineteenth centuries', in Hamilton, C. (ed.) *The Mfecane Aftermath: Reconstructive Debates in Southern African History*, Witwatersrand University Press, Johannesburg and University of Natal Press, Pietermaritzburg.

Hamilton, C. (ed.) (1995) *The Mfecane Aftermath: Reconstructive Debates in Southern African History*, Witwatersrand University Press, Johannesburg and University of Natal Press, Pietermaritzburg.

Hamilton, C. (1998) *Terrific Majesty: The Powers of Shaka Zulu and the Limits of Historical Invention*, Harvard University Press, Cambridge, Mass.

Hanlon, B. (1986) *Beggar Your Neighbours: Apartheid Power in Southern Africa*, CIIR, London.

Hanretta, S. (1998) 'Women, marginality and the Zulu state: women's institutions and power in the early nineteenth century', *Journal of African History*, **39**, 389–415.

Harries, P. (1982) 'Kinship, ideology and the nature of pre-colonial labour migration: labour migration from the Delagoa Bay hinterland to South Africa, up to 1895', in Marks, S. and Rathbone, R. (eds) *Industrialisation and Social Change in South Africa: African Class Formation, Culture and Consciousness, 1870–1930*, Longman, London and New York.

Harries, P. (1989) 'Exclusion, classification and internal colonialism: the emergence of ethnicity among the Tsonga-speakers of South Africa', in Vail, L. (ed.) *The Creation of Tribalism in Southern Africa*, James Currey, London; University of California Press, Berkeley and Los Angeles and David Philip, Claremont.

**Harries, P.** (1994) *Work, Culture and Identity: Migrant Labourers in Mozambique and South Africa, c. 1860–1910*, Heinemann, Portsmouth; Ravan Press, Johannesburg and James Currey, Oxford.

**Hart, C. and Parnell, S.** (1989) 'Church, state and the shelter of white working-class women in Johannesburg, 1920–1955', *South African Geographical Journal*, **71**, 25–31.

**Hart, D. M. and Pirie, G. H.** (1984) 'The sight and soul of Sophiatown', *Geographical Review*, **74**, 38–47.

**Harte, N. and Quinalt, R.** (eds) (1996) *Land and Society in Britain 1700–1914*, Manchester University Press, Manchester.

**Hartley, G.** (1995) 'The Battle of Dithakong and "Mfecane" theory', in Hamilton, C. (ed.) *The Mfecane Aftermath: Reconstructive Debates in Southern African History*, Witwatersrand University Press, Johannesburg and University of Natal Press, Pietermaritzburg.

**Harvey, D.** (1990) *The Condition of Postmodernity*, Blackwell, Oxford.

**Hennock, E. P.** (1994) 'Poverty and social reforms', in Johnson, P. (ed.) *Twentieth Century Britain: Economic, Social and Cultural Change*, Longman, London and New York.

**Hexham, I. and Poewe, K.** (1997) 'The spread of Christianity among Whites and Blacks in Transorangia', in Elphick, R. and Davenport, R. (eds) *Christianity in South Africa: A Political, Social and Cultural History*, James Currey, Oxford and David Philip, Cape Town.

**Heymans, C.** (1991) 'Privatization and municipal reform', in Swilling, M., Humphries, R. and Shubane, K. (eds) *The Apartheid City in Transition*, Oxford University Press, Cape Town.

**Hill, R. A. and Pirio, G. A.** (1987) '"Africa for the Africans": the Garvey movement in South Africa, 1920–1940', in Marks, S. and Trapido, S. (eds) *The Politics of Race, Class and Nationalism in Twentieth Century South Africa*, Longman, London and New York.

**Hilton, B.** (1988) *The Age of Atonement: The Influence of Evangelicalism on Social and Economic Thought, 1795–1865*, Clarendon Press, Oxford.

**Hilton, R.** (ed.) (1976) *The Transition From Feudalism to Capitalism*, Verso, London.

**Hindson, D.** (1987) *Pass Controls and the Urban African Proletariat*, Ravan Press, Johannesburg.

**Hindson, D., Byerley, M. and Morris, M.** (1994) 'From violence to reconstruction: the making, disintegration and remaking of an apartheid city', *Antipode*, **26**, 323–50.

**Hirsch, A.** (1991) 'Industrial decentralisation and the spatial economy of apartheid', in Konczacki, Z. A., Parpart, J. L. and Shaw, T. M. (eds) *Studies in the Economic History of South Africa, Vol. 2: South Africa, Lesotho and Swaziland*, Frank Cass, London.

**Hirson, B.** (1979) *Year of Fire, Year of Ash: The Soweto Revolt: Roots of a Revolution?* Zed Press, London.

Hobsbawm, E. (1968) *Industry and Empire*, Penguin, London.

Hobsbawm, E. (1975) *The Age of Capital, 1848–1875*, Charles Scribner's Sons, New York.

Hobsbawm, E. (1987) *The Age of Empire, 1875–1914*, Weidenfeld and Nicolson, London.

Hodgson, J. (1997) 'A battle for sacred power: Christian beginnings among the Xhosa', in Elphick, R. and Davenport, R. (eds) *Christianity in South Africa: A Political, Social and Cultural History*, James Currey, Oxford and David Philip, Cape Town.

Hofmeyr, I. (1987) 'Building a nation from words: Afrikaans language, literature and ethnic identity, 1902–1924', in Marks, S. and Trapido, S. (eds) *The Politics of Race, Class and Nationalism in Twentieth Century South Africa*, Longman, London and New York.

Hofmeyr, I. (1991) 'Popularising history: the case of Gustav Preller', in Clingman, S. (ed.) *Regions and Repertoires: Topics in South African Politics and Culture*, Ravan Press, Johannesburg.

Holness, S., Nel, E. L. and Binns, T. (forthcoming) 'The changing nature of informal street trading in post-apartheid South Africa: the case of East London's central business district', *Urban Forum*.

Holt, T. (1992) *The Problem of Freedom: Race, Labor, and Politics in Jamaica and Britain, 1832–1938*, Johns Hopkins University Press, Baltimore and London.

Houghton, D. H. (1967) *The South African Economy*, Oxford University Press, Cape Town.

Hughes, H. (1990) '"A lighthouse for African womanhood": Inanda Seminary, 1869–1945, in Walker, C. (ed.) *Women and Gender in Southern Africa to 1945*, David Philip, Cape Town and James Currey, London.

Humphries, R. and Shubane, K. (1991) 'Homelands and provinces; dynamics of change and transition', in Lee, R. and Schlemmer, L. (eds) *Transition to Democracy: Policy Perspectives, 1991*, Oxford University Press, Cape Town and Oxford.

Hyam, R. (1972) *The Failure of South African Expansion, 1980–1948*, Macmillan, London.

Hyslop, J. (1988) 'State education policy and the social reproduction of the urban African working class: the case of the Southern Transvaal 1955–1976', *Journal of Southern African Studies*, **14**, 446–76.

Hyslop, J. (1991) 'Food, authority and politics: student riots in South African schools, 1945–1976', in Clingman, S. (ed.) *Regions and Repertoires: Topics in South African Politics and Culture*, Ravan Press, Johannesburg.

Hyslop, J. (1993) '"A destruction coming in": Bantu education as response to social crisis', in Bonner, P., Delius, P. and Posel, D. (eds) *Apartheid's Genesis, 1935–1962*, Ravan Press and Witwatersrand University Press, Johannesburg.

Hyslop, J. (1995) 'White working-class women and the invention of apartheid: "purified" Afrikaner nationalist agitation for legislation against "mixed" marriages, 1934–9', *Journal of African History*, **36**, 57–81.

Hyslop, J. (1998) 'Why did apartheid's supporters capitulate? Whiteness, consumption and class in urban South Africa, 1985–1995', paper presented to The Societies of Southern Africa in the Nineteenth and Twentieth Centuries seminar, Institute of Commonwealth Studies, University of London, 20 November.

Iheduru, O. (1996) 'Post-apartheid South Africa and its neighbours: a maritime transport perspective', *Journal of Modern African Studies*, **34**, 1, 1–26.

Iliffe, J. (1987) *The African Poor: A History*, Cambridge University Press, Cambridge and Ravan Press, Johannesburg.

Innes, D. (1984) *Anglo-American and the Rise of Modern South Africa*, Heinemann, London.

Jeeves, A. H. (1975) 'The control of migratory labour on the South African gold mines in the era of Kruger and Milner', *Journal of Southern African Studies*, **2**, 3–29.

Jeeves, A. H. (1985) *Migrant Labour in South Africa's Mining Economy: The Struggle for the Gold Mines' Labour Supply, 1890–1920*, McGill-Queen's University Press, Kingston and Montreal.

Jeeves, A. H. (1991) 'Migrant labour in the industrial transformation of South Africa, 1920–1960', in Konczacki, Z. A., Parpart, J. L. and Shaw, T. M. (eds) *Studies in the Economic History of South Africa, Vol. 2: South Africa, Lesotho and Swaziland*, Frank Cass, London.

Jeeves, A. H. and Crush, J. (1997) 'Introduction', in Jeeves, A. H. and Crush, J. (eds) *White Farms, Black Labour:*

*The State and Agrarian Change in Southern Africa, 1910–1950*, Heinemann, Portsmouth N.H.; University of Natal Press, Pietermaritzburg and James Currey, Oxford.

Jochelson, K. (1995) 'Women, migrancy and morality: a problem of perspective', *Journal of Southern African Studies*, **21**, 2, 323–32.

Johnson, P. (1994) 'The role of the state in twentieth-century Britain', in Johnson, P. (ed.) *Twentieth Century Britain: Economic, Social and Cultural Change*, Longman, London and New York.

Johnstone, F. (1976) *Class, Race and Gold: a Study of Class Relations and Racial Discrimination in South Africa*, Routledge and Kegan Paul, London.

Joseph Rowntree Foundation (1995) *Inquiry into Income and Wealth*, JRF, York.

Kahn, B. (1991) 'The crisis and South Africa's balance of payments', in Gelb, S. (ed.) *South Africa's Economic Crisis*, David Philip, Cape Town and Zed Books London.

Keegan, T. (1979) 'The restructuring of agrarian class relations in a colonial economy: the Orange River Colony, 1902–1910', *Journal of Southern African Studies*, **5**, 234–54.

Keegan, T. (1986) *Rural Transformations in Industrialising South Africa: The Southern Highveld to 1914*, Ravan Press, Johannesburg.

Keegan, T. (1988) 'The making of the Orange Free State, 1846–1854: sub-imperialism, primitive accumulation and state formation', *Journal of Imperial and Commonwealth History*, **17**, 26–54.

Keegan, T. (1989) 'Race, class and economic development in South Africa: a review article', *Social Dynamics*, **15**, 111–31.

Keegan, T. (1991) 'The making of the rural economy: from 1850 to the present', in Konczacki, Z. A., Parpart, J. L. and Shaw, T. M. (eds) *Studies in the Economic History of South Africa, Vol. 2: South Africa, Lesotho and Swaziland*, Frank Cass, London.

Keegan, T. (1996) *Colonial South Africa and the Origins of the Racial Order*, Leicester University Press, London.

Keet, D. (1996) 'The European Union's proposed free trade agreement with South Africa: the implications and some proposals', *Development Southern Africa*, **13**, 4, 555–66.

**Kent State University, College of Business** (1996) *Southern African Development Community: Country Profiles/Trade Opportunities*. http://business.kent.edu/sabos/intro.htm

Kentridge, M. (1990) *An Unofficial War: Inside the Conflict in Pietermaritzburg*, David Philip, Cape Town and Johannesburg.

Kimble, H. (1996) Prospects for R and D in the New SA, *Journal fur Enttwicklungspolitik* **12**, 1, 75–106.

Kinghorn, J. (1997) 'Modernization and apartheid: the Afrikaner churches', in Elphick, R. and Davenport, R. (eds) *Christianity in South Africa: A Political, Social and Cultural History*, James Currey, Oxford and David Philip, Cape Town.

Kinsman, M. (1983) '"Beasts of burden": the subordination of Southern Tswana women, ca. 1800–1840', *Journal of Southern African Studies*, **10**, 39–54.

Kinsman, M. (1995) '"Hungry wolves": the impact of violence on Rolong life, 1823–1836', in Hamilton, C. (ed.) *The Mfecane Aftermath: Reconstructive Debates in Southern African History*, Witwatersrand University Press, Johannesburg and University of Natal Press, Pietermaritzburg.

Kirk, T. (1980) 'The Cape economy and the expropriation of the Kat River Settlement, 1846–53', in Marks, S. and Atmore, A. (eds) *Economy and Society in Pre-Industrial South Africa*, Longman, London.

Kotelo, K. (1995) 'The reshaping of trade relations in Southern Africa', *Bulletin of the Africa Institute, South Africa*, **35**, 6, 10–11.

Krikler, J. (1993) *Revolution From Above, Rebellion From Below: The Agrarian Transvaal at the Turn of the Century*, Oxford University Press, Oxford.

Krikler, J. (1996) 'Women, violence and the Rand Revolt of 1922', *Journal of Southern African Studies*, **22**, 349–72.

Krikler, J. (1998) 'War service and workers' insurrection in South Africa: commandos in 1922', paper presented at the Imperial History Seminar, Institute of Historical Research, University of London, 2 February.

Kruger, D. (1992) 'District Six, Cape Town: an apartheid landscape', *Landscape*, **13**, 9–15.

Kubicek, R. V. (1991) 'Mining: patterns of dependence and development 1870–1930', in Konczacki, Z. A., Parpart, J. L. and Shaw, T. M. (eds) *Studies in the Economic History of South Africa, Vol. 2: South Africa, Lesotho and Swaziland*, Frank Cass, London.

**Kuper, A.** (1988) 'Anthropology and apartheid', in Lonsdale, J. (ed.) *South Africa in Question*, African Studies Centre, Cambridge; James Currey, London and Heinemann, Portsmouth N.H.

**Kuper, L., Watts, H. and Davies, R.** (1958) *Durban: A Study in Racial Ecology*, Jonathan Cape, London.

**la Hausse, P.** (1989) 'The message of the warriors: the ICU, the labouring poor and the making of a popular political culture in Durban, 1925–1930', in Bonner, P., Hofmeyr, I., James, D. and Lodge T. (eds) *Holding Their Ground: Class, Locality and Culture in Nineteenth and Twentieth Century South Africa*, Witwatersrand University Press and Ravan Press, Johannesburg.

**la Hausse, P.** (1993) 'So who was Elias Kuzwayo? Nationalism, collaboration and the picaresque in Natal', in Bonner, P., Delius, P. and Posel, D. (eds) *Apartheid's Genesis, 1935–1962*, Ravan Press and Witwatersrand University Press, Johannesburg.

**Lacey, M.** (1981) *Working for Boroko: The Origins of a Coercive Labour System in South Africa*, Raven Press, Johannesburg.

**Laing, M.** (1999) 'An African renaissance', in *African Business Focus 1999*, British African Business Association, London.

**Lambert, J.** (1995) *Betrayed Trust: Africans and the State in Colonial Natal*, University of Natal Press, Pietermaritzburg.

**Lambert, R. and Webster, E.** (1988) 'The re-emergence of political unionism in contemporary South Africa', in Cobbett, W. and Cohen, R. (eds) *Popular Struggles in South Africa*, James Currey, London.

**Lambert, R. V.** (1993) 'Trade unionism, race, class and nationalism in the 1950s resistance movement', in Bonner, P., Delius, P. and Posel, D. (eds) *Apartheid's Genesis, 1935–1962*, Ravan Press, Witwatersrand University Press, Johannesburg.

**Lazar, J.** (1987) *Conformity and Conflict: Afrikaner Nationalist Politics in South Africa, 1948–1961*, unpublished D. Phil, Oxford University.

**Lazar, J.** (1993) 'Verwoerd versus the "visionaries": the South African Bureau of Racial Affairs (Sabra) and apartheid, 1948–1961', in Bonner, P., Delius, P. and Posel, D. (eds) *Apartheid's Genesis, 1935–1962*, Ravan Press, Witwatersrand University Press, Johannesburg.

**Le Cordeur, B.** (1981) *Eastern Cape Separatism, 1820–54*, Oxford University Press, Cape Town.

**Lea, J. P.** (1982) 'Government dispensation, capitalist imperative or liberal philanthropy? Responses to the black housing crisis in South Africa', in Smith, D. (ed.) *Living Under Apartheid: Aspects of Urbanization and Social Change in South Africa*, George Allen and Unwin, London.

**Lee, R., Sutherland, M., Phillips, M. and McLennan, A.** (1991) 'Speaking or listening? Observers or agents of change? Business and public policy: 1989/90', in Lee, R. and Schlemmer, L. (eds) *Transition to Democracy: Policy Perspectives, 1991*, Oxford University Press, Cape Town and Oxford.

**Legassick, M.** (1974) 'South Africa: capital accumulation and violence', *Economy and Society*, 3, 254–91.

**Legassick, M.** (1977) 'Gold, agriculture and secondary industry in South Africa, 1885–1970: from periphery to sub-metropole as a forced labour system', in Palmer, R. and Parsons, N. (eds) *The Roots of Rural Poverty in Central and Southern Africa*, Heinemann, London, Ibadan, Nairobi and Lusaka.

**Legassick, M.** (1988) 'The northern frontier to c. 1840: the rise and decline of the Griqua people', in Elphick, R. and Giliomee, H. (eds) *The Shaping of South African Society, 1652–1840*, Wesleyan University Press, Middletown.

**Legassick, M.** (1993) 'The state, racism and the rise of capitalism in the nineteenth century Cape Colony', *South African Historical Journal*, **28**, 329–68.

**Legassick, M.** (1995) 'British hegemony and the origins of segregation in South Africa, 1901–14', in Beinart, W. and Dubow, S. (eds) *Segregation and Apartheid in Twentieth Century South Africa*, Routledge, London and New York.

**Legum, C.** (1993) 'South Africa's potential role in the Organisation of African Unity', *South African Journal of International Affairs*, **1**, 1, 15–25.

**Leistner, E.** (1997) 'Regional co-operation in sub-Saharan Africa, with special reference to Southern Africa', *Africa Insight*, **27**, 2, 112–23

**Lemon, A.** (1987) *Apartheid in Transition*, Gower, Aldershot.

**Lemon, A.** (1991) 'The apartheid city', in Lemon, A. (ed.) *Homes Apart: South Africa's Segregated Cities*, Paul Chapman, London; Indiana University Press, Bloomington and David Philip, Cape Town.

**Lemon, A.** (1999) 'Shifting inequalities in South Africa's schools: some evidence from the Western Cape', *South African Geographical Journal*, **81**, 2, 96–105.

**Lemon, A.** (forthcoming) 'New directions in a changing global environment: foreign policy in post-apartheid South Africa', *South African Geographical Journal*, **82**.

**Lemon, A. and Stevens, L.** (1999) 'Reshaping education in the New South Africa', *Geography*, **84**, 3, 222–32.

**Lester, A.** (1997) 'The margins of order: strategies of segregation on the eastern Cape frontier, 1806–c. 1850', *Journal of Southern African Studies*, **23**, 635–53.

**Lester, A.** (1998a) '"Otherness" and the frontiers of empire: the eastern Cape Colony, 1806–c.1850', *Journal of Historical Geography*, **24**, 2–19.

**Lester, A.** (1998b) 'Settlers, the state and colonial power: the colonization of Queen Adelaide Province, 1834–37', *Journal of African History*, **39**, 221–46.

**Lester, A.** (1998c) *Colonial Discourse and the Colonization of Queen Adelaide Province, South Africa*, Royal Geographical Society/Institute of British Geographers, Historical Geography Research Series, 34, London.

**Lester, A.** (1998d) 'Reformulating identities: British settlers in early nineteenth-century South Africa', *Transactions of the Institute of British Geographers*, **23**, 515–31.

**Lester, A.** (2000) *From Colonization to Democracy: a New Historical Geography of South Africa*, I. B. Tauris, London and New York.

Lester, A. (2000) 'Historical geographies of imperialism', in Graham, B. and Nash, C. (eds) *Modern Historical Geographies*, Longman, London and New York.

Lester, A. (forthcoming) *The Eastern Cape: British Colonialism in the Nineteenth Century*, Routledge, London and New York.

Lewis, G. (1987) *Between the Wire and the Wall: a History of South African 'Coloured' Politics*, David Philip, Cape Town and Johannesburg.

Lewis, J. (1984) *Industrialisation and Trade Union Organisation in South Africa, 1924–55: The Rise and Fall of the South African Trades and Labour Council*, Cambridge University Press, Cambridge.

Lipton, M. (1986) *Capitalism and Apartheid: South Africa 1910–1986*, Wildwood House, Aldershot.

Livingstone, D. (1992) *The Geographical Tradition*, Blackwell, Oxford.

Lodge, T. (1983) *Black Politics in South Africa Since 1945*, Longman, London and New York.

Lodge, T. (1992) 'Rebellion: the turning of the tide', in Lodge, T. and Nasson, B. (eds) *All Here and Now: Black Politics in South Africa in the 1980s*, Hurst, London.

Lonsdale, J. (1998) 'Conclusion: South Africa in African History', in Greenstein, R. (ed.) *Comparative Perspective on South Africa*, Macmillan, London.

Loomba, A. (1998) *Colonialism/Postcolonialism*, Routledge, London and New York.

Lowe, P. (1999) 'Life after Lomé', in *African Business Focus 1999*, British African Business Association, London.

Luckhart, K. and Wall, B. (1980) *Organize ... or Starve! The History of the South African Congress of Trade Unions*, Lawrence and Wishart, London.

Lundahl, M. (1997a) 'The post-apartheid economy and after?', in Petersson, L. (ed.), *Post-Apartheid Southern Africa: Economic Challenges and Policies for the Future*, Routledge, London.

Lundahl, M. (1997b) *The South African Economy in 1996: From Reconstruction and Development to Growth, Employment and Redistribution*, Swedish International Development Co-operation Agency, Stockholm.

Mabin, A. (1984) *The Making of Colonial Capitalism: Intensification and Expansion in the Economic Geography of the Cape Colony, South Africa, 1854–1899*, unpublished PhD, Simon Fraser University.

Mabin, A. (1986) 'Labour, capital, class struggle and the origins of residential segregation in Kimberley, 1880-1920', *Journal of Historical Geography*, **12**, 4–26.

MacCrone, I. D. (1937) *Race Attitudes in South Africa: Historical, Experimental and Psychological Studies*, Oxford University Press, London.

Maclennan, B. (1986) *A Proper Degree of Terror: John Graham and the Cape's Eastern Frontier*, Ravan Press, Johannesburg.

Macmillan, H. (1989) 'A nation divided? The Swazi in Swaziland and the Transvaal, 1865–1986', in Vail, L. (ed.) *The Creation of Tribalism in Southern Africa*, James Currey, London; University of California Press, Berkeley and Los Angeles and David Philip Claremont.

**Macmillan, W. M.** (1927) *The Cape Colour Question: A Historical Survey*, Faber and Gwyer, London.

**Macmillan, W. M.** (1963) *Bantu, Boer and Briton: The Making of the South African Native Problem*, Clarendon Press, Oxford.

**Mager, A. K.** (1999) *Gender and the Making of a South African Bantustan: A Social History of the Ciskei, 1945–1959*, Heinemann, Portsmouth, NH; James Currey, Oxford and David Philip, Cape Town.

**Maharaj, B.** (1992) 'The "spatial impress" of the central and local states: the Group Areas Act' in Durban, in Smith, D. (ed.) *The Apartheid City and Beyond: Urbanization and Social Change in South Africa*, Routledge, London and New York and Witwatersrand University Press, Johannesburg.

**Maharaj, B.** (1995) 'The local state and residential segregation: Durban and the prelude to the Group Areas Act', *South African Geographical Journal*, **77**, 33–41.

**Mamdani, M.** (1996) *Citizen and Subject: Contemporary Africa and the Legacy of Late Colonialism*, Princeton University Press, Princeton.

**Mandela, N.** (1993) 'South Africa's foreign policy', *Foreign Affairs*, **Nov/Dec**, 88.

**Manicom, L.** (1992) 'Ruling relations: rethinking state and gender in South African history', *Journal of African History*, **33**, 441–65.

**Manson, A.** (1995) 'Conflict in the western highveld/southern Kalahari c. 1750–1820', in Hamilton, C. (ed.) *The Mfecane Aftermath: Reconstructive Debates in Southern African History*,

Witwatersrand University Press, Johannesburg and University of Natal Press, Pietermaritzburg.

**Manzo, K.** (1992) *Domination, Resistance and Social Change in South Africa: The Local Effects of Global Power*, Praeger, Westport and London.

**Marais, H.** (1996) 'South Africa: The popular movement in flux and the Reconstruction and Development Programme (RDP)', *Africa Development*, **21**, 2 & 3, 211–33.

**Marais, H.** (1998) *South Africa: Limits to Change: The Political Economy of Transformation*, Zed Books, London and University of Cape Town Press, Cape Town.

**Marais, J. S.** (1957) *The Cape Coloured People, 1652–1937*, Witwatersrand University Press, Johannesburg.

**Marcus, T.** (1989) *Modernizing Super-Exploitation: Restructuring South African Agriculture*, Zed Books, London and New Jersey.

**Maré, G.** (1993) *Ethnicity and Politics in South Africa*, Zed Books, London.

**Maré, G. and Hamilton, G.** (1987) *An Appetite for Power: Buthelezi's Inkatha and South Africa*, Indiana University Press, Bloomington and Indianapolis, Ravan Press, Johannesburg.

**Marks, S.** (1970) *Reluctant Rebellion: The 1906–8 Disturbances in Natal*, Clarendon Press, Oxford.

**Marks, S.** (1972) 'Khoisan resistance to the Dutch in the seventeenth and eighteenth centuries', *Journal of African History*, **13**, 55–80.

Marks, S. (1978) 'Natal, the Zulu royal family and the ideology of segregation', *Journal of Southern African Studies*, **4**, 172–94.

Marks, S. (1985a) 'Southern Africa, 1867–1886', in Oliver, R. and Sanderson, G. N. (eds) *The Cambridge History of Africa, Vol. 6, From 1870 to 1905*, Cambridge University Press, Cambridge.

Marks, S. (1985b) 'Southern and Central Africa, 1886–1910', in Oliver, R. and Sanderson, G. N. (eds) *The Cambridge History of Africa, Vol. 6, From 1870 to 1905*, Cambridge University Press, Cambridge.

Marks, S. (1986) *The Ambiguities of Dependence in South Africa: Class, Nationalism, and the State in Twentieth-Century Natal*, Ravan Press, Johannesburg.

Marks, S. (1987) 'White supremacy: a review article', *Comparative Studies in Society and History*, **29**, 385–97.

Marks, S. (1989) 'Patriotism, patriarchy and purity: Natal and the politics of Zulu ethnic consciousness', in Vail, L. (ed.) *The Creation of Tribalism in Southern Africa*, James Currey, London; University of California Press, Berkeley and Los Angeles and David Philip, Claremont.

Marks, S. and Atmore, A. (1980) 'Introduction', in Marks, S. and Atmore, A. (eds) *Economy and Society in Pre-Industrial South Africa*, Longman, London.

Marks, S. and Rathbone, B. (1982) 'Introduction', in Marks, S. and Rathbone, R. (eds) *Industrialisation and Social Change in South Africa: African Class Formation, Culture and Consciousness 1870–1930*, Longman, London and New York.

Marks, S. and Trapido, S. (1979) 'Lord Milner and the South African state', *History Workshop Journal*, **2**, 50–80.

Marks, S. and Trapido, S. (1987) 'The politics of race, class and nationalism', in Marks, S. and Trapido, S. (eds) *The Politics of Race, Class and Nationalism in Twentieth Century South Africa*, Longman, London and New York.

Marks, S. and Trapido, S. (eds) (1987) *The Politics of Race, Class and Nationalism in Twentieth Century South Africa*, Longman, London and New York.

Marks, S. and Trapido, S. (1991) 'Introduction', *Journal of Southern African Studies*, **18**, 1–18.

Mason, J. (1994) 'Paternalism under siege: slavery in theory and practice during the era of reform, *c*. 1825 through emancipation', in Worden, N. and Crais, C. (eds) *Breaking the Chains: Slavery and its Legacy in the Nineteenth-Century Cape Colony*, Witwatersrand University Press, Johannesburg.

Mathews, K. (1989) 'The Organisation of African Unity in world politics', in Onwuka, R. I. and Shaw T. M. (eds) *Africa in world politics*, Macmillan, London.

Matsetela, T. (1982) 'The life-story of Nkgono Mma-Pooe: aspects of sharecropping and proletarianisation in the northern Orange Free State, 1890–1930', in Marks, S. and Rathbone, R. (eds) *Industrialisation and Social Change in South Africa: African Class Formation, Culture and Consciousness, 1870–1930*, Longman, London and New York.

Matthews, Z. K. (1981) *Freedom for My People: The Autobiography of Z. K.*

*Matthews: Southern Africa 1901 to 1968*, ed. Monica Wilson, David Philip, Cape Town and Johannesburg.

**Mayall, J.** (1991) 'The hopes and fears of independence: Africa and the world 1960–90', in Rimmer, D. (ed.), *Africa 30 years on*, James Currey, London.

**Mayer, P.** (1971) *Townsmen or Tribesmen: Conservatism and the Process of Urbanization in a South African City*, 2nd edn., Oxford University Press, Cape Town.

**Maylam, P.** (1990) 'The rise and decline of urban apartheid in South Africa', *African Affairs*, **89**, 57–84.

**Mbeki, G.** (1964) *The Peasants' Revolt*, Penguin, Harmondsworth.

**McCarthy, C. L.** (1992) 'Revenue distribution and economic development in the Southern African Customs Union', *South African Journal of Economics*, **62**, 3, 167–87.

**McClintock, A.** (1995) *Imperial Leather: Race, Gender and Sexuality in the Colonial Contest*, Routledge, London and New York.

**MDB (Municipal Demarcation Board)** (1999) *An Integrated Framework of Nodal Points for Metropolitan and District Council Areas in South Africa*, MDB, Pretoria.

**Meintjes, S.** (1990) 'Family and gender in the Christian community at Edendale, Natal, in colonial times', in Walker, C. (ed.) *Women and Gender in Southern Africa to 1945*, David Philip, Cape Town and James Currey, London.

**Meltzer, L.** (1994) 'Emancipation, commerce and the role of John Fairbairn's *Advertiser*', in Worden, N. and Crais, C. (eds) *Breaking the Chains: Slavery and its Legacy in the Nineteenth-Century Cape Colony*, Witwatersrand University Press, Johannesburg.

**Michie, J. and Padayachee, V.** (1997) *The Political Economy of South Africa's Transition: Policy Perspectives in the Late 1990s*, Dryden Press, London.

**Mills, G.** (ed.) (1994) *From Pariah to Participant: South Africa's Evolving Foreign Relations, 1900–1994*, South African Institute of International Affairs, Johannesburg.

**Mills, G.** (1995) 'South Africa and Southern Africa', *The Courier*, **153**, 59–65

**Mills, G.** (1998a) 'South Africa's foreign policy: from isolation to respectability', in Simon, D. (ed.) *South Africa in Southern Africa: Reconfiguring the Region*, James Currey, Oxford.

**Mills, G.** (1998b) 'South Africa's foreign policy after 1994: a template for foreign policy integration?', paper presented at African Studies Association of the UK Conference, School of Oriental and African Studies, London, 14–16 September.

**Mills, G. and Baynham, S.** (1990) 'Changing the guard: South African foreign policy into the 1990s', *Africa Insight*, **20**, 3, 176–88.

**Mills, W. G.** (1997) 'Millennial Christianity, British imperialism, and African nationalism', in Elphick, R. and Davenport, R. (eds) *Christianity in South Africa: A Political, Social and Cultural History*, James Currey, Oxford and David Philip, Cape Town.

**Ministry of Finance** (1996) *Growth, Employment and Redistribution: A Macroeconomic Strategy*, Pretoria. http://www.polity.org.za/govdocs/polcy/growth.html

**Mitchell, M. and Russell, D.** (1989) 'Black unions and political change in South Africa', in Brewer, J. (ed.) *Can South Africa Survive?* Macmillan, Basingstoke.

**Moll, T.** (1990) 'From booster to brake? Apartheid and economic growth in comparative perspective', in Nattrass, N. and Ardington, E. (eds) *The Political Economy of South Africa*, Oxford University Press, Cape Town.

**Moodie, T. D.** (1975) *The Rise of Afrikanerdom: Power, Apartheid, and the Afrikaner Civil Religion*, University of California Press, Berkeley, Los Angeles and London.

**Moodie, T. D.** (1986) 'The moral economy of the black miners' strike of 1946', *Journal of Southern African Studies*, 13, 1–35.

**Moodie, T. D. with Ndatshe, V.** (1994) *Going for Gold: Men, Mines and Migration*, University of California Press, Berkeley, Los Angeles and London.

**Moon, B. P. and Dardis, G. F.** (eds) (1988) *The Geomorphology of Southern Africa*, Southern Book Publishers, Johannesburg.

**Morna, C. L.** (1990) 'SADCC's first decade', *Africa Report*, **May–June**, 49–52.

**Morrell, R.** (1998) 'Of boys and men: masculinity and gender in southern African studies', *Journal of Southern African Studies*, 24, 4, 605–30.

**Morris, M. L.** (1976) 'The development of capitalism in South African agriculture: class struggle in the countryside', *Economy and Society*, 5, 292–43.

**Mostert, N.** (1992) *Frontiers: The Epic of South Africa's Creation and the Tragedy of the Xhosa People*, Jonathan Cape, London.

**Munslow, B. and Fitzgerald, P.** (1997) 'Search for a development strategy: the RDP and beyond', in Fitzgerald, P., McLennan, A. and Munslow, B. (eds), *Managing Sustainable Development in South Africa*, Oxford University Press, Cape Town.

**Murray, C.** (1987) 'Displaced urbanization: South Africa's rural slums', *African Affairs*, **86**, 311–29.

**Murray, C.** (1992) *Black Mountain: Land, Class and Power in the Eastern Orange Free State, 1880s–1980s*, Witwatersrand University Press, Johannesburg.

**Murray, M. J.** (1997) 'Factories in the fields: capitalist farming in the Bethal District, *c*. 1910–1950', in Jeeves, A. H. and Crush, J. (eds) *White Farms, Black Labour: The State and Agrarian Change in Southern Africa, 1910–1950*, Heinemann, Portsmouth, N.H, University of Natal Press, Pietermaritzburg, James Currey, Oxford.

**Nagle, G.** (1994) 'Challenges for the new South Africa', *Geographical Magazine*, **5**, 45–7.

**Nash, C.** (2000) 'Historical geographies of modernity', in Graham, B. and Nash, C. (eds) *Modern Historical Geographies*, Prentice Hall, London.

**Nash, M.** (1982) *Bailie's Party of 1820 Settlers*, Balkema, Cape Town.

**Nasson, B.** (1991) *Abraham Essau's War: A Black South African War in the Cape: 1899–1902*, Cambridge University Press, Cambridge.

Nattrass, J. (1988) *The South African Economy: Its Growth and Change*, Oxford University Press, Cape Town.

Nattrass, N. (1990) 'Economic power and profits in post-war manufacturing', in Nattrass, N. and Ardington, E. (eds) *The Political Economy of South Africa*, Oxford University Press, Cape Town.

Nattrass, N. (1996) 'Economic reconstruction and development in South Africa', in Rich, R. B. (ed.) *Reaction and Renewal in South Africa*, Macmillan, Basingstoke.

Nel, E. L. (1991) *The Spatial Planning of Racial Residential Segregation in East London, 1948–1973*, unpublished MA thesis, University of the Witwatersrand, Johannesburg.

Nel, E. L. (1994) 'Local development initiatives and Stutterheim', *Development Southern Africa*, 11, 363–77.

Nel, E. L. (1999) *Regional and Local Economic Development in South Africa: The Experience of the Eastern Cape*, Ashgate, Aldershot.

Nel, E. L. (2000) 'Economics and Economic development', in Fox, R. and Rowntree, K. (eds) *The Geography of South Africa in a Changing World*, Oxford University Press, Cape Town.

Nel, E. L. and Binns, T. (1999) 'Changing the geography of apartheid education in South Africa', *Geography*, 84, 2, 119–28.

Nel, E. L., Hill, T. and Binns, T. (1997) 'Development from below in the "new" South Africa: the case of Hertzog, Eastern Cape', *Geographical Journal*, 163, 1, 57–64.

Neumark, S. D. (1957) *Economic Influences on the South African Frontier, 1652–1836*, Stanford University, Stanford.

Newton-King, S. (1992) *The Enemy Within: The Struggle for Ascendancy on the Cape Eastern Frontier, 1760–99*, unpublished PhD, University of London.

Newton-King, S. (1994) 'The enemy within', in Worden, N. and Crais, C. (eds) *Breaking the Chains: Slavery and its Legacy in the Nineteenth-Century Cape Colony*, Witwatersrand University Press, Johannesburg.

Newton-King, S. (1999) *Masters and Servants on the Cape Eastern Frontier*, Cambridge University Press, Cambridge.

Nyerere, J. K. (1996) 'Introduction', in Adedeji, A. (ed.) *South Africa: Within or Apart?*, Zed, London.

Nzo, A. (1994) Speech by the South African Minister of Foreign Affairs to the 48th session, 95th meeting of the United Nations General Assembly, 23 June.

O'Meara, D. (1976) 'The 1946 African mineworkers' strike in the political economy of South Africa', *Journal of Commonwealth and Comparative Politics*, 13, 146–73.

O'Meara, D. (1982) '"Muldergate" and the politics of Afrikaner nationalism', *Work in Progress*, Supplement, 22, 1–19.

O'Meara, D. (1983) *Volkskapitalisme: Class, Capital and Ideology in the Development of Afrikaner Nationalism, 1934–1948*, Cambridge University Press, Cambridge.

O'Meara, D. (1996) *Forty Lost Years: The Apartheid State and the Politics of the National Party, 1948–1994*, Ravan Press, Johannesburg and Ohio University Press, Athens.

**Oldfield, J. R.** (1998) *Popular Politics and British Anti-Slavery: The Mobilisation of Public Opinion Against the Slave Trade, 1787–1807*, Frank Cass, London and Portland.

**Omer-Cooper, J. D.** (1966) *The Zulu Aftermath: A Nineteenth Century Revolution in Bantu Africa*, Longman, Harlow.

**Omer-Cooper, J. D.** (1995) 'The Mfecane survives its critics', in Hamilton, C. (ed.) *The Mfecane Aftermath: Reconstructive debates in Southern African History*, Witwatersrand University Press, Johannesburg and University of Natal Press, Pietermaritzberg.

**Onimode, B.** (1992) *A Future for Africa*, Earthscan, London.

**Owoeye, J.** (1994) 'What can Africa expect from a post-apartheid South Africa?', *Africa Insight*, **24**, 1, 44–6.

**Packenham, T.** (1979) *The Boer War*, Weidenfeld and Nicolson, London.

**Parnell, S.** (1988) 'Public housing as a device for white residential segregation in Johannesburg', *Urban Geography*, **9**, 584–602.

**Parnell, S.** (1991) 'Sanitation, segregation and the Natives (Urban Areas) Act: African exclusion from Johannesburg's Malay Location, 1897–1925', *Journal of Historical Geography*, **17**, 271–88.

**Parnell, S.** (1996) 'Changes in the British Colonial Office's policy towards urban Africa, 1939–1945', paper presented to the Conference on Africa's Urban Past, School of Oriental and African Studies, University of London, 6 July.

**Parnell, S. and Mabin, A.** (1995) 'Rethinking urban South Africa', *Journal of Southern African Studies*, **21**, 39–61.

**Parsons, N.** (1995a) '"The time of troubles": difaqane in the interior', in Hamilton, C. (ed.) *The Mfecane Aftermath: Reconstructive Debates in Southern African History*, Witwatersrand University Press, Johannesburg and University of Natal Press, Pietermaritzburg.

**Parsons, N.** (1995b) 'Prelude to difaqane in the interior of southern Africa, *c*. 1600–*c*. 1822', in Hamilton, C. (ed.) *The Mfecane Aftermath: Reconstructive Debates in Southern African History*, Witwatersrand University Press, Johannesburg and University of Natal Press, Pietermaritzburg.

**Parsons, N. and Palmer, R.** (1977) 'Introduction: historical background', in Palmer, R. and Parsons, N. (eds) *The Roots of Rural Poverty in Central and Southern Africa*, Heinemann, London, Ibadan, Nairobi and Lusaka.

**Peires, J. B.** (1981) *The House of Phalo: A History of the Xhosa People in the Days of Their Independence*, University of California Press, Berkeley, Los Angeles and London.

**Peires, J. B.** (1988) 'The British and the Cape, 1814–1834', in Elphick, R. and Giliomee, H. (eds) *The Shaping of South African Society, 1652–1840*, Wesleyan University Press, Middletown.

**Peires, J. B.** (1989) *The Dead Will Arise: Nongqawuse and the Great Xhosa Cattle-Killing Movement of 1856–7*, Ravan Press, Johannesburg; Indiana University Press, Bloomington and Indianapolis and James Currey, London.

Peires, J. B. (1993) 'Paradigm deleted: the materialist interpretation of the Mfecane', *Journal of Southern African Studies*, **19**, 2, 295–313.

Peires, J. B. (1995) 'Matiwane's road to Mbholompo: a reprieve for the Mfecane?', in Hamilton, C. (ed.) *The Mfecane Aftermath: Reconstructive Debates in Southern African History*, Witwatersrand University Press, Johannesburg and University of Natal Press, Pietermaritzburg.

Penn, N. (1989) 'Land, labour and livestock in the Western Cape during the eighteenth century: the Khoisan and the colonists', in James, W. G. and Simons, M. (eds) *The Angry Divide: Social and Economic History of the Western Cape*, David Philip, Cape Town and Johannesburg.

Perry Curtis Jnr, L. (1997) *Apes and Angels: The Irishman in Victorian Caricature*, Smithsonian Institute Press, Washington and London.

Pillay, B. (1976) *British Indians in the Transvaal: Trade, Race Relations and Imperial Policy in Republican and Colonial Transvaal, 1885–1906*, Longman, London.

Pillay, V. (1996) 'Wanted: a new economic policy', *SA Labour Bulletin*, **20**, 2, 34–41.

Pirie, G. H. (1984) 'Race zoning in South Africa: board, court, parliament, public', *Political Geography Quarterly*, **3**, 207–21.

Pirie, G. H. (1991) 'Kimberley', in Lemon, A. (ed) *Homes Apart: South Africa's Segregated Cities*, Paul Chapman, London; Indiana University Press, Bloomington and David Philip, Cape Town.

Pityana, B., Ramphele, M., Mpumlwana, M. and Wilson, L. (eds) (1991) *Bounds of Possibility: The Legacy of Steve Biko and Black Consciousness*, Zed Books, London and David Philip, Cape Town.

Plaatje, S. (1916) *Native Life in South Africa*, reprinted 1982, Ravan Press, Johannesburg.

Platzky, L. and Walker, C. (1985) *The Surplus People: Forced Removals in South Africa*, Ravan Press, Johannesburg.

Porter, A. (1980) *The Origins of the South African War: Joseph Chamberlain and the Diplomacy of Imperialism, 1895–99*, Manchester University Press, Manchester.

Porter, A. (1990) 'The South African War (1899–1902): context and motive reconsidered', *Journal of African History*, **31**, 31–57.

Porter, A. (1997) '"Cultural imperialism" and Protestant missionary enterprise, 1780–1914', *Journal of Imperial and Commonwealth History*, **25**, 367–91.

Porter, B. (1996) *The Lion's Share: A Short History of British Imperialism, 1850–1995*, Longman, London and New York.

Posel, D. (1991) *The Making of Apartheid, 1948–1961: Conflict and Compromise*, Clarendon Press, Oxford.

Potter, R. B., Binns, T., Elliott, J. and Smith, D. (1999) *Geographies of Development*, Longman, Harlow.

Power, M. (1997) 'The dissemination of development', *Environment and Planning D: Society and Space* **16**, 577–98.

Pratt, M. L. (1992) *Imperial Eyes: Travel Writing and Transculturation*, Routledge, London and New York.

Preston-Whyte, R. A., and Tyson P. D. (1988) *The Atmosphere and Weather of Southern Africa*, Oxford University Press, Cape Town.

Price, R. (1991) *The Apartheid State in Crisis: Political Transformation in South Africa, 1975–1990*, Oxford University Press, Oxford.

Price, R. (1997) 'Race and reconciliation in the New South Africa', *Politics and Society*, 25, 2, 149–78.

Ramsamy, E. (1995) 'South Africa and SADC(C): a critical evaluation of future development scenarios', in Lemon, A. (ed.), *The Geography of Change in South Africa*, John Wiley, Chichester.

Ranger, T. (1983) 'The invention of tradition in colonial Africa', in Hobsbawm, E. and Ranger, T. (eds) *The Invention of Tradition*, Cambridge University Press, Cambridge.

Rapley, J. (1996) *Understanding Development: Theory and Practice in the Third World*, UCL Press, London.

Rapoo, T. (1996) *The Theory and Practice of the RDP*, Policy Issues and Action Series No. 45, Centre for Policy Studies, University of the Witwatersrand, Johannesburg.

Rayner, M. (1986) *Wine and Slaves: The Failure of an Export Economy and the Ending of Slavery in the Cape Colony, South Africa, 1806–34*, unpublished PhD, Duke University.

RDP (1995) *Key Indicators of Poverty in South Africa*, South African Communications Services, Pretoria.

Reader's Digest (1994) *Illustrated History of South Africa: The Real Story*, Reader's Digest, Capetown and London.

Rich, P. B. (1984) *Race and Empire in British Politics*, Cambridge University Press, Cambridge.

Rich, P. B. (1986) *White Power and the Liberal Conscience*, Manchester University Press, Manchester.

Rich, P. B. (1994a) The search for security in Southern Africa, in Rich, P. B. (ed.) *The Dynamics of Change in Southern Africa*, Macmillan, London.

Rich, P. B. (1994b) 'South Africa and the politics of regional integration in Southern Africa in the post-apartheid era', in Rich, P. B. (ed.), *The Dynamics of Change in Southern Africa*, Macmillan, London.

Rich, P. B. (1996) *State Power and Black Politics in South Africa, 1912–51*, Macmillan, London and St. Martin's Press, New York.

Richardson, P. (1986) 'The Natal sugar industry in the nineteenth century', in Beinart, W., Delius, P. and Trapido, S. (eds) *Putting a Plough to the Ground: Accumulation and Dispossession in Rural South Africa, 1850–1930*, Ravan Press, Johannesburg.

Richardson, P. and van-Helten, J. J. (1982) 'Labour in the South African gold mining industry, 1886–1914', in Marks, S. and Rathbone, R. (eds) *Industrialisation and Social Change in South Africa: African Class Formation, Culture and Consciousness, 1870–1930*, Longman, London and New York.

Riddell, J. B. (1992) 'Things fall apart again: structural adjustment programmes in sub-Saharan Africa', *Journal of Modern African Studies*, 30, 1, 53–68.

Robinson, J. (1992) 'Power, space and the city: historical reflections on apartheid and post-apartheid urban orders', in Smith, D. M. (ed.) *The Apartheid City and Beyond: Urbanization and Social Change in South Africa*, Routledge, London and New York, Witwatersrand University Press, Johannesburg.

Robinson, J. (1996) *The Power of Apartheid: State, Power and Space in South African Cities*, Butterworth-Heinemann, Oxford.

Robinson, J. (1999) 'Divisive cities: power and segregation in cities', in Pile, S., Brook, C. and Mooney, G. (eds) *Unruly Cities?*, Routledge, London and New York.

Robson, B. T. (1990) 'The years between', in Dodgshon, R. A. and Butlin, R. A. (eds) *An Historical Geography of England and Wales*, Academic Press, London and San Diego.

Rogerson, C. M. (1982) Apartheid, decentralisation and spatial industrial change, in Smith, D. M. (ed.) *Living Under Apartheid: Aspects of Urbanization and Social Change in South Africa*, George Allen and Unwin, London.

Rogerson, C. M. (1995) 'The employment challenge in a democratic South Africa', in Lemon, A. (ed.), *The Geography of Change in South Africa*, John Wiley, Chichester.

Rogerson, C. M. (1997) 'Restructuring the post-apartheid space economy', paper presented at the United Nations Centre for Human Settlements Conference on Regional Development Planning and Management of Urbanization, Nairobi, 26–30 May 1997.

Rogerson, C. M. and Pirie, G. H. (1979) 'Apartheid, urbanization and regional planning in South Africa', in Obudho, R. A. and El-Shakhr, S. (eds) *Development of Urban Systems in Africa*, Praeger, New York.

Ross, A. (1986) *John Philip (1775–1851): Missions, Race and Politics in South Africa*, Aberdeen University Press, Aberdeen.

Ross, R. (1981) 'Capitalism, expansion, and incorporation on the southern African frontier', in Lamar, H. and Thompson, L. (eds) *The Frontier in History: North America and Southern Africa Compared*, Yale University Press, New Haven and London.

Ross, R. (1982) 'Pre-industrial and industrial racial stratification in South Africa', in R. Ross (ed.) *Racism and Colonialism: Essays on Ideology and Social Structure*, Martinus Nijhoff for Leiden University Press, Leiden.

Ross, R. (1983) *Cape of Torments: Slavery and Resistance in South Africa*, Routledge and Keegan Paul, London.

Ross, R. (1986) 'The origins of capitalist agriculture in the Cape Colony: a survey', in Beinart, W., Delius, P. and Trapido, S. (eds) *Putting a Plough to the Ground: Accumulation and Dispossession in Rural South Africa, 1850–1930*, Ravan Press, Johannesburg.

Ross, R. (1988) 'The Cape of Good Hope and the world economy, 1652–1835', in Elphick, R. and Giliomee, H. (eds) *The Shaping of South African Society, 1652–1840*, Wesleyan University Press, Middletown.

Ross, R. (1993a) 'The developmental spiral of the white family and the expansion of the frontier', in Ross, R. *Beyond the Pale*: *Essays on the History of Colonial South Africa*, Witwatersrand University Press, Johannesburg.

Ross, R. (1993b) 'The rise of Afrikaner Calvinism', in Ross, R., *Beyond the Pale*: *Essays on the History of Colonial South Africa*, Witwatersrand University Press, Johannesburg.

Ross, R., with van Arkel, D. and Quispel, G. C. (1993) 'Going beyond the pale: on the roots of white supremacy in South Africa', in Ross, R. *Beyond the Pale*: *Essays on the History of Colonial South Africa*, Witwatersrand University Press, Johannesburg.

Ross, R. (1994) '"Rather mental than physical": emancipations and the Cape economy', in Worden, N. and Crais, C. (eds) *Breaking the Chains*: *Slavery and its Legacy in the Nineteenth-Century Cape Colony*, Witwatersrand University Press, Johannesburg.

Ross, R. (1999) *Status and Respectability in the Cape Colony, 1750–1870*, Cambridge University Press, Cambridge.

Rostow, W. W. (1960) *The Stages of Economic Growth*: *A Non-Communist Manifesto*, Cambridge University Press, Cambridge.

RSA (Republic of South Afria) (1990) *Annual Report of the Board of the Decentralization of Industry*, Government Printer, Pretoria.

RSA (1995a) *RDP White Paper*, RDP Ministry, Pretoria.

RSA (1995b) *Tourism Green Paper*, Department of Environmental Affairs, Pretoria.

RSA (1996a) *Growth, Employment and Redistribution*: *A Macro-Economic Strategy*, Ministry of Finance, Pretoria.

RSA (1996b) *The Constitution of the Republic of South Africa*, Typeface Media, Pretoria.

RSA (1997) *The Foundations for a Better Life are Laid*: *The Government's Mid-term Report to the Nation*, SA Communications Service, Pretoria.

RSA (1998a) *Local Government White Paper*, Department of Constitutional Development, Pretoria.

RSA (1998b) unpublished statistical tables, Department of Trade and Industry, Pretoria.

Ryall, D. (1997) 'Caught between two worlds: understanding South Africa's foreign policy options', *Third World Quarterly*, **18**, 397–402.

SA Communications Service (1995) *Key Indicators of Poverty in South Africa*, SA Communications Service, Pretoria.

Sapire, H. (1989) 'African settlement and segregation in Brakpan, 1900–1927', in Bonner, P., Hofmeyr, I., James, D. and Lodge T. (eds) *Holding Their Ground*: *Class, Locality and Culture in Nineteenth and Twentieth Century South Africa*, Witwatersrand University Press and Ravan Press, Johannesburg.

Sapire, H. (1993) 'African political organisations in Brakpan in the 1950s', in Bonner, P., Delius, P. and Posel, D. (eds) *Apartheid's Genesis, 1935–1962*, Ravan Press, Witwatersrand University Press, Johannesburg.

Saul, J. and Gelb, S. (1986) *The Crisis in South Africa*, Zed Books, London.

Saunders, C. (1995) 'Pre-Cobbing Mfecane historiography', in Hamilton, C. (ed.) *The Mfecane Aftermath: Reconstructive Debates in Southern African History*, Witwatersrand University Press, Johannesburg and University of Natel Press, Pietermaritzburg.

Savage, M. and Miles, A. (1994) *The Remaking of the British Working Class 1840–1940*, Routledge, London.

Schreiner, O. (1923) *Thoughts on South Africa*, T. Fisher Unwin, London.

Schreuder, D. M. (1976) 'The cultural factor in Victorian imperialism: a case study of the British "civilising mission' *Journal of Imperial and Commonwealth History*, 4, 283–317.

Schrire, R. (1992) *Adapt or Die: The End of White Politics in South Africa*, Hurst and Company, London.

Scully, P. (1994) 'Private and public worlds of emancipation in the rural Western Cape, *c.* 1830–42', in Worden, N. and Crais, C. (eds) *Breaking the Chains: Slavery and its Legacy in the Nineteenth-Century Cape Colony*, Witwatersrand University Press, Johannesburg.

Scully, P. (1997) *Liberating the Family: Gender and British Slave Emancipation in the Rural Western Cape, South Africa, 1823–1853*, Heinemann, Portsmouth, NH; James Currey, Oxford and David Philip, Cape Town.

Seekings, J. (1988) 'The origins of political mobilisation in the PWV townships, 1980–1984', in Cobbett, W. and Cohen, R. (eds) *Popular Struggles in South Africa*, James Currey, London.

Seekings, J. (1991) 'Township resistance in the 1980s', in Swilling, M., Humphries,

R. and Shubane, K. (eds) *The Apartheid City in Transition*, Oxford University Press, Cape Town.

Shain, M. (1994) *The Roots of Antisemitism in South Africa*, University Press of Virginia, Charlottesville and London.

Shell, R. (1994) *Children of Bondage: A Social History of the Slave Society at the Cape of Good Hope, 1652–1838*, University of the Witwatersrand Press, Johannesburg.

Shillington, K. (1982) 'The impact of diamond discoveries on the Kimberley hinterland: class formation, colonialism and resistance among the Tlhaping of Griqualand West in the 1870s', in Marks, S. and Rathbone, R. (eds) *Industrialisation and Social Change in South Africa: African Class Formation, Culture and Consciousness, 1870–1930*, Longman, London and New York.

Shillington, K. (1985) *The Colonisation of the Southern Tswana, 1870–1900*, Ravan Press, Johannesburg.

Shuters-Macmillan, (1995) *New Secondary School Atlas for South Africa*, Shuter and Shooter, Pietermaritzburg.

Sibley, D. (1995) *Geographies of Exclusion: Society and Difference in the West*, Routledge, London and New York.

Sidaway, J. D. and Gibb, R. (1998) 'SADC, COMESA, SACU: Contradictory formats for regional "integration" in Southern Africa?', in Simon D. (ed.) *South Africa in Southern Africa: Reconfiguring the region*, James Currey, Oxford.

Sidaway, J. D. (1998) 'The (geo) politics of regional integration: the example of the

Southern African Development Community', *Environment and Planning D: Society and Space*, **16**, 549–76.

Simkins, C. (1981) 'Agricultural production in the African reserves of South Africa, 1919–1969', *Journal of Southern African Studies*, **7**, 256–83.

Simon, D. (1998) 'Introduction: shedding the past, shaping the future', in Simon, D. (ed.) *South Africa in Southern Africa: Reconfiguring the Region*, James Currey, Oxford.

Simons, J. and Simons, R. (1983) *Class and Colour in South Africa, 1850–1950*, International Defence and Aid Fund for Southern Africa, London.

Slater, D. (1993) 'The geopolitical imagination and the enframing of development theory', *Transactions of the Institute of British Geographers* **18**, 419–37.

Slater, H. (1980) 'The changing pattern of economic relationships in rural Natal, 1838–1914', in Marks, S. and Atmore, A. (eds) *Economy and Society in Pre-Industrial South Africa*, Longman, London.

Smets, C. (1996) 'Power to the people', *The Courier*, **160**, 66.

Smit, P., Olivier, J. J. and Booysen, J. J. (1982) 'Urbanization in the homelands', in Smith, D. M. (ed.) *Living Under Apartheid: Aspects of Urbanization and Social Change in South Africa*, George Allen and Unwin, London.

Smith, D. M. (ed.) (1982) *Living Under Apartheid: Aspects of Urbanization and Social Change in South Africa*, George Allen and Unwin, London.

Smith, D. M. (1994) *Geography and Social Justice*, Blackwell, Oxford.

Smith, I. R. (1996) *The Origins of the South African War, 1899–1902*, Longman, London and New York.

Solomon, H. (1995) 'A region of migration ... and refugees', *The Courier*, **153**, 75

Solomon, H. (1996) 'South Africa and the UN', *Africa Insight*, **26**, 1, 65–76.

Stadler, A. (1987) *The Political Economy of Modern South Africa*, David Philip, Cape Town and Croom Helm, London.

Stadler, A. H. (1979) 'Birds in the cornfield: squatter movements in Johannesburg, 1944–47', *Journal of Southern African Studies*, **6**, 93–123.

Stasiulis, D. and Yuval-Davis, N. (1995) 'Introduction: beyond dichotomies – gender, race, ethnicity and class in settler societies', in Stasiulis, D. and Yuval-Davis N. (eds) *Unsettling Settler Societies: Articulations of Gender, Race, Ethnicity and Class*, Sage, London.

StatsSA (1996) *Census in Brief*, Statistics South Africa, Pretoria.

StatsSA (1998) *Census in Brief*, Statistics South Africa, Pretoria.

StatsSA (1999a) *Census in Brief*, Statistics South Africa, Pretoria.

StatsSA (1999b) *Rural Survey, 1997*, Statistics South Africa, Pretoria.

Stedman Jones, G. (1984) *Outcast London: A Study in the Relationship Between Classes in Victorian Society*, Penguin, London.

Stepan, N. (1982) *The Idea of Race in Science: Great Britain, 1800–1960*, Macmillan, London.

**Stepan, N.** (1990) 'Race and gender: the role of analogy in science', in Golberg, D. (ed.) *Anatomy of Racism*, University of Minnesota Press, Minneapolis and London.

**Stock, R.** (1995) *Africa South of the Sahara: A Geographical Interpretation*, Guilford Press, New York.

**Stoler, A. L.** (1989) 'Rethinking colonial categories: European communities and the boundaries of rule', *Comparative Studies in Society and History*, 13, 134–61.

**Stoler, A. L.** (1995) *Race and the Education of Desire: Foucault's History of Sexuality and the Colonial Order of Things*, Duke University Press, Durham.

**Stoler, A. L. and Cooper, F.** (1997) 'Between metropole and colony: rethinking a research agenda', in Cooper, F. and Stoler, A. L. (eds) *Tensions of Empire: Colonial Cultures in a Bourgeois World*, University of California Press, Berkeley, Los Angeles and London.

**Straker, G. with Moosa, F., Becker, R. and Nkwale, M.** (1992) *Faces in the Revolution: The Psychological Effects of Violence on Township Youth in South Africa*, David Philip, Cape Town and Ohio University Press, Athens.

**Stutterheim Development Foundation** (1995) *Annual Report*, Stutterheim Development Foundation, Stutterheim.

**Swanson, M.** (1977) 'The sanitary syndrome: bubonic plague and urban native policy in the Cape Colony, 1900–1910', *Journal of African History*, 18, 387–410.

**Swanevelder, C. J., van Huyssteen, M. K. R. and Kotzé, J. C.** (1987) *Senior Geography Standard: New Syllabus 1987*, Nasou Ltd, Goodwood.

**Swilling, M. and Shubane, K.** (1991) 'Negotiating urban transition: the Soweto experience', in Lee, R. and Schlemmer, L. (eds) *Transition to Democracy: Policy Perspectives, 1991*, Oxford University Press, Cape Town and Oxford.

**Taylor, P. J.** (1996) *The Way the Modern World Works: World Hegemony to World Impasse*, John Wiley, Chichester.

**Taylor, R. and Shaw, M.** (1998) 'The dying days of apartheid', in Howarth, D. R. and Norval, A. (eds) *South Africa in Transition: New Theoretical Perspectives*, Macmillan, London and St. Martin's Press, New York.

**Thomas, D. S. G. and Middleton, N. J.** (1994) *Desertification: Exploding the Myth*, John Wiley, Chichester.

**Thomas, N.** (1994) *Colonialism's Culture: Anthropology, Travel and Government*, Polity Press, Cambridge.

**Thompson, E. P.** (1980) *The Making of the English Working Class*, Penguin, London.

**Thompson, F. M. L.** (1988) *The Rise of Respectable Society: A Social History of Victorian Britain 1830–1900*, Fontana Press, London.

**Thompson, L.** (1960) *The Unification of South Africa, 1902–1910*, Oxford University Press, Oxford.

**Thompson, L.** (1982a) 'Co-operation and conflict: the high veldt', in Wilson, M. and Thompson, L. (eds) *A History of South Africa to 1870*, Croom Helm, London and Canberra.

**Thompson, L.** (1982b) 'Co-operation and conflict: the Zulu kingdom and

Natal', in Wilson, M. and Thompson, L. (eds) *A History of South Africa to 1870*, Croom Helm, London and Canberra.

Thompson, L. (1985) *The Political Mythology of Apartheid*, Yale University Press, New Haven and London.

Thompson, L. (1990) *A History of South Africa*, Yale University Press, New Haven and London.

Thorne, S. (1997) '"The conversion of Englishmen and the conversion of the world inseparable": missionary imperialism and the language of class in early industrial Britain', in Cooper, F. and Stoler, A. L. (eds) *Tensions of Empire: Colonial Cultures in a Bourgeois World*, University of California Press, Berkeley, Los Angeles and London.

Todorov, T. (1984) *The Conquest of America: The Question of the Other*, Harper and Row, New York.

Tomaselli, K. G. (1988) 'The geography of popular memory in post-colonial South Africa: a study of Afrikaans cinema', in Eyles, J. and Smith, D. M. (eds) *Qualitative Methods in Human Geography*, Polity Press, Cambridge.

Tomlinson, R. (1990) *Urbanization in Post-Apartheid South Africa*, Unwin Hyman, London.

Trapido, S. (1963) 'Natal's non-racial franchise, 1856–1863', *African Studies*, **22**, 22–32.

Trapido, S. (1964) 'The origins of the Cape franchise qualifications of 1853', *Journal of African History*, **5**, 37–54.

Trapido, S. (1971) 'South Africa in a comparative study of industrialisation', *Journal of Development Studies* **7**, 309–20.

Trapido, S. (1978) 'Landlord and tenant in a colonial economy: the Transvaal, 1880–1910', *Journal of Southern African Studies*, **5**, 26–58.

Trapido, S. (1980a) '"The friends of the natives": merchants, peasants and the political and ideological structure of liberalism in the Cape, 1854–1910', in Marks, S. and Atmore, A. (eds) *Economy and Society in Pre-Industrial South Africa*, Longman, London.

Trapido, S. (1980b) 'Reflections on land, office and wealth in the South African Republic, 1850–1900', in Marks, S. and Atmore, A. (eds) *Economy and Society in Pre-Industrial South Africa*, Longman, London.

Trapido, S. (1986) 'Putting a plough to the ground: a history of tenant production on the Vereeniging Estates, 1896–1920', in Beinart, W., Delius, P. and Trapido, S. (eds) *Putting a Plough to the Ground: Accumulation and Dispossession in Rural South Africa, 1850–1930*, Ravan Press, Johannesburg.

Trapido, S. (1990) 'From paternalism to liberalism: the Cape Colony, 1800–1834', *The International History Review*, **12**, 76–104.

Trollope, A. (1878) *South Africa*, reprinted 1973, A. A. Balkema, Cape Town.

Turley, D. (1991) *The Culture of Antislavery, 1780–1860*, Routledge, London and New York.

Turok, B. (1993) 'South Africa's skyscraper economy: growth or development', in Hallowes, D. (ed.) *Hidden Face: Environment, Development, Justice: South Africa and the Global Context*, Earthlife Africa, Scottsville.

Turrell, R. (1982) 'Kimberley: labour and compounds, 1871–1888', in Marks, S. and Rathbone, R. (eds) *Industrialisation and Social Change in South Africa: African Class Formation, Culture and Consciousness, 1870–1930*, Longman, Harlow and New York.

Turrell, R. (1984) 'Kimberley's model compounds', *Journal of African History*, 25, 59–76.

Turrell, R. (1987) *Capital and Labour on the Kimberley Diamond Fields, 1871–1890*, Cambridge University Press, Cambridge.

UNDP (United Nations Development Programme) (1994) *Human Development Report, 1994*, Oxford University Press, Oxford.

UNDP (United Nations Development Programme) (1997) *Human Development Report, 1997*, Oxford University Press, Oxford.

UNDP (United Nations Development Programme) (1998) *Human Development Report, 1998*, Oxford University Press, Oxford.

Unterhalter, E. (1987) *Forced Removal: The Division, Segregation and Control of the People of South Africa*, International Defence and Aid Fund for Southern Africa, London.

Unterhalter, E. (1995) 'Constructing race, class, gender and ethnicity: state and opposition strategies in South Africa', in Stasiulis, D. and Yuval-Davis, N. (eds) *Unsettling Settler Societies: Articulations of Gender, Race, Ethnicity and Class*, Sage, London.

Vail, L. (ed.) (1989) *The Creation of Tribalism in Southern Africa*, James Currey, London; University of California Press, Berkeley and Los Angeles and David Philip, Claremont.

Vale, P. (1997) 'The African Renaissance', *Election Talk* 2, 1–2.

van der Merwe, P. J. (1995) *The Migrant Farmer in the History of the Cape Colony, 1657–1842*, translated by Roger Beck, Ohio University Press, Athens.

van Duin, P. and Ross, R. (1987) *The Economy of the Cape Colony in the Eighteenth Century*, Centre for the History of European Expansion, Leiden.

van Heerden, S. (1996) 'The marginalisation of Africa?', unpublished address to conference, Coventry University, Coventry, 17 December.

van Jaarsveld, F. A. (1964) *The Afrikaner's Interpretation of South African History*, Simondium, Cape Town.

van Onselen, C. (1972) 'Reactions to rinderpest in Southern Africa, 1896–97', *Journal of African History*, 13, 473–88.

van Onselen, C. (1976) *Chibaro: African Mine Labour in Southern Rhodesia, 1900–1933*, Pluto Press, London.

van Onselen, C. (1982a) *Studies in the Social and Economic History of the Witwatersrand, 1886–1914, Vol. 1, New Babylon*, Ravan Press, Johannesburg.

van Onselen, C. (1982b) *Studies in the Social and Economic History of the Witwatersrand, 1886–1914, Vol. 2, New Nineveh*, Ravan Press, Johannesburg.

van Onselen, C. (1996) *The Seed is Mine: The Life of Kas Maine, a South African Sharecropper, 1894–1985*, James Currey, Oxford.

van Onselen, C. (1997) 'Paternalism and violence on the maize farms of the south-western Transvaal, 1900–1950', in Jeeves, A. H. and Crush, J. (eds) *White Farms, Black Labour: The State and Agrarian Change in Southern Africa, 1910–1950*, Heinemann, Portsmouth, NH; University of Natal Press, Pietermaritzburg and James Currey, Oxford.

van Tonder, D. (1993) '"First win the war, then clear the slums": the genesis of the Western Areas Removal Scheme, 1940–1949', in Bonner, P., Delius, P. and Posel, D. (eds) *Apartheid's Genesis, 1935–1962*, Ravan Press, Witwatersrand University Press, Johannesburg.

Vaughan, M. (1994) 'Colonial discourse theory and African history, or has postmodernism passed us by?', *Social Dynamics*, **20**, 1–23.

Venter, D. (1994) 'An evaluation of the OAU on the eve of South Africa's accession', *Africa Insight*, **24**, 1, 47–59.

Venter, L. (1997) *When Mandela Goes: The Coming of South Africa's Second Revolution*, Transworld Publishers, London.

Vogel, C. (1994) 'People and drought in South Africa: reaction and mitigation', in Binns, T. (ed.) *People and Environment in Africa*, John Wiley, Chichester.

Wagner, R. (1980) 'Zoutpansberg: the dynamics of a hunting frontier, 1848–67', in Marks, S. and Atmore, A. (eds) *Economy and Society in Pre-Industrial South Africa*, Longman, London.

Waldmeir, P. (1997) *Anatomy of a Miracle: The End of Apartheid and the Birth of the New South Africa*, Viking, London.

Walker, C. (1990a) 'Women and gender in southern Africa to 1945: an overview', in

Walker, C. (ed.) *Women and Gender in Southern Africa to 1945*, David Philip, Cape Town and James Currey, London.

Walker, C. (1990b) 'Gender and the development of the migrant labour system c. 1850–1930', in Walker, C. (ed.) *Women and Gender in Southern Africa to 1945*, David Philip, Cape Town and James Currey, London.

Walker, C. (1990c) 'The women's suffrage movement: the politics of gender, race and class', in Walker, C. (ed.) *Women and Gender in Southern Africa to 1945*, David Philip, Cape Town and James Currey, London.

Walker, C. (1995) 'Conceptualising motherhood in twentieth-century South Africa', *Journal of Southern African Studies*, **21**, 417–37.

Walvin, J. (ed.) (1982) *Slavery and British Society, 1776–1846*, Louisiana State University Press, Baton Rouge.

Warwick, P. (1983) *Black People and the South African War, 1899–1902*, Longman, London and Ravan Press, Johannesburg.

Watts, M. (1993) 'Development 1: power, knowledge, discursive practice', *Progress in Human Geography* **17**, 257–71.

Webster, A. (1995) 'Unmasking the Fingo: the war of 1835 revisited', in Hamilton, C. (ed.) *The Mfecane Aftermath: Reconstructive Debates in Southern African History*, Witwatersrand University Press, Johannesburg and University of Natal Press, Pietermaritzburg.

Wellings, P. and Black, A. (1986) 'Industrial decentralisation under apartheid: the relocation of industry to the South African periphery', *World Development*, **14**, 1, 1–38.

**Wells, J.** (1993) *We Now Demand! The History of Women's Resistance Against the Pass Laws in South Africa*, Witwatersrand University Press, Johannesburg.

**Wells, J.** (1998) 'Eva's men: gender and power in the establishment of the Cape of Good Hope, 1652–74', *Journal of African History* **39**, 417–37.

**Welsh, D.** (1971) *The Roots of Segregation: Native Policy in Natal, 1845–1910*, Oxford University Press, Cape Town.

**West, M.** (1988) 'Confusing categories: population groups, national states and citizenship', in Boonzaier, E. and Sharp, J. (eds) *South African Keywords: The Uses and Abuses of Political Concepts*, David Philip, Cape Town.

**West Africa** (1997) 'A gripping sideshow', *West Africa*, **4173**, 10–16 November, 1753–4.

**West Africa** (1998) 'Clinton's historic visit', *West Africa*, **4189**, 6–26 April, 386–9.

**Western, J.** (1996) *Outcast Cape Town*, University of California Press, Berkeley, Los Angeles and London.

**Wickens, P. L.** (1978) *The Industrial and Commercial Workers' Union of Africa*, Oxford University Press, Cape Town.

**Willan, B.** (1982) 'An African in Kimberley: Sol Plaatje, 1894–1898', in Marks, S. and Rathbone, R. (eds) *Industrialisation and Social Change in South Africa: African Class Formation, Culture and Consciousness, 1870–1930*, Longman, London and New York.

**Willan, B.** (1996) *Sol Plaatje: Selected Writings*, Witwatersrand University Press, Johannesburg and Ohio University Press, Athens.

**Wilson, F. and Ramphele, M.** (1989) *Uprooting Poverty: The South African Challenge*, W. W. Norton, New York and London.

**Wittenberg, M.** (1997) 'Growth, demand and redistribution: economic debate, rhetoric and some food for thought', in Michie, J. and Padayachee, V. (eds), *The Political Economy of South Africa's Transition: Policy Perspectives in the Late 1990s*, Dryden Press, London.

**Wolpe, H.** (1972) 'Capitalism and cheap labour power in South Africa: from segregation to apartheid', *Economy and Society*, **1**, 425–56.

**Wolpe, H.** (1988) *Race, Class and the Apartheid State*, UNESCO, London.

**Worden, N.** (1994) 'Between slavery and freedom: the apprenticeship period, 1834–8', in Worden, N. and Crais, C. (eds) *Breaking the Chains: Slavery and its Legacy in the Nineteenth-Century Cape Colony*, Witwatersrand University Press, Johannesburg.

**Worden, N. and Crais, C.** (1994) 'Introduction', in Worden, N. and Crais, C. (eds) *Breaking the Chains: Slavery and its Legacy in the Nineteenth-Century Cape Colony*, Witwatersrand University Press, Johannesburg.

**Worger, W.** (1987) *South Africa's City of Diamonds: Mine Workers and Monopoly Capitalism in Kimberley, 1867–1895*, Yale University Press, New Haven.

**Worger, W.** (1997) 'Historical setting', in Byrnes, R. M. (ed.) *South Africa: A Country Study*, Library of Congress, Washington.

**World Bank** (1994a) *Adjustment in Africa: a World Bank Policy Research Report*, Oxford University Press, Oxford.

World Bank (1994b) *Reducing Poverty in South Africa*: *Options for Equitable and Sustainable Growth*, The World Bank, Washington.

World Bank (1997) *World Development Report 1997*, Oxford University Press, New York.

World Bank (1998) *World Development Report 1998/99*, Oxford University Press, Oxford.

Wright, J. B. (1995) 'Beyond the concept of the "Zulu explosion": comments on the current debate', in Hamilton, C. (ed.) *The Mfecane Aftermath: Reconstructive Debates in Southern African History*, Witwatersrand University Press, Johannesburg and University of Natal Press, Pietermaritzburg.

Wright, J. B. and Hamilton, C. (1989) 'Traditions and transformations: the Phongolo–Mzimkulu region in the late eighteenth and early nineteenth centuries', in Duminy, A. and Guest, B. (eds) *Natal and Zululand From Earliest Times to 1910*: *A New History*, University of Natal Press and Shuter and Shooter, Pietermaritzburg.

Wylie, D. (1995) 'Language and assassination: cultural negations in white writers' portrayal of Shaka and the Zulu', in Hamilton, C. (ed.) *The Mfecane Aftermath: Reconstructive Debates in Southern African History*, Witwatersrand University Press, Jonannesburg and University of Natal Press, Pietermaritzburg.

Young, C. (1994) *The African Colonial State in Comparative Perspective*, Yale University Press, New Haven.

Young, R. (1995) *Colonial Desire*: *Hybridity in Theory, Culture and Race*, Routledge, London and New York.

Yudelman, D. (1984) *The Emergence of Modern South Africa: State, Capital, and the Incorporation of Organized Labour on the South African Gold Fields, 1902–1939*, David Philip, Cape Town and Johannesburg.

Zituta, H. M. (1997) *The Spatial Planning of Racial Residential Segregation in King William's Town, 1826–1991*, unpublished MA, Rhodes University.

# INDEX